SCHRÖDINGER'S PHILOSOPHY OF QUANTUM MECHANICS

BOSTON STUDIES IN THE PHILOSOPHY OF SCIENCE

VOLUME 188

SCHRÖDINGER'S PHILOSOPHY OF QUANTUM MECHANICS

by

MICHEL BITBOL
Centre National de la Recherche Scientifique,
Paris, France

KLUWER ACADEMIC PUBLISHERS
DORDRECHT / BOSTON / LONDON

A C.I.P. Catalogue record for this book is available from the Library of Congress

ISBN-13: 978-94-010-7291-5 e-ISBN-13: 978-94-009-1772-9
DOI: 10.1007/978-94-009-1772-9

Published by Kluwer Academic Publishers,
P.O. Box 17, 3300 AA Dordrecht, The Netherlands.

Kluwer Academic Publishers incorporates
the publishing programmes of
D. Reidel, Martinus Nijhoff, Dr W. Junk and MTP Press.

Sold and distributed in the U.S.A. and Canada
by Kluwer Academic Publishers,
101 Philip Drive, Norwell, MA 02061, U.S.A.

In all other countries, sold and distributed
by Kluwer Academic Publishers Group,
P.O. Box 322, 3300 AH Dordrecht, The Netherlands.

Printed on acid-free paper

Softcover reprint of the hardcover 1st edition 1996

CONTENTS

PREFACE

This book is the final outcome of two projects.

My first project was to publish a set of texts written by Schrödinger at the beginning of the 1950's for his seminars and lectures at the Dublin Institute for Advanced Studies. These almost completely forgotten texts contained important insights into the interpretation of quantum mechanics, and they provided several ideas which were missing or elusively expressed in Schrödinger's published papers and books of the same period. However, they were likely to be misinterpreted out of their context. The problem was that current scholarship could not help very much the reader of these writings to figure out their significance. The few available studies about Schrödinger's interpretation of quantum mechanics are generally excellent, but almost entirely restricted to the initial period 1925-1927. Very little work has been done on Schrödinger's late views on the theory he contributed to create and develop. The generally accepted view is that he never really recovered from his interpretative failure of 1926-1927, and that his late reflections (during the 1950's) are little more than an expression of his rising nostalgia for the lost ideal of picturing the world, not to say for some favourite traditional picture. But the content and style of Schrödinger's texts of the 1950's do not agree at all with this melancholic appraisal; they rather set the stage for a thorough renewal of accepted representations. In order to elucidate this paradox, I adopted several strategies. To begin with, I wished to understand the historical roots of the widespread view according to which Schrödinger was somehow "conservative" in his approach of quantum mechanics. The results of this inquiry can be found in paragraphs 1-2 to 1-6. Then, I attempted a rational reconstruction of Schrödinger's interpretation of the quantum theory in the long term, from 1924 to 1958. In the light of this reconstruction, it appeared that the whole chronological perspective had to be *reverted* in order to capture the internal logic of his successive positions during this period (see paragraph 1-9). Instead of considering that Schrödinger's interpretation of the 1950's is a late reminiscence of that of 1926, I thus tried to work out the idea that the interpretations of 1926 are an early and simplistic way of coming close to the interpretation of the 1950's while bypassing the careful epistemological analysis which eventually established it on firm grounds. Once this perspective has been adopted, studying Schrödinger's interpretation of quantum mechanics is tantamount to following the development by fits and starts of a coherent methodological program. True, some elements of this program were not explicitly formulated at the beginning; but evidence that part of it was nevertheless being carried out by Schrödinger very early, and that the conceptual background he needed to carry it out was already at his disposal in the 1920's and 1930's, is provided in chapters 2 and 3. The historical inaccuracy of the reconstruction is then only partial, and it is compensated by an insight into the philosophical motivations of Schrödinger's mature interpretation of quantum mechanics.

The key point in order to grasp the subtleties of Schrödinger's thought process is to make sense of his apparently conflicting attitudes towards realism. Whereas his philosophical writings reveal strong affinities with Mach's positivism and with German idealism of the beginning of the nineteenth century[1], Schrödinger expressed an almost naïve realist position in his seminal papers on wave mechanics of 1926, and then a more sophisticated scientific realist attitude in his writings of the 1950's. However, these positions are not as incompatible as they seem, provided naïve realism and sophisticated scientific realism are not mixed up with *metaphysical* realism. In paragraphs 1-5, 2-1, 4-2 and 5-9, this point is developed at length by reference to modern minimal versions of realism such as Putnam's internal realism, or Blackburn's quasi-realism. But here, I would like to show more rapidly how a proper appraisal of Schrödinger's realism may account for his slight but decisive interpretative shift between 1926 and 1950.

In order to do so, commenting on some remarks of Hume's and of Kant's will prove very useful. Hume and Kant were both willing to take into account the recent birth of Newtonian physics in their philosophical system. And both of them rejected (though in a different way and for different reasons) any dogmatically realist interpretation of classical mechanics. But they both recognized at the same time that for a man-in-the-street or a working physicist, keeping on with realism or adopting their (empiricist or transcendentalist) position would make very little difference.

In a well-known paragraph of his *Treatise of human nature,* Hume emphasizes that, according to him, there is no "radical cure"[2] to our skeptical doubts about the existence of material bodies in an external world beyond the sense impressions. However, he says, in everyday life we use a substitutive "remedy", that is *"carelessness and inattention"*. With this unrefined but powerful remedy, we very soon forget our doubts and come back to the standard situation where the existence of material bodies is a presupposition of discourse. This existence "(...) is a point which we must take for granted in all our reasonings". Therefore, even from an empiricist standpoint, one must recognize that in the course of his careless and inattentive life, the man-in-the-street can (or even *should* in many circumstances) behave and speak as if he were a realist about material bodies.

[1]E. Schrödinger, *My view of the world,* Cambridge University Press, 1964

[2]D. Hume, *A Treatise of Human Nature,* (ed. L.A. Selby-Bigge), Oxford University Press, 1960; Book I, part IV, section II.

As for Kant, he insists in the *Transcendental aesthetic* section of his *Critique of pure reason* that space is not a concept abstracted from our outer experiences, but rather the *a priori* form of all outer intuitions. It is only this way that one can understand how it is possible to have a knowledge of the *necessary* propositions of geometry. But in paragraph 13 of his Prolegomena, Kant also accepts that, with respect to any possible experience, everything remains exactly as if ("als ob") space were an intrinsic feature of things and of their relations. Here again, the critical attitude, which is essential from the meta-standpoint of the philosopher, proves mostly irrelevant from the ordinary standpoint of the man-in-the-street or the scientist. The latter can (or *should* in certain circumstances) adopt a realist attitude about space, time, substance and causality, and forget that these are only *a priori* forms of intuition and thought.

Now, this succession of reflexivity and objectifying directedness, of philosophical critique and realist posture, is exactly what one witnesses in Schrödinger's writings of the years 1925-1926. In spite of his philosophical affinities with Mach's positivism, he did not see why, as a scientist, he could not behave towards the new theoretical entities (especially ψ-waves) exactly *as if* they were real; so much so that nothing prevented the "as if" itself from being completely wiped out. After all, Schrödinger thought, it was exactly this path that physics and science in general had always followed in the past. But, as he soon realized, the structure of quantum phenomena is not such that this traditional strategy can be adopted without harm.There were then two alternatives left. Either one retreated to a purely instrumentalist conception of theories, or one invented a new variety of *self-conscious* realist attitude towards theoretical entities: a variety of realism in which the *process* of objectification is still recognized as such, even when a set of objectified entities has been picked out. After a short hesitation in 1930, Schrödinger chose the second alternative, and he developed it more and more systematically until the 1950's. But few thinkers were able to follow him in this highly elaborated approach, and they generally misunderstood his position.

This has been, in outline, the contents of the first four chapters of this book, which were initially intended to be an introductory essay to Schrödinger's unpublished writings of the 1950's. The essay was entitled: *Schrödinger's interpretation of quantum mechanics in the 1950's.* But its excessive length, and various editorial ups and downs, led me to cut it off from Schrödinger's texts. These texts were published separately by Ox Bow Press[1], with a much shorter introduction. The former introductory essay then became part of a new and more ambitious project, which was no longer restricted to providing a reconstruction of Schrödinger's route towards his mature interpretation of quantum mechanics. It rather aimed

[1]E. Schrödinger, *The interpretation of quantum mechanics (Dublin seminars 1949-1955 and other unpublished texts)*, edited and with introduction by M. Bitbol, Ox Bow Press, 1995

at gaining a better appraisal of the philosophical motivations of Schrödinger's views on quantum physics[1]. The first step in this direction consists in elucidating the structure of these views from a modern standpoint, by confronting them with some recent interpretations of quantum mechanics. In paragraph 4-4 and 4-5, a systematic comparison between Schrödinger's approach of the measurement problem and some current no-collapse conceptions (especially Everett's and Van Fraassen's) is undertaken. The second step consists in putting Schrödinger's interpretation of quantum mechanics in a broader philosophical context. In chapter 5, Schrödinger's complete rejection of corpuscularian entities, and his adoption of wave entities instead, is seen to be a natural byproduct of his phenomenalistic conception of material bodies. Then, in chapter 6, his combination of thorough criticism of the idea of "wave packet collapse", and growing skepticism towards any attempt at making macroscopic experimental events an emergent feature of some universal ψ-function, is analyzed. Schrödinger's mature position on the measurement problem is characterized as a "parallelism" between the continuous and uninterrupted time-development of a holistic wave-function and the succession of discontinuous experimental events. The only link which is assumed between these two series is Born's probabilistic rule. Schrödinger's and Bohr's positions are systematically compared at this stage (Paragraphs 6-1 to 6-3). It is found that their disagreement stems from a slight but crucial difference between Bohr's *dualistic* analysis of the measurement process (functional cut between a classical and a quantum fraction of the measurement chain), and Schrödinger's *parallelist* conception of the relation between the stochastic evolution of experimental oucomes and the unitary evolution of ψ-functions. This parallelism is finally confronted (in paragraphs 6-4 to 6-8) with another kind of parallelism which is currently the subject of intense philosophical discussions: parallelism between the intentional and causal accounts of action. It is shown that both parallelisms reflect the same fundamental and well-documented difference between two attitudes, neither of which is reducible to or completely intertranslatable with the other: the attitude of intersubjective understanding and the attitude of objectifying explanation.

* * *

I am especially indebted to Robert S. Cohen for permanent encouragements and lasting patience. At many stages of this work, the impetus came from him. As my manuscript took shape, my friend Michael Lockwood made many fruitful criticisms and helped me to edit several chapters. His permanent and warm interest in Schrödinger's ideas

[1]This work was prepared by my previous introductory essays (in French) on two philosophical writings of Schrödinger's. See M. Bitbol, *L'élision*, in: E. Schrödinger, *L'esprit et la matière*, Seuil, 1990 and M. Bitbol, *La Clôture de la représentation*, in: E. Schrödinger, *La nature et les grecs*, Seuil, 1992.

and in mine were an invaluable support. Mara Beller read very carefully the first four chapters and offered significant suggestions about them, with her typically lucid and original approach of the history of quantum mechanics. Bernard d'Espagnat, Rom Harré, and Mioara Mugur-Schächter, have provided many basic reflections and discussions which were the departure point of several parts of my work during the past five years. I am grateful to Rachel Zahn and Roland Sypel for having carefully edited my manuscript, at various stages of its completion. They have helped me clarify my text and develop some important points. I also owe gratitude to Ruth Braunizer, for her permanent hospitality and friendship, and for her permission to consult as frequently as I wish the remarkable archive in her possession about her father Erwin Schrödinger.

Work on this volume has been performed partly in the libraries of the university of Oxford, and partly in *Linacre college,* Oxford, of which I am a visiting senior member. The effectiveness of my thought and writing have been much enhanced by this busy, friendly and highly stimulating environment. Other parts of the book were written in connection with my institution in Paris: The *Institut d'Histoire et Philosophie des sciences et des Techniques,* which depends both of the CNRS and the University Paris I / Panthéon-Sorbonne. There, I was helped and motivated by the high standard and great expectations of my best graduate students, especially Léna Soler.

Finally, I wish to aknowledge my indebtedness to Annie and Anne-Florence, my wife and our daughter, who bore patiently and usually cheerfully the "carelessness and inattention" which is too often the correlate of my academic work.

CHAPTER 1

THE CONTROVERSY BETWEEN SCHRÖDINGER AND THE GÖTTINGEN-COPENHAGEN PHYSICISTS IN THE 1950's

When surveying the literature, one often gets the impression that Schrödinger held, in succession, *four* distinct interpretations of quantum mechanics, and that, except for the one he *borrowed* from the Copenhagen group, these interpretations all fell into a complete and deserved oblivion. People generally recognize the great importance of his contributions to the interpretation of quantum mechanics. But what they regard as important here are, as a rule, only the lines of argument and ingenious thought-experiments by which Schrödinger challenged the current orthodoxy, thus forcing his contemporaries to clarify their positions. On the face of it, none of Schrödinger's own positive suggestions appear to have had any lasting influence. Let us then begin with a brief statement of this widespread view of Schrödinger's philosophy of quantum mechanics, especially as originally stated in the writings of such contemporaries as Heisenberg and Born, before we subject it to critical scrutiny.

1-1 *Schrödinger's successive interpretations of quantum mechanics according to the current views*

Schrödinger's first interpretation of quantum mechanics was sketched out in January and February of 1926, in the pioneering papers entitled "Quantization as a problem of proper values I and II"[1]. It amounted to taking the ψ-function at face-value and treating it as a direct description of wave-like processes occurring within the boundaries of atoms. However, in the early spring of 1926, Schrödinger realized that this way of dealing with the wave-mechanical formalism could yield only the *proper modes* of the vibrating system, providing no hint of a satisfactory account of the radiative processes (especially the line intensities). As soon as he had demonstrated the mathematical equivalence of his wave mechanics with Heisenberg's, Born's and Jordan's matrix mechanics, Schrödinger came to the conclusion that the ψ-function must be thought of merely as an intermediate-level concept *(hilfbegriff)*, and that the correct description of the atomic processes is actually given by the product $-e\psi\psi^*$, considered as an electric charge density. This new approach was only partially successful, however, and it did not remove all the difficulties that plagued the original wave interpretation. The onset of Born's probabilistic interpretation of ψ, the strong criticisms coming from the Göttingen-Copenhagen physicists, their elaboration of a full-

[1]E. Schrödinger, "Quantization as a problem of proper values" (I and II), in: *Collected papers on wave mechanics,* Blackie and son, 1928. For a valuable assessment on the possibility of holding this Schrödinger's initial position nowadays, see J. Dorling, "Schrödinger's original interpretation of the Schrödinger equation: a rescue attempt", in: C.W. Kilmister (ed.), *Schrödinger, centenary celebration of a polymath,* Cambridge University Press, 1987

blown synthetic apprehension of quantum mechanics whose two corner stones were Heisenberg's "uncertainty relations" and Bohr's complementarity principle, put an end to these initial attempts of Schrödinger to make sense of the wave function. From 1928 on, Schrödinger resigned himself to teaching quantum mechanics according to the mainstream "Copenhagen interpretation". The year 1935 marked a noticeable turning point. A few weeks after the publication of the Einstein-Podolsky-Rosen paper and the burst of correspondence with Einstein which followed, Schrödinger published both his "cat-paper"[1] and a more technical article concerning the "entanglement" of wave-functions[2]. In these two papers, Schrödinger expressed a skepticism about the current interpretation of quantum mechanics, which was grounded in a deep understanding of the issues involved, even though he admitted his inability to offer any satisfactory alternative. Finally, in the late forties and the early fifties, he became increasingly self-assertive, and declared his attachment to an idiosyncratic conception of quantum mechanics which, to many, appeared as a mere revival of his first 1926 wave-interpretation.

It is no exaggeration to say that this later development was unanimously rejected by the scientific community. True, some isolated physicists of renown, such as Einstein and de Broglie, welcomed both Schrödinger's renewed "realist" approach and the valuable support which he was now giving to their own struggle against the current "dogma" (as they called the cluster of Copenhagen-like interpretations commonly accepted at that time). But they did not approve, let alone accept, Schrödinger's wave-interpretation. Moreover, the apparent convergence of the three thinkers on the issue of "realism" was mostly epistemological, partly verbal, and certainly *not* metaphysical; for they did not even agree about the meaning of the word "reality". As for the former Göttingen-Copenhagen physicists, their reception of Schrödinger's late interpretation of quantum mechanics was quite hostile. Pauli went as far as denouncing Schrödinger's "neurotic" regression[3], and accusing him of entertaining the "(...) dream of a way back, back to the classical style of Newton-Maxwell, that is hopeless, off the way, bad taste (...) and not even a lovely dream"[4]. Many physicists rejected Schrödinger's proposals, without even bothering to examine his arguments with any care. Viewing them as a futile attempt to resuscitate the original 1926 wave conception, which had been convincingly refuted by the end of 1927, they thought it sufficient to

[1] The two papers are: A. Einstein, B. Podolsky and N. Rosen, "Can quantum-mechanical description of physical reality be considered complete?" Phys. Rev. 47, 777-780, 1935; E. Schrödinger, "Die gegenwärtige Situation in der Quantenmechanik", Naturwissenschaften, 23, 807-812, 823-828, 844-849, 1935. They are both reprinted, in English translation, in: J. A. Wheeler and W.H. Zurek, *Quantum mechanics and measurement,* Princeton University Press, 1983.

[2] E. Schrödinger, "Discussion of probability relations between separated systems", Proc. Cambridge Phil. Soc., 31, 555-563, 1935

[3] Quoted by: K.V. Laurikainen, *Beyond the atom,* Springer-Verlag, 1988, p. 31

[4] W. Pauli, letter to M. Born, quoted by M. Born, "The interpretation of quantum mechanics", Brit. J. Phil. Sci., 4, 95-106, 1953. Reprinted in: M. Born, *Physics in my generation,* Pergamon Press, 1956

rehearse the former objections and quickly return their attention to more fruitful tasks.

1-2 *Born's and Heisenberg's criticism of Schrödinger's late interpretation of quantum mechanics*

Max Born was more closely involved than anyone else in the debate with Schrödinger. In November 1952, he was due to hold a series of lectures at the university of London, and he expected Schrödinger to be one of the main participants in the public discussion. As it turned out, Schrödinger was unable to attend, due to ill health, but the elements of the controversy were recorded in two articles published by the British Journal of Philosophy of Science[1], as well as in Born's edition of the Born-Einstein letters[2]. Let me first try to summarize Born's account of Schrödinger's position:

1) It is an essentially "conservative attitude towards quantum mechanics"[3]; an attempt to recover the "classical physics of clearly comprehensible events"[4],

2) It tends to dismiss the "statistical concept of quantum mechanics"[5] and to reinstate determinism, in agreement with Einstein's views,

3) It leads one to the discarding of the very concept of a particle, to asserting that "there are no particles and there are no energy quanta"[6],

4) Schrödinger considers that "particles are narrow wave packets"[7],

5) Schrödinger "(...) insists that *there is* something *behind* the phenomena, the sense impressions, namely waves moving in a still scantily explored medium"[8]; he tends to forget the multi-dimensional character of the ψ-functions and to insist on waves in ordinary 3-dimensional space, which are supposed to rescue the "*Anschaulichkeit*" (picturability) of the theoretical description; he "believes that his waves constitute the final deterministic solution"[9].

With such a simplified account of Schrödinger's views in mind, it was not very difficult to refute them. Here are Born's main arguments:

1') Schrödinger's regressive trend towards classical pictures is just a matter of philosophical incapacity, an absence of "willingness to sacrifice traditional concepts and to accept new ones, like Bohr's principle of complementarity"[10].

[1]E. Schrödinger, "Are there quantum jumps?", Brit. J. Phil. Sci., 3, 109-123, 233-242, 1952; M. Born, "The interpretation of quantum mechanics", loc. cit.
[2]M. Born (ed.), *The Born-Einstein letters,* Mac Millan, 1971
[3]M. Born (ed.), *The Born-Einstein letters,* op. cit. p. 196
[4]M. Born, in: Physicalische Blätter, 17, 85-87, 1961
[5]M. Born (ed.), *The Born-Einstein letters,* op. cit. p. 195
[6]M. Born, "The interpretation of quantum mechanics", loc. cit. p. 96
[7]M. Born (ed.), *The Born-Einstein letters,* op. cit. p. 202
[8]M. Born, "The interpretation of quantum mechanics", loc. cit. p. 100
[9]M. Born (ed.), *The Born-Einstein letters,* op. cit. p. 195
[10]M. Born, "The interpretation of quantum mechanics", loc. cit. p. 105

2') By his insistence on the deterministic evolution of the ψ-function according to his own equation, Schrödinger tries to sweep under the carpet the last stage of any experiment, namely the detection which "produces countable events"[1].

3') The complete elimination of the particle notion is precluded, for one still has to *speak* in terms of particles in order to account for certain experimental situations (such as scattering experiments, photo-ionisation, etc.). One has to fall back on a particulate description in order "to describe the discontinuous recording of a Geiger counter"[2]. At any rate, the reference to particles is essential in order to *interpret* the number *n* in 3n-dimensional ψ-functions; namely to interpret it as the number *of* particles in the system under investigation.

Moreover, a radical elimination of the atomistic picture would break the continuity[3] of our concepts, both from a historical standpoint and from a methodological one.

-Historical: quantum mechanics arose from the modern revival of the atomistic frame of thought, and it would be "presumptuous" to overthrow it without having a powerful substitute.

-Methodological: "(...) if (Schrödinger) wants to connect his results with experimental facts, he has to describe them in terms of physical apparatus. These consist of bodies, not of waves. Thus, at some point, the wave description, even if it were possible, would have to be connected with ordinary bodies"[4]

These are the reasons why Born, in spite of recognizing that they could be assigned no precise trajectories and that they lacked individuality, still considered particles to be one of the two main (complementary) "aspects of phenomena". His paper "physical reality"[5], published in 1953, the same year as his rebuttal of Schrödinger's position, was partly intended as a strong defence of the concept of a particle, as one may recollect from the following sentence: "I maintain that we are justified in regarding these particles as real in a sense not essentially different from the usual meaning of the word".

4') "(W)ave packets representing solutions of the Schrödinger equations do not propagate without change of shape, but disperse"[6].

5') Born agreed with Schrödinger that one cannot be content with just describing experimental results, but he did not believe that what is "behind the phenomena" can be reduced to a wave-like process. Schrödinger's attempt at recovering the *"Anschaulichkeit"* of the theoretical description by means of the wave-concept seemed to him

[1]M. Born, "The interpretation of quantum mechanics", loc. cit. p. 104

[2]M. Born, "The interpretation of quantum mechanics", loc. cit. p. 102

[3]This argument was obviously intended to be ironical. It mimics Schrödinger's insistence on historical and representational continuity.

[4]M. Born, "The interpretation of quantum mechanics", loc. cit. p. 99

[5]M. Born, "Physical reality", Phil. Quart., 3, 139-49, 1953. Reprinted in: M. Born, *Physics in my generation*, Pergamon Press, 1956

[6]M. Born (ed.), *The Born-Einstein letters*, op. cit. p. 202

especially hopeless: "one generally (...) needs waves in spaces of many dimensions, which are something entirely different from the waves of classical physics, and impossible to visualise"[1]; "(...) this means that the claim of simplicity and of *'Anschaulichkeit'* , the possibility of seeing the process in space, is illusory"[2]. Born recognized in a footnote[3] that Schrödinger did not baldly assert the 3-dimensional visualisability of the original ψ-functions of wave mechanics, and that 3-dimensionality was saved "with the help of second quantization", but he emphasized the abstract character of the latter formalism and pointed out that "Anschaulichkeit" could not therefore be preserved in this way either.

This represents the most detailed, if not the most accurate, attempt at refutation of Schrödinger's late interpretation of quantum mechanics. Heisenberg took essentially the same line in the eighth chapter of his book *Physics and Philosophy* [4], but he restricted himself to a criticism of points 3) and 5) of Schrödinger's alleged conception:

3") Schrödinger cannot account for the "element of discontinuity that is found everywhere in atomic physics; any scintillation screen or Geiger counter demonstrates this element at once"[5]. At this point, Heisenberg introduced a personal view into the debate. To Schrödinger's contention that wave functions represent "objective reality", he opposed the idea that they are only meant to describe some kind of Aristotelian *potentia.* The appearance of a discontinuous event should thus be construed as a "transition from the possible to the actual".

5") Schrödinger "(...)overlooks the fact that only the waves in configuration space (...) are probability waves in the usual interpretation, while the three-dimensional matter waves or radiation waves are not".

These arguments were supposed to put an end to Schrödinger's late attempt at interpreting quantum mechanics. Unfortunately, the conception they tended to refute had actually little in common with the one Schrödinger was trying to promote, and they were therefore mostly irrelevant. Of course, in order to provide a precise demonstration of the latter statement, we shall have to go much farther in our historical and philosophical analysis of the debate on the interpretation of quantum mechanics. But the major reasons for which Born's and Heisenberg's criticisms of Schrödinger's views were almost entirely misplaced may easily be outlined. In providing a preliminary account of these reasons, we shall also be indicating the lines along which our examination of Schrödinger's late interpretation of quantum mechanics will proceed.

[1]ibid.
[2]M. Born, "The interpretation of quantum mechanics", loc. cit. p. 98
[3]ibid.
[4]W. Heisenberg, *Physics and Philosophy,* Harper and Brothers, 1958
[5]W. Heisenberg, *Physics and Philosophy,* op. cit. p. 143

1-3 *Historical flaws in the Born-Heisenberg critique of Schrödinger's late interpretation of quantum mechanics*

To begin with, Born's and Heisenberg's own texts display symptoms of their inability to perceive (or to recognize) the changes that occurred between the original wave interpretation of 1926 and Schrödinger's position in the 1950's. This can be seen most clearly when one compares points 3) and 4), in Born's account of Schrödinger's conception. According to point 4), Schrödinger's interpretation of quantum mechanics assumes that particles *are* something, since they *are* at least narrow wave packets moving in ordinary 3-dimensional space. But according to point 3), the same interpretation is based on a complete *dismissal* of the concepts of particle and of energy quanta. Schrödinger can hardly have held both views simultaneously. Rather, he was obliged to choose between two distinct approaches: (i) trying to provide the corpuscular representation (especially its continuous trajectories in ordinary space) with some counterpart in his wave picture, and (ii) eliminating it from the outset. The first approach is typical of Schrödinger's initial work on wave mechanics, whereas the second one was arrived at after a long period of reflection (which began in the late thirties and culminated in the fifties), about the ontological inconsistency of entities which are *in principle* incapable of being individuated, and whose identity through time is doubtful. This transformation was recorded by Schrödinger himself in his 1952 paper "Are there quantum jumps?". There, he began with some loose statements expounding what *would* happen "(...) *if* we picture the particle as a composite wave-phenomenon (usually called a 'wave-parcel')"; and he then listed some traditional objections to this way of "picturing" particles. He finally gave the following reply to widely expressed doubts about the ability of pure wave models to represent particles: "I am aware of these questions. They are no longer as embarassing as they were, before we had gained the insight we have now gained into what a particle certainly is *not;* it is *not* a durable little thing with individuality"[1]. In other texts of the same period, Schrödinger went even further, avoiding any mention of wave packets and claiming that what we call "particles" reduce to series of events which sometimes give us the *illusion* of continuous trajectories (see paragraph 4-2). Born was not completely unaware of this situation, for he did not try to criticize the model of wave packets in his 1953 paper[2]. But his inability to understand how one could merely *reject* one of the two "complementary aspects of phenomena" led him to overlook the important transition from *particles as wave-crests* to *no particle at all.* In the later comment he added to his correspondence with Einstein, he thus reverted to an irrelevant criticism of the original wave packet notion.

[1] E. Schrödinger, "Are there quantum jumps?", loc. cit.

[2] M. Born, "The interpretation of quantum mechanics", loc. cit.

Another element of chronological confusion can be detected in Born's and Heisenberg's arguments against the attempt to recover a picture of events in ordinary 3-dimensional space. Both authors seem to think that Schrödinger was willing simply to ignore the necessity of using a 3n-dimensional ψ-function when dealing with composite systems. But Schrödinger had actually been one of the most clear-headed exponents of what, in his own terminology, we now call the "entanglement" of the sub-systems in a composite system[1]. As far back as the beginning of 1927, he was fully aware of the far-reaching holistic implications of the multi-dimensional character of ψ-functions[2]. How could anyone think that he just reverted to a naive representation of phenomena as comprised of waves propagating and meeting in 3-dimensional space? Well, on a closer analysis, there may seem to be some good reason for such a misinterpretation. When they criticized Schrödinger's views, Heisenberg and Born were relying exclusively on the paper of 1952, entitled "Are there quantum jumps?"; and in this paper one can read some loose statements in which collision problems are dealt with *as if* they concerned two (or more) plane waves in ordinary space. True, Schrödinger tries to correct this elementary picture by admitting that "this model is obsolete", or by explaining that one has to rely on the "auxiliary concept, familiar to quantum physicists, of wave parcels in more that three dimensions"; but his insistence on the idea that the 3-dimensional wave picture is less inappropriate than the corpuscular one, and that "the means used now to depict the physical situation still follow the pattern to which it belonged", could create a confusion in the mind of his readers.

This is why it is very important to compare the article of 1952 with other texts of the same period. Only in this way can we keep clear the distinction between what was merely intended by Schrödinger as a partially acceptable metaphor and what was regarded by him as an up to date theoretical advance.

It is when he refers to another paper of 1953[3], where Schrödinger explicitly adopts the second quantization point of view, that Born makes the transition to a comparative study of Schrödinger's writings. Properly conducted, such a study would suggest that, when he makes assertions about three-dimensional theoretical entities, Schrödinger is not just dreaming of a way back to the 1926 original wave picture. Rather, he has a precise idea according to which any statement couched in the n-"particle" 3n-dimensional ψ-function language may be translated into a statement belonging to the second quantization language (n^{th} level of excitation of a three-dimensional "vacuum" state). But still, Born's account remains historically inaccurate. For Schrödinger's approval of

[1]E. Schrödinger, "Discussion of probability relations between separated systems", Proc. Camb. Phil. Soc., 31, 555-563, 1935

[2]E. Schrödinger, "The exchange of energy according to wave mechanics", in: *Collected papers on wave mechanics,* Blackie & son, 1928, p. 137-146

[3]E. Schrödinger, "The meaning of wave mechanics", in: A. George (ed.), *Louis de Broglie physicien et penseur,* Albin Michel, 1953

the second quantization scheme was by no means new in 1953. It was already expressed in the first 1944 edition of his *Statistical Thermodynamics* [1], and it dates back to 1927, namely to the publication of Jordan's first paper on *quantized matter waves*. This is attested to by a letter Schrödinger wrote to Jordan during the summer 1927: "I am particularly interested in your remark [on coupling quantized light and quantized matter waves]. Indeed, as far as I understand it, it is also my opinion"[2]. Schrödinger's approval of second quantization is not surprising since, as O. Darrigol[3] rightly pointed out, this approach was only a conceptual development of his own early theory of gases. Such a kinship was recognized by Jordan himself, in a letter to Schrödinger: "the ideas that I set out in the last paragraph of my work on Fermi gas [quantized matter waves coupled with light waves] have, as I claimed, an earlier origin (...) At that time I had given a lot of thought to Einstein's gas theory and I had specified the representation in a way similar to your work (...): the number of atoms in a cell corresponds to the quantum number of a cavity-mode oscillator"[4]. The only aspect of the quantized matter waves theory which did not suit Schrödinger was stylistic: it was its exclusive use of the same type of non-commutative algebra as the one which pervaded matrix mechanics[5]. A slight mathematical transformation, a simple shift of attention from the commutation relations to the state vectors in Fock space, proved sufficient to win Schrödinger's warm and unconditional acceptance of the second quantization formalism. This being granted, it becomes obvious that when Born[6] insists on the "abstract" character of second quantization, and on the idea that *"anschaulichkeit"* cannot be rescued by these means, he completely misses the point. It was not second quantization in itself but its algebraic formulation which Schrödinger considered as excessively abstract and *"unanschauliche"*. One could reasonably assume (see paragraph 2-2 for more details) that, according to him, it was just as unproblematic to provide second quantization with a continuous formulation as it had been to provide matrix mechanics with a continuous equivalent, namely wave mechanics.

1-4 *Misunderstandings about the concept of particle*

Let us provisionally set aside the historical issues, and focus on the deep philosophical misunderstanding that is revealed by some of the criticisms Born and Heisenberg tried to formulate against Schrödinger's late interpretation of quantum mechanics. As we saw earlier, Born blamed

[1] E. Schrödinger, *Statistical Thermodynamics*, Cambridge University Press, 1944, chapter VII.
[2] E. Schrödinger, letter to P. Jordan, July 28, 1927; quoted by: O. Darrigol, "The origin of quantized matter waves", H.S.P.S., 16, 197-239, 1986
[3] O. Darrigol, "The origin of quantized matter waves", loc. cit.
[4] Letter of P. Jordan to E. Schrödinger, summer 1927, quoted by: O. Darrigol, "The origin of quantized matter waves", loc. cit.
[5] E. Schrödinger, in: *Electrons et photons*, 5th Solvay conference (1927), Paris 1928, p. 208
[6] M. Born, "The interpretation of quantum mechanics", loc. cit.

Schrödinger for refusing to account for the *corpuscular aspect of phenomena*. But Schrödinger had retorted in advance by arguing that there is nothing in experimental physics which *has* to be interpreted in terms of *corpuscular* entities: "it is fair to state that we are not *experimenting* with single particles, any more than we can raise Ichtyosauria in the zoo. We are scrutinizing records of events long after they have happened"[1]. This remark was not completely ignored by Born. However, according to him, to consider it as a decisive rebuttal of the concept of particle would amount to adopting a positivistic trend of thought. It would amount to considering any attempt at going beyond elementary experimental event as a pure "construct" of the mind. The charge of positivism, which had so often been used by Schrödinger and Einstein against the prominent figures of the Göttingen-Copenhagen group was thus ironically sent back by Born. As we shall see later, this awkward reciprocity in philosophical accusation is symptomatic of the fact that almost no creator of quantum mechanics could dispense with positivist-like arguments, at one stage or another of his investigation. Therefore, what makes the difference is not the fact of making some use or *no* use of such arguments, but the way one clings to them or further elaborates upon them; and for the physicists who make further elaborations, it is the strategy by which they surmount the initial *tabula rasa* in order to obtain either an innovative picture of the situation or a complementary set of mutually exclusive classical representations with but a symbolic status. The disagreement between Born and Schrödinger about the concept of particle is precisely of the latter sort. Born accepted that "the particles have no individuality", that "their position can be determined only with a restricted inaccuracy"[2]. But he thought this was not exclusive of a corpuscularian representation with restricted validity. Such a representation was at least helpful in accounting for the language physicists still use, and *must* use according to Born, when they speak of scattering phenomena. Schrödinger, on the other hand, believed that there was no possibility of further elaborating on the ruins left by the concept of particle. According to him, a particle which has no well-defined trajectory and no individuality is no particle at all, for reasons I shall discuss later (in chapter 4 and 5). His own attempt at surmounting the positivist or operationalist *tabula rasa* thus radically rules out the corpuscularian categories.

It is now time to distentangle two criticisms: the first one concerns the relinquishment of the particle concept, and the second one is directed towards the inability of Schrödinger's wave interpretation to deal with the discontinuous aspect of phenomena. These criticisms were neatly distinguished by Heisenberg, but they were handled more or less as a single one by Born. For Born insisted that, in order to account for scintillations on screens or discontinuous recordings of a Geiger counter,

[1]E. Schrödinger, "Are there quantum jumps?", loc. cit.
[2]M. Born, "The interpretation of quantum mechanics", loc. cit.

one has imperatively to use a corpuscularian mode of description. And when he commented about Schrödinger's "omission" to interpret the square modulus of the ψ-function as a *probability,* Born could not help adding that this was the "probability of the appearance *of particles*". He acknowledged Schrödinger's refusal to make use of the particulate concepts and to consider that the source of the discontinuity on a screen is already contained in the impinging beam. However, since Born could not see how it was then possible to figure out the "countable events"[1] which are displayed by the instruments, he considered that elimination of any reference to particles meant the renouncement at tackling the experimental discontinuities. He thus mixed two elements in his reasonings: the bare fact of discontinuity, and the Copenhagen *interpretation* of this discontinuity in terms of a *particulate aspect of phenomena.* Having proved himself unable to avoid such an intermingling, he once more missed his target. Schrödinger's interpretation can indeed be accused of being unable to afford any account of the experimental discontinuities, but the charge must at least be formulated in terms which preserve its self-consistency and *not* according to a ready-made Bohrian complementary mode of using ordinary language. The self-consistent formulation of the charge is well-known nowadays: how can pure unitary quantum mechanics (namely wave mechanics) cope with the *measurement problem?* How do its macroscopic superpositions connect themselves with well-defined experimental events? This is actually a very delicate question, and Schrödinger recognized he had no solution at hand, and no precise idea about how he could find a satisfactory one. But his attitudes, his suggestions, as well as his first sketches, paved the way towards many modern approaches to the problem. The very agnosticism he manifested on that respect proved quite fruitful. The "mystical attitude" with which Born reproached Schrödinger, the hope "that the future will solve this riddle in a satisfactory way"[2], at least prevented any premature burial of the problem. But it did more: it prompted Schrödinger and his successors to formulate the program according to which that any physical situation, macroscopical as well as microscopical, could be tackled *entirely* by means of pure unitary quantum mechanics. That at no stage of the description *must* a discontinuous change be imposed upon ψ-function, provided one incorporates enough degrees of freedom within it (or provided one encompasses a sufficient number of sub-systems within the system under study). That at no intermediate level of the processes, between the initial preparation and the moment the rules of empirical correspondence are worked out, *has* the concept of a discontinuous change any role to play. The connection with facts does *not* have to appear within the field of the formal developments; the theoretician is only asked to show how it can be made *compatible* with the formalism. This newly

[1]ibid.
[2]ibid.

established distance between the description and the experimental facts, which Schrödinger[1] so strongly insisted upon, is the basis of every contemporary no-collapse interpretation of quantum mechanics. But Schrödinger's contribution to a satisfactory analysis of the measurement problem was not only negative. He also developed the quantum theory of measurement pioneered by Von Neumann[2] for its own sake, and he emphasized very convincingly the significance of considering macroscopic superpositions as conjunctions of statements, rather than disjunctions[3] (see chapter 4 for further developments).

This is enough to anticipate the implicit answer Schrödinger gave to one of Born's strongest objections against his interpretation of quantum mechanics, namely its inability to ensure a continuity between the wave-like properties of the atomic domain and the body-like structure of the measuring apparatuses. Whereas Born suggested that this continuity could not be established without extrapolating the body-like mode of description down to the atomic scale, a strategy which, when taken at face value, comes very close to Bohm's hidden variable theory, Schrödinger preferred to extrapolate his wave-like model up to the macroscopic scale. He decided to perform an upward extension of the new theoretical picture, rather than a downward extension of the familiar presuppositions of everyday speech and action. In order to reach this aim, Schrödinger went very far in his ontological inquiry. On the one hand he undertook such an extreme phenomenalist deconstruction of the concept of ordinary thing (or "body")[4] that he considerably weakened the obligation to consider localized and permanent bodies as cornerstones of our ontology, even at the macroscopical level (see chapter 5). And on the other hand he tried to figure out how the *appearance* of ordinary things could be accounted for by the wave mechanical formalism[5], adopting a strategy which has some features in common with the one which is carried out nowadays under the name of "decoherence"[6].

[1]E. Schrödinger, *Science and Humanism*, Cambridge University Press, 1951, p. 39-41

[2]E. Schrödinger, "Probability relations between separated systems", loc. cit.; E. Schrödinger, "The present situation in quantum mechanics", in: J.A. Wheeler & W.H. Zurek (eds.), *Quantum theory and measurement*, op. cit.; E. Schrödinger, *Transformation and interpretation in quantum mechanics*, (Dublin seminar 1952), in: E. Schrödinger, *The interpretation of quantum mechanics (Dublin seminars 1949-1955 and other unpublished texts)*, edited and with introduction by M. Bitbol, Ox Bow Press, 1995. The relevant chapters of J. Von Neumann's book *Mathematical foundations of quantum mechanics* are included in Wheeler's and Zurek's collection.

[3]E. Schrödinger, Dublin seminar, *July 1952 colloquium* (in: E. Schrödinger, *The interpretation of quantum mechanics (Dublin seminars 1949-1955 and other unpublished texts)*, op. cit. p. 19-20), and E. Schrödinger, "Are there quantum jumps?", loc. cit.

[4]See E. Schrödinger, *Nature and the Greeks*, Cambridge University Press, 1954, Chapter VII; E. Schrödinger, *William James lectures (c. 1954)*, 3rd lecture, (in: E. Schrödinger, *The interpretation of quantum mechanics (Dublin seminars 1949-1955 and other unpublished texts)*, op. cit. p.145-149).

[5]E. Schrödinger, *The problem of matter in quantum mechanics*, Notes for seminar 1949, (in: E. Schrödinger, *The interpretation of quantum mechanics (Dublin seminars 1949-1955 and other unpublished texts)*, op. cit. p. 97).

[6]See W.H. Zurek, "Environment-induced superselection rules", Phys. Rev. D26, 1862-1880, 1982; B. d'Espagnat, "Towards an empirical separable reality?" Found. Phys. 20, 1147-1172, 1990. As we shall see later on (paragraphs 4-3, 4-5, 6-8), Schrödinger's final views on the measurement problem were however quite unlike the theories of decoherence.

1-5 *Misunderstandings about the concept of "reality"*

Finally, we have to assess the criticisms Born and many other members of the former Göttingen-Copenhagen directed against the general epistemological attitude of Schrödinger, as well as against his metaphysical outlook. Did Schrödinger adopt an essentially "conservative attitude" towards physical theories? Did he attempt to recover a classical picture of the world? Did he tend to come back to determinism? Did he try to discover something "behind the phenomena" and to identify this something with his ψ-waves? Let us proceed from the last question to the first, for this is the order of increasing intricacy. To begin with, it is not very difficult to rule out a statement according to which Schrödinger held some variety of metaphysical realism[1]. No conception was more alien to his thought than that of a hidden "reality" underlying the phenomena. In his essay "What is real?", which was written in 1960, he rejects from the outset the assertion that, in addition to our experience, "there must also, externally to it, or alongside it, exist an object of which it is the idea and by which it is caused"[2]. The reason for such a position is that the distinction between a Lockean idea and a real object seems to him "a completely superfluous duplication which offends against Occam's razor"[3]. The same remark can be transposed to the duplication between the internal agreement of a linguistic community and the real object which such an agreement is about. Accounting for the mutual understanding between distinct human subjects by invoking "a real world of bodies which are the causes of sense impressions and produce roughly the same impression on everybody" is considered by Schrödinger as no "explanation at all; it is simply to state the matter in different words"[4]. It is only translating the intentional statement "everybody agrees on something" into the metaphysical statement "there exists something which causes everybody's agreement". And moreover, this translation is inappropriate, because it means extrapolating incorrectly the concept of a causal link from the field of phenomena, where it can be established by transcendental deduction, to the limbos which separate the thing-in-itself and the phenomena. Accordingly, his strong commitment *against* Kant's "sublime, but empty, idea of thing-in-itself"[5], very akin to the post-

[1]See, for instance: Y. Ben-Menahem, "Struggling with realism: Schrödinger's case", in: M. Bitbol and O. Darrigol (eds.), *Erwin Schrödinger, Philosophy and the birth of quantum mechanics,* Editions Frontières, 1993. M.F Melgar ("The philosophy of Erwin Schrödinger: a diachronic view of Schrödinger's thought", Found. Phys., 18, 357-371, 1988) has underlined the disarray of most commentators when they are confronted to the question "was Schrödinger a realist or an idealist?". See also chapter II-2 of M. Bitbol, *L'élision,* in: E. Schrödinger, *L'esprit et la matière,* Seuil, 1990.

[2]E. Schrödinger, *My view of the world,* Cambridge University Press, 1964

[3] E. Schrödinger, *My view of the world,* op. cit. p. 64

[4]ibid. p. 68

[5]E. Schrödinger, *Mind and Matter,* in: *What is life? and Mind and matter,* Cambridge University Press, 1967, p. 137.

kantian trend of thought, had been life-long. As soon as 1925[1], and even earlier[2], he elaborated his own blend of idealistic monism, by joining Mach's influence and Schopenhauer's elaboration of vedantic spirituality.

True, his vocabulary, especially in his scientific papers and his correspondence, was pervaded by a realist tinge. For instance he uses repeatedly the word "real" in the original series of papers about quantum mechanics: "we have often previously spoken in such an intuitive concrete way of the 'ψ-vibrations' as something quite real. But there is something tangibly real behind this conception also, namely the very real electrodynamically effective fluctuations of the electric space-density"[3]. See also his insistence on the reality of what quantum mechanical statements refer to, in a letter to Einstein of November 18, 1950: "A probabilistic assertion presupposes the full reality of its subject"[4]. However, before claiming that there is an internal incoherence in Schrödinger's thought, or a gap between his philosophical writings and his scientific papers, one has to examine the definition he gave of the word "reality", especially in the texts wherein he expressed his *realism* in physics. As Schrödinger further explained to Einstein (in the above-mentioned letter): "the metaphysical significance of this reality does not matter to us at all. It comes about for us as, so to speak, the intersection pattern of the determinations of many - indeed of all conceivable - individual observers. It is a condensation of their findings for economy of thought"[5]. The "real world around us", the reality of the physicist, is scarcely considered by Schrödinger as any more than a *construct* out of Machian elements, namely "sense perceptions, memory images, imagination, thought"[6]. A construct which has been hypostasized as an object, after Heraclitus introduced the *principle of objectivation* as a definite feature of the western way of understanding nature[7]. The "real world around us" is *only* a construct, then; but the importance of this construct must not be "questioned in the least" in so far as "without (it) we cannot achieve a single step in practical life"[8]. Indeed, "(I)f we wanted to give up this mode of thought before we have found an equivalent that at least gives the same thing", the findings of the individual subjects "would fall apart without any connections"[9]. The two faces of Schrödinger's attitude towards the concept of "reality" can thus be characterized as follows. Fully recognize that the "real objects which

[1]E. Schrödinger, *The seek of the road,* in: *My view of the world,* op. cit.

[2]M. Bitbol, *L'élision,* in: E. Schrödinger, *L'esprit et la matière,* précédé de *L'élision* par M. Bitbol, Seuil, 1990.

[3]E. Schrödinger, "Quantization and proper values, IV", in: *Collected papers on wave mechanics,* Blackie and son, 1928

[4]E. Schrödinger to A. Einstein, November 18 1950, in: K. Przibram, *Letters on wave mechanics,* (Tr. M.J. Klein), Philosophical Library, 1967 p. 37

[5]ibid. p. 38

[6]E. Schrödinger, *Nature and the Greeks,* op. cit. p. 92

[7]E. Schrödinger, *Nature and the Greeks,* op. cit.

[8]E. Schrödinger, *My view of the world,* op. cit. p. 64

[9]E. Schrödinger to A. Einstein, November 18 1950, in: K. Przibram, *Letters on wave mechanics,* (Tr. M.J. Klein), op. cit. p. 38

surround us" are nothing else than constructs, but take these constructs *very* seriously, since they are a precondition for our life. And conversely, when you have found a clear and adequate theoretical construct, do not diminish its significance by calling it *just a product of our minds*, or *a mere symbolic pattern;* think it is exactly the same type of structure as the one you are accustomed to call "a real object".

Therefore, according to Schrödinger, the disagreement between the Göttingen-Copenhagen group and himself did not bear at all on the metaphysical issue of the "existence" of some external reality, but on the necessity to provide quantum mechanics with at least a *full "equivalent"* [1] of the everyday life *construct* called "reality". His problem was just to find a good candidate for this position. For several reasons I shall expound later on, and in spite of his explicit renouncing the consideration of the ψ-function as a flat description of observed facts[2], the latter theoretical entity was still considered by him as the best possible choice in the 1950's: "as long as the state vector plays the role it does, it must be taken to represent the 'real world in space and time'"[3].

At this point, one can see most clearly that it was just Schrödinger's extreme *anti-(metaphysical) realism* which provided him with such a high capacity to accept awkward theoretical constructs as proper substitutes for the ordinary *concept* of reality. It was his very anti-(metaphysical) realism that led him to adopt norms of clarity and of completeness which are overtly *intellectual* when he had to state the requirements of his strong epistemological (or "methodological"[4]) realism. Would it be absurd to think that, by contrast, some spurious remnants of metaphysical (macro)-realism have been the actual reason for Göttingen-Copenhagen physicist's reluctance towards Schrödinger's light-hearted tendency to endow new theoretical constructs (ψ-functions) with the status of "real entities"? Think for example of Born's assertion that (macroscopic) physical apparatus "consist of bodies, not of waves": *consist,* and not *appear* to consist. Think of Heisenberg's remark that "the ontology of materialism rested upon the illusion that the kind of existence, the direct 'actuality' of the world around us, can be extrapolated into the atomic range. This extrapolation is impossible, however"[5]: the *extrapolation* into the atomic range is impossible, but the "kind of existence" of the objects of daily life remains almost unquestioned. Of course, the members of the Göttingen-Copenhagen group were very far from being just naive realists, even about the macroscopic bodies. However, the type of pragmatic analysis to which they submitted the concept of "real objects" of daily life prevented them from going very far in their ontological inquiry. Heisenberg's

[1]ibid.

[2]E. Schrödinger, *Science and Humanism,* op. cit. p. 40

[3]E. Schrödinger, "Might perhaps energy be merely a statistical concept?", Nuovo cimento, 9, 162-170, 1958

[4]L. Wessels, *Schrödinger's interpretations of wave mechanics,* Ph. D. dissertation, Indiana University, 1975

[5]W. Heisenberg, *Physics and philosophy,* op. cit. p. 145

Bohrian views according to which the usual concepts of "object" and of "actual occurrence of an event" are unsurpassable tools for unambiguous communication[1] did not prompt him to assess the *constitution* of the concept of "real object", nor did it suggest him putting ordinary bodies and scientific constructs on the same footing, as Schrödinger did[2]. And Born's stress on the priority of the observational invariant (or *Gestalt*)[3] over the variable perceptual elements, rather than the other way round, progressively led him to endow the invariant with a sort of unconditioned existence, and to extend the latter towards the micro-world. Surprisingly enough for those who are accustomed to consider Schrödinger as a "realist" and Born (as well as Heisenberg and Bohr) as "positivists"[4], Born's reference to something "behind the phenomena" thus proves much closer to his own position than to Schrödinger's.

1-6 *Misunderstandings about "causality"*

Now what about Schrödinger's alleged dream of coming back to determinism? At the beginning of the twenties, Schrödinger was known as a very strong supporter of F. Exner's ideas according to which macroscopic causal laws may perfectly well arise from statistical regularities of large numbers of stochastic micro-events. In his Zurich conference of 1922 entitled "What is a natural law?", he pointed out that Boltzmann's statistical mechanics had already showed that "chance is the common root of all the rigid conformity to Law that has been observed, at least in the overwhelming majority of natural processes, the regularity and invariability of which have led to the establishment of the postulate of universal causality"[5]. This being granted, there is no decisive argument for thinking that the micro-world is actually ruled by deterministic laws. Since every regularity observable in nature *can* be explained statistically, the assertion that the basic microscopical processes underlying these regularities are ruled by causal law clearly "goes beyond the reach of experience"[6]. The conclusion of these remarks is not that micro-causality is impossible, but that it is "not at all very probable", for it involves a kind of unnecessary duplication of the type of law (causal or statistical). According to the kind of empiricist perspective Schrödinger explicitly adopts, the focus then shifts from the laws to our biased perception of them. The relevant questions to be asked, Schrödinger says, are about our lasting *preference* for causality, rather than on causality itself: "Whence

[1]ibid. p. 144
[2]E. Schrödinger, *William James lectures (c. 1954)*, 3rd lecture (in: E. Schrödinger, *The interpretation of quantum mechanics (Dublin seminars 1949-1955 and other unpublished texts)*, op. cit. p. 149). See chapter 5.
[3]M. Born, *Natural philosophy of cause and chance,* Oxford University Press, 1949, p. 125, 209; M. Born, "Physical reality", Phil. Quart., 3, 139-149, 1953
[4]These were exactly the terms they used in order to qualify each other, during their life-long controversy.
[5] E. Schrödinger, *Science and the Human temperament,* G. Allen & Unwin, 1935, p. 109
[6]ibid. p. 118

arises the widespread belief that the behaviour of molecules is determined by absolute causality; whence the conviction that the contrary is *unthinkable?* Simply from the *custom,* inherited through thousands of years, of thinking *causally,* which makes the idea of undetermined events, of absolute, primary casualness, seem complete nonsense, a *logical* absurdity"[1]. In so far as micro-causality is just a belief, and moreover a belief which is quite unlikely to be true, "the burden of proof falls on those who champion absolute causality, and not those who question it"[2].

In order to account for Schrödinger's later rejection of Born's probabilistic interpretation of quantum mechanics, some prominent historians of science[3] thus attempted to demonstrate that Schrödinger changed his mind completely in 1926, about the issue of determinism. They based their statement on a letter to Wien (August 25, 1926), where Schrödinger appears to acknowledge his own intellectual shift: "today I would no longer care to assume, together with Born, that such an individual event is 'absolutely random' that is, completely undetermined. I no longer think that there is much to be gained by means of this assumption (which four years ago I staunchly started to defend)"[4]. But, quite recently, Y. Ben-Menahem[5] has argued very convincingly against the idea that this letter represents the landmark of a radical and lasting change in Schrödinger's views about determinism. She especially based her statement on the numerous post-1926 texts (going from two conferences of 1929 and 1931 to his last paper of 1958[6]) where Schrödinger stuck to his ideas of 1922. In 1929, for instance, Schrödinger still insisted that: "(...) from the point of view of the physicist, chance lies at the root of causality"[7]. The argument in favour of continuity of thought is all the more compelling since Schrödinger's adherence to Exner's conceptions survived several changes of attitude towards the interpretation of quantum mechanics. Ben-Menahem's case can even be strengthened if one adds to her list a short paper of 1944[8], wherein Schrödinger urged physicists to promote statistical thought from an ancillary status to the role of leader, and a very detailed philosophical investigation of 1948[9] in which he reiterated his agnosticism about the underlying determinism of micro-events.

[1]ibid. p. 115
[2]ibid. p. 118
[3]W.T. Scott, *Erwin Schrödinger,* University of Massachussetts Press, 1967 p. 50; V.V. Raman & P. Forman, "Why was it Schrödinger who developed de Broglie's ideas?", HSPS, 1, 291-314, 1969
[4]In: K. Przibram, *Letters on wave mechanics,* (Tr. M.J. Klein), op. cit.
[5]Y. Ben-Menahem, "Struggling with causality: Schrödinger's case", Stud. Hist. Phil. Sci., 20, 307-334, 1989
[6]E. Schrödinger, "The law of chance" (1929) and "Indeterminism in physics" (1931), in: *Science and the Human temperament,* op. cit.; E. Schrödinger, "Might perhaps energy be a merely statistical concept?", Nuovo cimento, 9, 162-170, 1958
[7]E. Schrödinger, "The law of chance" in: *Science and the Human temperament,* op. cit. p. 36
[8]E. Schrödinger, "The statistical law of nature", Nature, 153, 704, 1944
[9]E. Schrödinger, "Die Besonderheit des Weltbilds der Naturwissenschaft", Acta Physica Austriaca, 1, 201-245, 1948

However, this is not sufficient. Once one has realized there is no actual discontinuity in Schrödinger's very open attitude towards indeterminism, the reasons for his initial rejection of Born's probabilistic interpretation of quantum mechanics still remain to be understood. Ben-Menahem faces the difficulty by having recourse to the very text which was considered by other historians (and which appears at first sight) as the clearest proof of Schrödinger's change of position. The relevant sentence is "I no longer think there is much to be gained by means of this assumption". It is echoed in a paper written some months later: "I am averse to this (Born's probabilistic) conception, not so much on account of its complexity as on account of the fact that a theory which demands our assent to an absolute primary probability as a law of nature should at least repay us by freeing us from the old 'ergodic difficulties' and establishing us to understand the one-way course of natural processes without further supplementary assumptions"[1]. Schrödinger then rejected Born's interpretation not because he disliked indeterminism as such, but because he thought it did not "repay" us with any further understanding of the "one way course of natural processes", and there was accordingly little "to be gained" by this mean.

But why did he think so? Let us suppose, as Born did in his original papers of 1926[2], that the fundamental entities of the theory are *particles* as well as individual processes called *'quantum jumps'*. If it is taken at face value, this view implies that the only significance of the ψ–function is that its square modulus $|\psi|^2$ represents the density of probability for a particle to be[3] within a certain element of volume, and that the values $|c_{nm}|^2$ obtained from the coefficients c_{nm} of the linear superposition $\psi_n=\sum_m c_{nm}\psi_m$ represent the probability of a particle performing a 'quantum jump' from a well-defined energy-level E_n to another level E_m. However, if one strictly adheres to this view, if one considers that the probabilities calculated from the ψ-functions only reflect our (compensable or incompensable) *ignorance* about discontinuous processes[4], no possibility is

[1]E. Schrödinger, "The exchange of energy according to wave mechanics" (1927), in *Collected papers on wave mechanics,* op. cit. p. 146

[2]M. Born, "On the quantum mechanics of collisions" in: J.A. Wheeler and W.H. Zurek, *Quantum theory and measurements,* op. cit.; M. Born, "Das Adiabatenprinzip in der Quantenmechanik", Z. Phys. 40, 167-192, 1926. Mara Beller ("Born's probabilistic interpretation: a case study of 'concepts in flux'", Stud. Hist. Phil. Sci., 21, 563-588, 1990) however pointed that, in 1926, Born did not insist so much on particles as on quantum jumps. His strong corpuscularian commitment was developed much later (see his paper "Physical reality" op. cit., 1953).

[3]In his original paper on collisions, Born refers to the probability for an electron *"to be thrown out"* in a certain direction. Later on (in his *Atomic physics,* Blackie & son, 1944), he maintains a certain ambiguity about the epistemological status of quantum probabilities. Are they probabilities of a particle to *be* or to be *found* in a given element of volume? At page 139 of his book, we read that " $|\psi_E|^2$dv is the probability that the electron (regarded as a corpuscle) *is* in the volume element dv"; and at page 140, that " $|\psi_n|^2$dv is the probability that the electron will be *found* in the volume element dv". See discussion in §2-4

[4]M. Jammer, *The philosophy of quantum mechanics,* op. cit. p. 43

left to account for the diffraction and interference patterns which are so widely observed in Davisson and Germer's-like experiments. True, one can modify slightly Born's initial interpretation (as Born himself did) by insisting that $|\psi|^2$ is not the only computational tool to be used, and that the *"probability amplitude"* ψ retains an importance through the cross-product terms which appear when one takes the square of its modulus. But the question now becomes: can one still speak of a *probabilistic theory* in the usual sense of the expression?

Schrödinger did not think so. As he wrote to Planck in a letter of July 1927, "What seems most questionable to me in Born's probability interpretation is that (...) the probabilities of events that a naive interpretation would consider to be independent do not simply multiply when combined, but instead the 'probability amplitudes interfere' in a completely mysterious way (namely, just like my wave amplitudes, of course)"[1]. In order to contend that quantum mechanics is a probabilistic theory of individual events occurring in nature, one would have first to show that quantities that have the mathematical properties of ordinary probabilities can be associated to *any* kind of event. This is not nevertheless true, as long as these probabilities are taken to refer to physical events taken *in abstracto,* before an effective experimental arrangement is even mentioned; for in such case one has to take interference of ψ-amplitudes into account. This is only true in so far as one restricts attention to the measurement results themselves, namely to *experimental events* rather than to putative events occurring before an experiment or between two experiments; for in these circumstances additivity of $|\psi|^2$ values holds. In other terms (borrowed from Bohr and Reichenbach[2]), the rules of the standard theory of probabilities apply to *phenomena,* not to *inter-phenomena.*

One may of course single out the concept of "actual experimental result" (of a position measurement, for instance) and say that ψ has no other significance than providing the density of probability $|\psi|^2$ that such *results* are effectively obtained; but then the very assertion that quantum mechanics is a probabilistic theory of individual bodies called "particles" and of the discontinuous individual changes they undergo becomes void, for it goes beyond the strict level of the experimental results to which the quantum description is admittedly confined. This argument lies at the core of Schrödinger's later interpretation of quantum mechanics[3], but it was already implicit in his initial rejection of Born's views.

Let us summarize: even though Schrödinger did not eliminate from the outset the possibility that the individual processes are basically

[1]E. Schrödinger, Letter to Planck, July 4, 1927, in: K. Przibram, *Letters on wave mechanics,* op. cit. p. 20

[2]H. Reichenbach, *Philosophic foundations of quantum mechanics,* California University Press, 1946

[3]See for instance the appendix to the 1952 edition of E. Schrödinger's "Statistical thermodynamics", op. cit.

indeterministic, he thought that the development of quantum mechanics did not favour this view. Not only for the general reason that "however considerable finally the success achieved by the employment of an indeterministic picture may be, it is unlikely that we shall ever be able to demonstrate the impossibility of finding any deterministic model of nature capable of doing justice to the facts"[1], which is but the mirror image of Schrödinger's previous remark that it is unlikely that we shall ever be able to demonstrate the impossibility that the ultimate microscopic laws of nature are *indeterministic*. But also because quantum mechanics can be called an "indeterministic theory" *only* in the restricted epistemological sense of lack of predictibility of experimental results[2] given a maximal set of information about the experimental *preparation*. Quantum mechanics is only indeterministic in so far as the link between two successive experiments is concerned, but it gives no univocal indication about what happens between these two experiments.

Now, can we supplement quantum mechanics by attempting to describe what Reichenbach calls the "interphenomena"? Even if we try to do this, we are bound to recognize that indeterminism provides no clue by itself. No attempt at filling the gaps between two subsequent experiments with a local model of particles undergoing stochastic 'quantum jumps' or travelling along single well-defined trajectories (be they randomly determined by the initial conditions) can lead to a successful calculation of the probability of the final observation. No attempt at picturing the intermediate processes as indeterministic and discontinuous rather than deterministic and continuous has proved able *as such* to account for certain (wave-like) aspects of the phenomena. In other words, it is by no means *sufficient* to acknowledge the stochastic character of the micro-events in order to recover all the effects predicted by quantum mechanics.

The latter remarks can also be expressed more explicitly, in terms of "hidden variables". It is widely acknowledged that local *stochastic* hidden variable schemes are just as able to yield Bell-type inequalities as local *deterministic* hidden variable models are. Both types of models are incompatible with quantum predictions. It is therefore by no means *sufficient* to introduce *stochasticity* alone in hidden variable models for accounting for all the quantum predictions. One has to accept in addition a version "non-locality"[3] which is most conveniently described in terms of some wave formalism.

On the other hand, as Schrödinger progressively succeeded to demonstrate in example after example, it is not even *necessary* to retain the idea that the intermediate events, if any, are stochastic. Taking the quantum formalism at face value and figuring out the processes which

[1]E. Schrödinger, "Indeterminism in physics", in: *Science and the human temperament*, op. cit. p. 45.
[2]ibid. p. 43
[3]One must evoke at this point Bohm's theory. It accounts perfectly for all quantum predictions, but it does so at the cost of accepting a contextuality of trajectories which make them inaccessible *in principle* to any experimental assessment. Schrödinger's insistence on operational criteria for the continuity of trajectories is definitely incompatible with such an extreme option.

take place between two successive observations as a *continuous* wave which propagates in a 3n-dimensional space according to a *deterministic* law of evolution, always provides one with the proper probabilistic link between the initial and the final experiments (or between a preparation and an experiment).

The indeterminism of the micro-processes being thus neither *sufficient* nor *necessary* to account for the distinctively non-classical phenomena which are manifest at the atomic scale, it can be dispensed with. It can be dispensed with because there is little (not to say nothing) "to be gained by means of this assumption". This was, in substance, the major reason Schrödinger had to be so reluctant towards the probabilistic interpretation of quantum mechanics. Such a reason did not prevent him from accepting eventually that the ψ-function can be used as a tool to calculate the probability of *experimental events,* but it definitively disqualified the idea according to which the ψ-function is only a statistical description of some underlying random processes which concern corpuscular entities. So much for Schrödinger's "determinism".

1-7 *Schrödinger's "over-revolutionary" attitude*

Last but not least, we must address the accusation of "conservatism" which was so often issued against Schrödinger. Did he value so much the classical modes of thinking that he was unable adopt the new ideas of the Göttingen-Copenhagen group? Some of his statements at the very beginnings of wave mechanics, as well as his age (he was 39 years old in 1926), may support the portrait of an old-fashioned physicist struggling to restore a classical field-like theory against Heisenberg's revolutionary algebra of observables. Wasn't it Schrödinger who wrote in his 1926 paper "Quantization as a problem of proper values II": "(...)we cannot really alter our manner of thinking in space and time, and what we cannot comprehend within it we cannot understand at all"[1]? Wasn't it he who expressed his preference for the wave-mechanical ideas over the matrix-mechanical scheme in these terms: "To me it seems extraordinarily difficult to tackle problems of the above kind, as long as we feel obliged on epistemological grounds to repress intuition in atomic dynamics, and to operate only with such abstract ideas as transition probabilities, energy levels, etc."[2]? Wasn't it Schrödinger again who was praised by the entire establishment of classical physics for his ability to bring back atomic processes within the boundary of formerly accepted theoretical models[3]? All this is true, of course, but does not demonstrate in the least that

[1]E. Schrödinger, "Quantization as a problem of proper values II", in: *Collected papers on quantum mechanics,* op. cit. p. 27

[2]E. Schrödinger, "On the relation between the quantum mechanics of Heisenberg, Born and Jordan, and that of Schrödinger", in: *Collected papers on quantum mechanics,* op. cit., p. 59

[3]See the letter of July 28, 1926, where W. Heisenberg reported to W. Pauli the outcome of Schrödinger's conference in Munich. In: N. Bohr, *Collected works,* E. Rüdinger (gen. ed.), vol. 6, J. Kalckar (ed.), North-Holland, 1985, p. 10.

Schrödinger was personally attached to classical concepts, let alone to classical theories *as such.* For reasons that we shall discuss at length in the next chapter (section 2-5), he was a strong supporter of some methodological requirements of clarity and spatio-temporal representation. These requirements happen to be fulfilled by most parts of classical physics, but they exceed by far the strict domain of classical theories and even the general principles on which they are grounded. They allow an impressive range of variations which may go well beyond the traditionally accepted types of theories. Therefore we should not be surprised if, except for this nucleus of epistemological ideals, Schrödinger proved to be extremely open-minded, and even sometimes revolutionary, as to the conceptual content of the newly formulated quantum theory. In 1924, for instance, he immediately supported the ideas expressed in the Bohr-Kramer-Slater paper, finding no obstacle in its indeterministic features, and not even in its renouncement of the principles of conservation of energy and momentum[1]. Later on, from 1926 to 1935, he often stressed some features of quantum mechanics which, in spite of their being scarcely recognized as such due to the insistence of the Göttingen-Copenhagen group on epistemological changes, appeared radically new to him. Let us give some prominent examples of these new characteristics.

One of them was the *superposition principle,* which Schrödinger first expressed in terms of multiple excitation of the proper vibrating modes of the atoms[2], but that he soon came to single out and to recognize as one of the most puzzling distinctive features of quantum mechanics[3]. Some of its consequences carried quantum mechanics very far away from *both* classical particle mechanics *and* classical vibration theories:

(i) The principle of superposition does not agree with classical particle mechanics, for (in terms which are strongly reminiscent of Feynman's path integral concept): "the true laws of quantum mechanics do not consist of definite rules for the *single path,* but that in these laws the elements of the whole manifold of paths of a system are bound together by equations, so that apparently a certain reciprocal action exist between the different paths"[4].

(ii) The particular form taken by the principle of superposition in relation to the Schrödinger equation does not agree with classical vibration theories *either,* for, in quantum mechanics, the proper frequencies become additive when two or more systems are combined into one composite system[5]. According to Schrödinger, then, quantum

[1]M. Jammer, *The conceptual development of quantum mechanics,* Mc Graw Hill, 1966, p. 184

[2]E. Schrödinger, "Quantization and proper values-I", in: *Collected papers on quantum mechanics,* op. cit., p. 11

[3]E. Schrödinger, "The fundamental ideal of wave mechanics" (Nobel lecture, 1933), in: *Science and the Human temperament,* op. cit. p. 152

[4]E. Schrödinger, "Quantization and proper values-II", in: *Collected papers on quantum mechanics,* op. cit., p. 26

[5] E. Schrödinger, "The exchange of energy according to wave mechanics", in: *Collected papers on quantum mechanics,* op. cit., p.140; the fact that this additivity of frequencies does not fit with classical vibration theories was stressed much later in: E. Schrödinger, "Are there quantum jumps?", loc. cit. p.

mechanics does not only depart from classical particles theories, but *also from classical wave theories.*

Another very important new characteristic of the theory is highlighted by the previous reference to composite systems: quantum mechanics has a *holistic* descriptive structure. Schrödinger had already caught a glimpse of it in his 1926 paper about Einstein's theory of gas[1]; he then began to provide it with a precise mathematical form (namely the appearance of linear superpositions of products of eigenfunctions) in 1927[2]; and finally, in 1935, he stated: "I would not call that *one* but rather *the* characteristic trait of quantum mechanics"[3] (see sections 2-2 and 2-3 for more details).

With these examples in mind, Schrödinger claimed more strongly than any other physicist that the gap between quantum mechanics and classical theories hindered any partial rescue of the classical concepts as well as any attempt at preserving the remnants of the *natural ontological attitude* [4] towards material bodies which are incorporated in classical thought. The situation appeared to him "exciting, novel, revolutionary"[5] in a way that few of his colleagues were able to appreciate. As soon as 1928, he explained to Bohr that one of the major reasons of his persistent skepticism towards the current interpretation of quantum mechanics was that, even though the classical corpuscular concepts were altered beyond any possibility of recognition, they were not completely abandoned: "If you want to describe a system, e.g. a mass point, by giving its p and q, then [according to the uncertainty relations] you find that the description is possible only with a limited degree of exactness. This is very interesting, for it seems to me a limitation on the applicability of the *old* phenomenal concepts. But it seems to me imperative to require the introduction of *new* concepts, in which these restrictions *no longer* exist"[6]. This argument was reproduced at the end of the series of lectures he gave in London in 1928: "It seems to me that this statement [i.e. that the result of measuring a classical variable will be one of the proper value of the corresponding observable] contains a rather vague conception, namely that of measuring a quantity (...) which relates to the classical picture of the atom, i.e. to an obviously wrong one. Is it not rather bold to interpret measurements according to a picture which we know to be wrong? May they not have quite another meaning according to the picture

116, and also in: E. Schrödinger, Dublin seminar, *July 1952 colloquium* (in: E. Schrödinger, *The interpretation of quantum mechanics (Dublin seminars 1949-1955 and other unpublished texts)*, op. cit. p. 25)

[1]E. Schrödinger, "Zur Einsteinschen Gastheorie", Phys. Z., 27, 95-101, 1926

[2]E. Schrödinger, "The exchange of energy according to wave mechanics", in: *Collected papers on quantum mechanics*, op. cit., p. 141

[3]E. Schrödinger, "Discussion of probability relations between separated systems", Proc. Camb. Roy. Soc., 31, 555-563, 1935

[4]The expression is borrowed from A. Fine (*The shaky game*, University of Chicago Press, 1986)

[5]E. Schrödinger, *Science and humanism*, op. cit. p. 11

[6]E. Schrödinger, letter to N. Bohr, May 5, 1928, (Tr. L. Wessels, *Schrödinger's interpretations of wave mechanics*, op. cit. p. 340), in: N. Bohr, *Collected works*, vol. 6, op. cit., p. 463; Einstein warmly approved Schrödinger's call to relinquish the classical concepts of position and momentum in a letter of May 31, 1928; N. Bohr, *Collected works*, vol. 6, op. cit., p. 51.

which will finally be forced upon our minds?"[1]. The same trend of thought was then developed at lengh in a paper of 1934[2] and in the second and in the third paragraph of the "cat-paper". There, Schrödinger noticed how strange was an interpretation of quantum mechanics wherein the old classical concepts were still used as a guideline for defining observables, in spite of the fact that the *relations* classical physics establish between these observables are admittedly inadequate[3]. He then concluded: "Does one not have the impression that at issue here are fundamentally new characteristics, which have only their name in common with the classical properties?"[4].

The names of these characteristics are retained, due to their relation to certain experimental procedures which must eventually be described in classical terms, as Bohr emphasized repeatedly and as Schrödinger accepted with qualifications (see §4-4, point (8), and §6-3). But in order for them to be the very *same* concepts as in classical mechanics, they would also have to be nodes into the same network of lawlike relations as in classical physics. And this is obviously not true.

The previous texts, as well as many others, strongly support the non-conventional position of the few scholars who have carefully studied Schrödinger's views on quantum mechanics. These historians of physics rightly claim that, far from being conservative, Schrödinger adopted an over-revolutionary attitude towards the conceptual content of quantum mechanics. As Y. Ben-Menahem puts it, "the Copenhagen interpretation appeared to him as a rather contrived effort to retain the classical foundations while confining the revolutionary element of the theory to its probabilistic interpretation"[5]. Or, according to L. Wessels' nice metaphor[6], Schrödinger proved the only physicist who, confronted with the clear impossibility of pouring the new wine of atomic phenomena into the old bottles of classical physics, suggested *blowing entirely new bottles*. Schrödinger's extremely innovative line of thought was not very easy to hold because, as we already noticed, the experimental devices and results have still, at some stage, to be described in agreement with the natural ontological attitude, or even in agreement with those classical concepts which incorporate the relevant elements of the natural ontological attitude. But it had both the merit of coherence and the aptitude to circumscribe most precisely the difficulties *within the framework of the*

[1]E. Schrödinger, *Four lectures on wave mechanics*, Blackie & son, 1928, p. 52

[2]E. Schrödinger, "Über die Unanwendbarkeit der geometrie im Kleinen", Naturwissenschaften, 22, 518-520, 1934:
"Current quantum mechanics erroneously keeps the concepts of the classical mechanics of mass points, e.g. energy, momentum, position (...) but the price it has to pay for this is that a system in an exctly defined state cannot be ascribed definite values of those quantities. This shows that the concepts are insufficient. The concepts have to be given up, not their exact definitions."

[3]E. Schrödinger, "The present situation in quantum mechanics", in: J.A. Wheeler & W.H. Zurek (eds.), *Quantum theory and measurement*, op. cit. §2

[4]ibid. § 3

[5]Y. Ben-Menahem, "Struggling with causality: Schrödinger's case", loc. cit.

[6]L. Wessels, *Schrödinger's interpretations of wave mechanics*, op. cit. p. 113

new theory, by identifying them with the well-known *measurement problem.*

1-8 *Modernity and post-modernity*

Our answer to the first-order question: "did Schrödinger adopt a conservative attitude in the interpretation of quantum mechanics?" is thus a definite *"no"*. But we are left with a second-order question: "why were the Göttingen-Copenhagen physicists so eager to claim Schrödinger's reluctance towards conceptual changes?". The reasons for this claim can now be more easily understood. Schrödinger's conception of the radical changes imposed by quantum mechanics did not fit at all with the current views. He therefore began many of his papers by a strong criticism of the way his colleagues conceived these changes:

(i) Schrödinger rejected Bohr's complementarity from the outset[1] and never abandoned his extreme skepticism about this concept[2], even though (see paragraph 6-2) he also felt the necessity to find a proper equivalent to it in his own framework of thought.

(ii) After the very short period (at the beginning of the thirties) when he accepted teaching the dominant conceptions, he criticized more and more vehemently Heisenberg's idea that quantum mechanics has in some way forced us to break down the "barrier between subject and object"[3]. In Schrödinger's metaphysics, subject and object are not two pre-given entities facing each other, which had to wait for quantum mechanics in order to merge. They were One *from the outset,* and every branch of science, including quantum mechanics, is based on a procedure called *objectivation* which consists in withdrawing every subjective element (*qualia* and values) from the primeval Unity. It is only after this withdrawal that one can isolate a common structure which deserves the name of "object".

Moreover, talking of an interaction between subject and object when interaction between apparatus and object is at stake, seemed to him utterly inappropriate. Something like Gilbert Ryle's *category mistake.* "For the observing mind is not a physical system, it cannot interact with any physical system. And it might be better to reserve the term 'subject' for the observing mind"[4]. In modern terms[5], one would say that Schrödinger felt very strongly the need of a clear distinction between the "Cartesian cut" (between *res cogitans* and *res extensa)* and the "Heisenbergian cut" (between apparatus and object).

[1]E. Schrödinger, Letter to N. Bohr, October 23, 1926, N. Bohr, *Collected works,* vol. 6, op. cit., p. 459

[2]see for instance E. Schrödinger, *Science and humanism,* op. cit. p. 66. For more details, see § 6-1

[3]E. Schrödinger, *Science and humanism,* op. cit. p. 47; E. Schrödinger, "Der Geist der Naturwissenschaft", Eranos Jahrbuch, 14, 491-520, 1946

[4]E. Schrödinger, *Science and humanism,* op. cit. p. 53

[5]H. Atmanspacher "Is the ontic/epistemic distinction sufficient to describe quantum systems exhaustively?", in: K.V. Laurikainen, C. Montonen, & K. Sunnaborg, *Symposium on the foundations of modern physics 94,* Editions Frontières 1994

(iii) He did not accept at all Jordan's contention that free-will was rescued by the inderministic features of quantum mechanics, not because he did not like indeterminism as such, but because, in good agreement with Cassirer's Kantian analysis[1], he thought that the determination of natural processes has absolutely nothing to do with the practical concept of free will. He insisted particularly on what he called the "moral objection" to Jordan's thesis: "(...) this haphazard side of the goings-on in the material world is certainly (says Cassirer) the very last to be invoked as the physical correlate of man's ethical behaviour"[2].

(iv) Finally, he did not think that "the impossibility of a continuous, gapless, uninterrupted description in space and time (is) really founded in incontrovertible facts"[3]. The relations between the said continuous description and the facts had only to be redefined, as we have already suggested, and as we shall explain at length in chapter 4 of this essay.

Being so alien to the Göttingen-Copenhagen conception of the conceptual changes brought about by quantum mechanics, it is not surprising that Schrödinger was considered by the physicists of that group as hostile to the very idea of change.

But there is another way to apprehend the reasons for the deep and persistent misunderstanding between Schrödinger and the Göttingen-Copenhagen physicists about the issue of "conservatism". A way which goes well beyond the quarrels on particular points, and parallels the major trends of the *Weltanschauung* of our century. My thesis is that the creators of quantum mechanics can be divided not in *two* groups (the classicists and the modernists), but in *three* groups (the classicists, the modernists and the post-modernists). Whereas the usual classification puts de Broglie, Einstein and Schrödinger on the classical side, and the members of the Göttingen-Copenhagen group on the modern side, my classification is both more refined and more time-sensitive: de Broglie, Einstein (mostly) and Schrödinger (in 1926 *only)* played the role of the classicists; Heisenberg, Pauli, Dirac, Born, Jordan and Bohr (mostly) were the modernists; finally Schrödinger *after 1926* (as well as Einstein and Bohr for some scattered aspects of their later thought) adopted a post-modernist attitude. I shall advocate the idea that the background reason why the Göttingen physicists accused Schrödinger of conservatism lies in the obvious difficulty for any modernist thinker to acknowledge a distinction between true old-fashioned classicists and somehow over-revolutionary post-modernists. Now, I have of course to justify this reordering and to explain the use of a vocabulary which is borrowed to the historians of contemporary art. The easiest strategy consists in

[1]E. Cassirer, *Determinismus und Inderterminismus in der modernen Physik,* Götheborgs Högskolas Arsskrift 42, 1937; *Determinism and indeterminism in modern physics,* Yale University Press, 1956
[2]E. Schrödinger, *Science and humanism,* op. cit. p. 62
[3]ibid. p. 49

focusing our attention on painting, for, after all, the problem of *representation* is at stake.

Classical (or "Raphaëlite") painting has two distinctive features: the figurative ideal, and the use of linear perspective in order to organize the figuration of three-dimentional sceneries on two-dimensional surfaces. According to one of the first theoreticians of perspective, Leon-Battista Alberti, the aim was to transform the painted surface into "(...)an open window through which one can see"[1]. Nothing was to be seen *on the surface,* but the surface had rather to be made transparent in order to give access to the represented part of the (mythological or real) world. By contrast, modern painting, from impressionism to abstract art, tended more and more insistently to revert the attention from what is represented to the painting itself, to its internal structure, to its patches of colour. One may even detect symptoms of over-reaction in modern pictural art: not only did the artists renounce representing an object, but they often claimed that their work tended to express their subjectivity, that its meaning was conditioned by an external written commentary, or that the painting had no other significance than being the trace of the performance of the painter (J. Pollock's action painting). They condensed and expressed in their way some of the most important shifts which affected the western outlook during the past three centuries. According to the usual classifications of philosophical doctrines, we could say that they first documented a transition from unproblematic realism to idealism (be it transcendental), and then from a philosophy of consciousness to a philosophy of language or to pragmatism. They first performed something like the cartesian discovery of the subject, and then took either the linguistic turn or the pragmatic turn. But the theories of painting did not stop at this point. Almost at the same time as the climax of deconstructive undertaking associated with abstract painting, an opposite trend began to develop; a trend which resulted in a bundle of (sometimes very strange) *reconstructive* attempts. The surrealists resumed representing fragments of worlds on their canvas, but the disposition of the fragments was intended to prevent any assimilation of the whole work to a mere imitation of the coherent universe of common experience. Magritte's most characteristic paintings were precisely aimed at showing that no representation could ever naively revert to the previous ideal of mimicry. His demonstration sometimes used pure pictorial techniques (painted pictures within the frame of the actual painting), or an association of pictorial and verbal elements (for instance his very famous "this is not a pipe" written beneath a coloured shape that anyone would identify with a pipe). Later on, some painters could even revert to a *quasi*-classical style, in so far as the confusion with genuine "Raphaëlite" painting was not to be feared any longer. This is how the so-called "post-

[1]L.B. Alberti, *De Pictura* (1435), Latin text and French translation by J.L. Schefer, Macula Dédale, 1992, p. 115

modern" trend arose. Here are then some of the characteristic features of post-modern painting:

(i) It makes use of the methods and conventions of classical art, except for a host of little details which reveal that it has fully integrated the teachings of the modern *deconstructive* process.

(ii) Its theoretical framework is akin to that of modern (impressionist, cubist or abstract) art, for here also the representational content of the work is considered as less important than its *presentational* aesthetic qualities and also sometimes less important than its expressive value.

(iii) Its effects on the inattentive spectator are quite close to that of classical painting, for the representational *rules and functions* are retained. But on the other hand, the link between the representation and the represented content, as well as between the represented content and what appears in the world, is broken, or just not taken seriously (see Magritte's "this is not a pipe").

(iv) In short, the post-modern *reconstructive* endeavour is a perfectly self-aware process. It makes extensive use of the material of colours and shapes left by the *tabula rasa* of modern painting, as well as of its theoretical understanding of the polyvalent status of art. It can by no means be mistaken for the (semi-unconsciously) *constructive* procedures of classical figurative art.

This definition of post-modernism and of its relations with both modern art and Raphaëlite painting can be used, almost without modification, to characterize Schrödinger's interpretation of quantum mechanics and its relations to both Göttingen-Copenhagen views and classical physics. The *modern* trend in physics arose from a progressive loss of confidence in the spatio-temporal pictures, which appeared as the main outcome of the crisis of Bohr's early theory of quanta from 1920 to 1925[1]. It culminated in 1925 with matrix mechanics, which relied upon Heisenberg's "reduction to the observables", and which consisted in a generalized use of the correspondence principle. Its mottos were: do not try to provide physical theories with any representational content; concentrate on the operational referents of the symbols; and try to bring out what the instrumental procedures reveal about (or express of) the scale and the linguistic-pragmatic features of the community of observers. In a word, physical theories were not supposed to retain their *figurative* role any longer, but to focus on their *expressive* content. In comparison to such an extremely coherent deconstructive line of thought, Bohr's later reconstructive attempts appear both fragmentary and utterly incomplete. The old corpuscular and undulatory *pictures* were both retained by Bohr from the second half of 1926 on, but only as mutually exclusive symbolic tools within the generalized scheme of complementarity.

As for Schrödinger, his reconstructive attitude manifested itself very early, in 1924 when he approved the Bohr-Kramers-Slater paper, and in

[1]See for instance W. Heisenberg, *Physics and Beyond, encounters and conversations,* George Allen & Unwin, 1971 (chapters 4, 5)

1925 when he began to recast de Broglie's concept of matter waves for the sake of the statistical theory of gases. At this point, however, very few details could distinguish Schrödinger's undertaking from a mere rehearsal of the classical ideal. True, one could guess that his early taste for representation was *not* underpinned by a naive realist doctrine which would amount to assimilate the picture to what is "really happening out there"; such a simplistic view is ruled out by those chapters of the essay "The seek of the road" (written in 1925)[1] where Schrödinger acknowledged his adherence to Mach's doctrine of elements-sensations and to a blend of idealistic monism, thus showing that he had no metaphysical reason to reject the *tabula rasa* of the modern physicists, and no *a priori* reluctance to use it as a methodological premise. But even though he was by no means committed to a primitive conception of physical theories as "reflection of nature", he left a very important series of epistemological questions unadressed, at least until the last months of 1926. And by leaving them unadressed he favoured the view that he was not fully aware of the implications of his reconstructive attempt, thus coming very close to a "classicist".

Here is a short list of these questions. Will it prove acceptable, after the "reconstruction" has been completed, to revert to the forgetful attitude which was so often adopted by the classical physicists? Will it be possible to hide the circumstance that these pictures are *primarily intellectual constructs?* Will it appear as unproblematic as during the past centuries to speak *as if* the physical theories were faithful (and even "true") models of some *reality-in-itself?*

It is clear, from the series of bare claims he made about the "reality" of ψ-waves (or of the density of electric charge $-e\psi\psi^*$), that Schrödinger would have bet it was the case in the beginning of 1926. He believed that, exactly as in the time of classical mechanics, physical theories could perfectly well be worked out *as if* they were just mimicking natural processes, irrespective of the metaphysical foundations one had previously ascribed to them, and also irrespective of Kant's critical assessment. However, from October 1926[2] on, he became gradually aware that things were not so simple, namely that the preliminary intellectual and operational scaffoldings of physical theories would not be as easy to sweep under the carpet as they had been previously. His reconstructive attempts then became increasingly self-conscious. They acquired more and more features in common with post-modernism and went farther and farther from the old classical ideals. By 1929, he had acknowledged explicitly an irreducible distance between representation and appearances which is typical of the post-modern trend of thought: "(...) very likely, the wave-corpuscle contradiction is the manifestation of an important new

[1]in: E. Schrödinger, *My view of the world,* op. cit.

[2]After his discussions with Bohr in Copenhagen in September 1926; Letter of E. Schrödinger to N. Bohr, October 23, In: N. Bohr, *Collected works,* E. Rüdinger (gen. ed.), vol. 6, J. Kalckar (ed.), North-Holland, 1985

fundamental principle: the non-identity of what is detailed in space-time, on the one hand, and of what is observable on the other"[1]. Bohr's strategy, which involved *couples* of complementary symbolic (wave-like and corpuscle-like) pictures, was replaced by a clear and "new" distinction between the picture and the events, between the (wave-like) content of an *unique* continuous representation and the discontinuous observable events. The representative *function* was fully retained, as against the modern trend of the Göttingen-Copenhagen physicists, but the link between the represented *content* and the experimental phenomena was seriously altered, far from the classical ideal and in good agreement with the post-modern prescriptions. The concept of *interpretation,* in the minimal sense of *set of correspondence rules between the formalism and the experiments,* accordingly became a central concern of Schrödinger's later work on quantum mechanics[2]. In so far as no isomorphism between the theoretical constructs and their experimental counterparts could be invoked any longer, the correspondence rules were to be made explicit. In 1928-1929, Schrödinger had thus adopted one of the most important features of the post-modern turn, namely the dissociation between the representative function, the represented content and the appearances. He would retain it until the end of his life. In 1950, he gave this dissociation its most lucid exposition: "we do give a complete description, continuous in space and time without leaving any gaps, conforming to the classical ideal - a description of something. But we do not claim that this 'something' is the observed or observable facts; and still less do we claim that we thus describe what nature (matter, radiation, etc.) really is. In fact, we use this picture (the so-called wave picture) in full knowledge that it is neither"[3].

Unfortunately, this major change in his attitude was generally not recognized by his colleagues. In 1935, Einstein still criticized his initial "realist" interpretation of the wave function, as if Schrödinger were still supporting it. Schrödinger had to explain to him at length that he did not cling to his old views any longer, namely that he had already abandoned several years earlier the idea according to which *the ψ-function is a direct representation of reality* [4]. He even expounded his cat-paradox as a device to show that one cannot "(...) naively accept as valid a 'blurred model' [namely the ψ-function itself, taken at face value] for representing reality"[5]. Still, this did not prevent many other misunderstandings. Rather, this promoted misunderstandings of an opposite kind. One or two

[1]E. Schrödinger, "Neue Wege in der Physik", *elektrotechnische Zeitschrift*, 50, 15-16, 1929

[2]E. Schrödinger, *Transformation and interpretation in quantum mechanics,* (in: E. Schrödinger, *The interpretation of quantum mechanics (Dublin seminars 1949-1955 and other unpublished texts),* op. cit. p. 50)

[3]E. Schrödinger, *Science and Humanism,* op. cit. p. 40

[4]E. Schrödinger, letter of august 19th 1935 to A. Einstein, quoted and translated by A. Fine, *The shaky game,* op. cit.

[5]E. Schrödinger, "The present sitation in quantum mechanics", in: J.A. Wheeler & W.H. Zurek (eds.), *Quantum theory and measurement,* op. cit., §5

years later, commenting on Schrödinger's "cat-paper", several authors[1] thought that Schrödinger had simply joined the *modernist* program, since he appeared to claim that the ψ-function was to be considered as a mere catalog of information, and not as "representing reality". But here again, things were much more intricate. Schrödinger did not just assimilate ψ-functions to catalogs of information; he rather tried to use this conception as a minimal interpretative skeleton, and then to work it out up to the point where it would clearly display its deficiencies. Later on, in 1952[2], in texts which are currently supposed to reflect his move back to realism of the ψ-functions after the skeptical period of the 1930's, he would resume his analysis of *ψ-functions as catalogs of information* exactly in the same spirit and often with the same wordings as in 1935.

This being recognized, the difficulties so many physicists and philosophers experienced in grasping Schrödinger's position are not entirely negative. We can learn something from them. Their very mutual discrepancies provide us with further indirect evidence of Schrödinger's post-modern attitude after 1928. These authors were confronted with a thinker who, on the one hand, formally went on making an extensive use of the methods of classical physics and who, on the other hand, demonstrated his perfect assimilation of Heisenberg's original *tabula rasa,* using it systematically as a preliminary step of his own further elaborations. When he insisted on the first aspect, he was considered as "classical" (by the modernists), whereas when he developed the second aspect, he was regarded as a "modernist" (by the exponents of classicism). His re-constructive undertaking was identified by some of his colleagues with the constructive procedure used unproblematically by classical physicists, whereas his insistence on the phenomenal material of the re-construction was considered by others as a sign of allegiance to the Copenhagen interpretation. No one could figure out Schrödinger's ideas in their full extent. No one could make sense of them all without invoking a series of complete about-faces in his attitude; something like: "Schrödinger was first an exponent of classicism (in 1926), then of modernism (from 1928 to1935), then of classicism once more (especially in the 1950's)".

1-9 *The continuity of Schrödinger's attitude towards quantum mechanics (an outline)*

As we indicated at the beginning of this chapter (section 1-1), this chronological division of Schrödinger's interpretation(s) of quantum mechanics has become common wisdom among most physicists and

[1]V. Lentzen, "The interaction between subject and object in observation", Erkenntnis, 6, 326-333, 1936; H. Margenau, "Critical points in modern physical theory", Philos. sci., 4, 337-370, 1937
[2]E. Schrödinger, *Transformation and interpretation in quantum mechanics,* (in: E. Schrödinger, *The interpretation of quantum mechanics (Dublin seminars 1949-1955 and other unpublished texts),* op. cit., p. 78)

historians of physics. But our previous analysis suggest a very different view. The idea of a series of about-faces in Schrödinger's attitude after all arose from a complete inability to understand the bonds which unite the two sides of his thought: the monistic idealism in metaphysics and the methodological realism in epistemology; the deconstructive phenomenalist analysis and the reconstructive undertaking; the extreme audaciousness about conceptual content and the demand of permanent intellectual standards. If these various aspects of Schrödinger's philosophy of physics were properly articulated, the apparent necessity of ascribing each one to a particular period in order to avoid contradictions would disappear. Nothing could then prevent one from perceiving the continuity and coherence of Schrödinger's conception of quantum mechanics from 1925 to the end of his life. That there is such a continuity has already been demonstrated by Y. Ben-Menahem, in connection with the problem of causality[1]. Our task is then to generalize her analysis in order to show that it extends to the whole of Schrödinger's interpretation of quantum mechanics.

Of course, we shall by no means try to demonstrate that there has been no change *at all* in Schrödinger's attitude towards quantum mechanics. Continuity does not mean immobility. Continuity here means that the successive steps of Schrödinger's thought about quantum mechanics can quite easily be considered as successive statements of, partial retreat from, reorganizations and further elaborations of, a *single* epistemological project. Such an approach may well be taken at first sight as a kind of rational *post-factum* reconstitution, but even if so, it will prove its usefulness and likelihood as the discussion proceeds.

Along with this line, we shall divide our study of Schrödinger's interpretation of quantum mechanics into three parts. In the first part (chapter 2), we shall try to identify the various components of Schrödinger's epistemological project, as they were stated very early (especially from 1924 to 1926 in so far as atomic physics physics is concerned). In the course of expounding this project and its earlier statements, we shall also bring out its permanent import on Schrödinger's thought. In the second part (chapter 3), we shall study the analytical and skeptical aspect of Schrödinger's approach of quantum mechanics. This aspect was of course dominant during the years 1928-1935, but it was not completely absent from earlier texts, and it was still fully acknowledged in the texts of the 1950's. The characteristic style of the analytical trend of thought consisted in pushing the instrumentalist (deconstructive) interpretation of quantum mechanics to its ultimate consequences; the skeptical element intervened when some loose current interpretations of these ultimate consequences were found to yield inconsistencies. Finally, in the third part (chapter 4), we shall describe at length the major shifts which gave Schrödinger enough confidence to overcome his purely

[1]Y. Ben-Menahem, "Struggling with causality: Schrödinger's case", loc. cit.

skeptical attitude and to undertake the reconstruction he was so eager to promote. These shifts were already foreseen in 1928 (and sometimes even earlier), some of them were worked out in the 1930's, but it was only in the 1950's that they were gathered and clearly perceived as such. They include: the complete relinquishment of the concept of particle (rather than its identification with wave packets); the *full* acceptance of the consequences of the holistic features of quantum mechanics; the establishment of a post-modern distance between the experimental results and the represented content of theoretical representations (rather than identifying the theoretical representation to a direct description of events occurring in space-time); and, last but not least, the correlative strategy consisting to postponing indefinitely the solution of the measurement problem, or pushing it towards the edges of theoretical thought (as we shall see, Schrödinger also formulated some possible solutions of the measurement problem, but their interest lies more in their general guiding principles than in their admittedly unconvincing details).

With this half-historical, half-methodological, framework in mind, it will prove quite easy to delineate the actual relationship between Schrödinger's original wave interpretation of quantum mechanics as it was formulated in January 1926, and his renewed wave interpretation of the 1950's. It will be easy to show that, despite the striking superficial similarity between the initial and the final wave-interpretations, the final one is by no means a rehearsal of the initial one. The wave interpretation of the 1950's has fully incorporated all the teachings of the analytical-skeptical period and, in almost every part of it, its meaning is deeply altered by the shifts which were listed above. Conversely, taking into account Schrödinger's ever-lasting metaphysical anti-realist commitment, his original 1926 wave interpretation will appear as a (too) hasty attempt at fulfilling his methodological realist demands, or alternatively as a (too) early realization of his reconstructive project, several years before he had completed his thorough analysis of the phenomenal material which was to be used for the (re)-construction. As we have already mentioned, Schrödinger's attempt of 1926 was pervaded by the implicit assumption that, quite apart from any remnant of metaphysical realism, it will prove acceptable, as in classical mechanics, to hide the instrumentalistic scaffoldings of the theoretical construct. By contrast, his late interpretation of quantum mechanics can be construed as a recognition that this creed was illusory. The reconstruction was still considered possible, but the scaffoldings had to be retained, made apparent, and given the status of a permanent element of the finished building, through the *interpretative scheme* (or, more precisely, through an explicit statement of the empirical correspondence rules).

This being admitted, we shall be able to challenge the current appraisal of Schrödinger's interpretation of quantum mechanics, which was so deeply conditioned by the criticisms of the Göttingen-Copenhagen group. It will appear in chapter 4 that, far from being the last conservative attempt of an old-fashioned physicist, Schrödinger's frame of thought paved the way towards several of the most advanced attempts at interpreting quantum mechanics. Schrödinger's radical criticism of the corpuscularian concepts, together with his motivated preference for a second-quantized Fock space representation over the usual trick of conceptually labelable but experimentally indistinguishable particles, is just beginning to receive the philosophical attention it deserves[1]. His strategy of providing the wave-mechanical formalism with its full development, of postponing indefinitely the solution of the measurement problem, of refusing to consider any interruption of the unitary evolution of the ψ-function, and of ascribing the state vector the status of a conjunction of occurrences rather than of a disjunction, is closely akin, as we shall see (paragraphs 4-4 and 4-5), to many contemporary no-collapse interpretations including Everett's and Van Fraassen's. Finally, his insistence on descriptive holism will be found to share many important features with modern quantum cosmological studies[2].

But actually, Schrödinger's conception afforded something more than a mere outline of several subsequent interpretations of quantum mechanics. It provided some of them in advance with a proper philosophical ground. To appreciate the importance of this contribution, we must remind ourselves that one of the recurrent criticisms some contemporary no-collapse interpretations such as Everett's interpretation have to face is their being oblivious of the extensive philosophical work which was performed by the physicists of the Göttingen-Copenhagen school. They are accused of entertaining a very straightforward and somewhat primitive version of realism, which does not take fully into account the series of difficulties which were documented by the epistemologically-inclined creators of quantum mechanics. In short, they are described as philosophically naive. Now, as we have already begun to show, Schrödinger was by no means a newcomer in philosophy. His own *methodological* realism even appears as epistemologically more advanced and more self-conscious, in some respects, than most blends of the Copenhagen interpretation. Having pushed the deconstructive step associated with the major scientific revolution he had contributed to promote to its ultimate consequences, Schrödinger was fully prepared to go one step further and to look for a way to recovering the use of the *concept* of real entities in physics. The present-day no-collapse "realist" interpretations of the (universal) ψ-function are then entitled to avail themselves of Schrödinger's preliminary foundational work in order to show that they are not doomed to be associated with a *metaphysical*, or pre-reflective, version of realism.

[1]M. Redhead & P. Teller, "Particle labels and the theory of indistinguishable particles in quantum mechanics", Brit. J. Phil. Sci., 43, 201-218, 1992; M.L. Dalla Chiara and G. Toraldo di Francia, "Individual, Kinds and Names in Physics", in: Corsi et al. (eds.) *Bridging the Gap: Philosophy, mathematics and physics*, Kluwer, 1993

[2]D. Deutsch, "Quantum theory as a universal physical theory", Int. J. Theor. Phys., 24, 1-41, 1985; J.D. Barrow and F. Tipler, *The anthropic cosmological principle*, Oxford University Press, 1986, p. 458 f.

CHAPTER 2

SCHRÖDINGER'S THEORETICAL PROJECT

Even though this essay is primarily devoted to an assessment of the most sophisticated version of Schrödinger's interpretation of quantum mechanics, namely at the beginning of the 1950's, we cannot avoid analyzing in some detail the ideas he defended during the mid-twenties. It was indeed during this early period that Schrödinger first formulated his life-long methodological requirements for a theory of atomic processes. But of course, the perspective we shall adopt is quite different from that of most historians of the beginnings of quantum mechanics. Our task goes beyond identifying an initial version of Schrödinger's requirements and inserting it within the intellectual context of the time. We also have to track successive statements of these requirements in later texts, and to comment retrospectively on their significance. The very vocabulary we use when we speak of a "theoretical project", and of the "methodological requirements" which are constitutive of it, has a retrospective tinge. For, after all, little had to be said about the project as long as it appeared to be immediately realized by wave mechanics, and few of the methodological requirements had to be made explicit when they were considered as unproblematically fulfilled by the current theory. It was only when the initial wave (and electrodynamic) interpretations of quantum mechanics happened to be seriously challenged that both the project and the requirements were isolated from their first theoretical embodyment, and that they began an independent career as guiding principles for an anticipated new interpretation.

2-1 *Reality and virtuality (1924)*

A few months after the issuing of the celebrated Bohr, Kramers and Slater paper[1] about the quantum theory of interaction between radiation and matter, Schrödinger published an article in which he expressed his warm approval[2]. The main purpose of the BKS paper was to show, at least programmatically, that it is possible to reconcile the quantized properties of atoms with the continuity of electromagnetic fields. Or, in other terms, that it is conceivable to avoid using Einstein's notion of light-quanta, which "in its most extreme form denies the wave constitution of light" and is thus unable to account for the interference phenomena. But one of the most important advantages of Einstein's light quanta is their ability to carry discrete amounts of energy that can then be directly transferred to atoms during a 'quantum jump'. By contrast, it was quite difficult to figure out how the continuous flux of energy of a classical radiation field can possibly be connected with discontinuous atomic

[1]N. Bohr, H.A. Kramers and J.C. Slater, "The quantum theory of radiation", Phil. Mag. 47, 785-802, 1924; in: B.L Van der Waerden (ed.), *Sources of quantum mechanics,* North Holland, 1967

[2]E. Schrödinger, "Bohrs neue Strahlungshypothese und der Energiesatz", Die Naturwissenschaften, 12, 720-724, 1924

transitions; at some moment during the process, the sum total of the energies of the emitter, the absorber, and the field, had to diverge from the initial energy of the emitter and the absorber. BKS thus decided to give up the detailed energy and momentum balance, and to work out the resulting ideas up to their ultimate consequences. They considered that the atoms (or rather their "virtual oscillators") communicate with one another through a spatio-temporal mechanism which is "virtually equivalent with the field of radiation" of classical electromagnetism. The virtual radiation field did not bear energy by itself but it was supposed to "induce" transitions between the stationary states of the atoms. The induction, in turn, was not individual, but statistical. In other words, the radiation field did not provoke directly a transition; it just generated a non-zero probability of transition. On the one hand, the energy and momentum conservation laws were violated by the individual processes, but on the other hand the values of the probabilities of transition were chosen in such a way that the overall balance of energy and momentum was in good statistical agreement with the conservation laws.

Almost every element of the BKS program was greeted by Schrödinger. He fully approved the relinquisment of Einstein's light-quanta, and he was especially interested, from a philosophical point of view, by the idea that the conservations laws were only statistical. As he wrote to Bohr, the renunciation of causality "touches me extraordinarily sympathetically. As pupil of the venerable Franz Exner, I have been on intimate terms for a long time with the idea that probably no microscopic lawfullness, but perhaps 'absolute accidents' forms the foundation of our statistics, and that perhaps even the energy and momentum principles are only statistically valid"[1]. The notion of a purely statistical status of the energy and momentum conservations law, as well as the criticism of Einstein's concept of light quanta, exerted a strong attraction on him. Such an attraction is made manifest by the recurrence of these ideas throughout his life. His last paper, in 1958, was devoted to an analysis of the idea that energy is conserved only statistically[2]. And, in a letter of 1957 to B. Bertotti[3], he wrote: "(...) in 1922, Einstein was given the (Nobel) prize, (...) for a rather trifling story about photoelectric effects, based on I daresay the only mistake he made in his life, viz. that the energy of electromagnetic radiation is lumped up in 'energy quanta', now termed photons". With such a deep and lasting adhesion to the general trends of the BKS paper, no wonder that Schrödinger defended it so eagerly. Even the original remark he made in his paper of 1924, that "the average squared fluctuation of the energy of a gas of Bohr atoms interacting in the BKS manner via the virtual fields would increase

[1]E. Schrödinger to N. Bohr, May 24, 1924, AHQP; quoted and translated by O. Darrigol, "Schrödinger's statistical physics and some related themes", in: M. Bitbol & O. Darrigol (eds.), *Erwin Schrödinger, Philosophy and the birth of quantum mechanics*, op. cit.

[2]E. Schrödinger, "Might perhaps energy be merely a statistical concept?", loc. cit.

[3]E. Schrödinger to B. Bertotti, December 27, 1957, in: B. Bertotti & U. Curi (eds.), *Erwin Schrödinger scienziato e filosofo*, Il poligrafo, 1994, p. 156

linearly in time"[1], which could have been taken as an unpleasant, not to say self-defeating, feature of the BKS theory, was not considered by him as a serious reason to abandon it. The linear increase of fluctuation could indeed, as he remarked, be distributed over an ever increasing number of atoms.

In this context, it is especially important to realize that there was nevertheless *one* point on which Schrödinger did *not* agree with Bohr, Kramers and Slater. This point was mentioned in his letter to Bohr: "I cannot really go along with you when you keep calling these waves 'virtual' (...) For what is the 'real' radiation if it is not that which 'causes' transitions, i.e. which conveys the transition probabilities?"[2]. In view of this criticism, Schrödinger did not mention in his own paper that the BKS radiation field is supposed to be "virtual"; he even assumed, unlike BKS, that this field is able to carry energy. The latter feature, namely the ability to carry energy, seems to introduce a very large difference between Schrödinger's ideas and the content of the BKS paper. I personally take it as circumstantial rather than fundamental. A field which transports energy is generally considered as "real" because, according to the laws of conservation, it may have *effects*. Its energy can in principle be transferred to an experimental device, and thereby detected. But the converse is not true. One can perfectly conceive, as BKS did, a field which has energetical effects without carrying energy. Now, according to Schrödinger, the crucial criterion for calling a theoretical entity "real" is its being ascribed the capability of "causing" effects (be it in a restricted probabilistic sense), and *not its being energetically homogeneous with the effects it produces*. The circumstance that Schrödinger associated a certain amount of energy with the BKS radiation field is therefore symptomatic of his reasonably strong confidence in the epistemological status of this field, rather than of his incapacity to conceive a non-energetically loaded *effective* field. It is symptomatic of his *preference* for theoretical entities which are both "effective" and "factual"[3] (namely isomorphic with the effects they are able to produce), rather than of his definite rejection of entities which are *only* "effective".

As Max Jammer rightly pointed out, the ideas developed in the BKS paper set the stage for the subsequent debate on the status of the ψ-function. "According to Heisenberg, Born's statistical interpretation of the Schrödinger's wave function had its ultimate root in the Bohr-Kramers-Slater paper"[4].

BKS indeed provided the model of a statistical linkage between a continous theoretical entity (the radiation field), and a discontinuous process (the atomic transition). A model which Born found very easy to

[1]Quoted and translated by O. Darrigol, "Schrödinger's statistical physics and some related themes", loc. cit.

[2]E. Schrödinger to N. Bohr, May 24, 1924, AHQP; quoted and translated by L. Wessels, *Schrödinger's interpretations of wave mechanics,* op. cit. p. 73

[3]See M. Jammer, *The conceptual development of quantum mechanics,* op. cit. p. 184 (footnote)

[4]ibid. p. 187 (footnote 137)

project onto the problem of the linkage between the continuous ψ-function and the discontinuous outcomes of particle collisions. Just as in the BKS situation there was no light-quantum which in any individual case caused an atomic transition, "from the standpoint of quantum mechanics there is no quantity which in any individual case causally fixes the consequence of a collision"[1]. Just as in the BKS situation, the intensity of the radiation field did not provide any answer to the question 'which atomic transition occurs, and at which moment' but only to the question 'what is the probability for such transition to occur', wave mechanics does not provide any answer to "the question 'what is the state after the collision', but only to the question, 'how probable is a specified outcome of the collision'"[2]. The articulation between the continuous and discontinuous concepts is thus identical in both cases. But this does not solve, as such, the problem of the epistemological status of the ψ-wave. For, after all, the epistemological status of the BKS radiation field was itself unclear. Concerning the status of the ψ-wave, one could choose between three options (at least). Let us list them according to an order of increasing "reality":

(i) The ψ-wave is but a conceptual tool allowing one to deal statistically with ensembles of particles. Both Einstein and Born took this option[3]. The only true difference between them was about the prospect of a future "completion" of the quantum-mechanical statistical description[4].

(ii) Just as in the BKS paper the radiation field is considered as "virtual", the ψ-wave has "some intermediate kind of reality"[5]; it plays the role of a "virtuality", or of "potentia". This is basically Heisenberg's late position: "(...) The paper of Bohr, Kramers and Slater revealed one essential feature of the correct interpretation of quantum theory. The concept of probability wave was something entirely new in theoretical physics since Newton. Probability in mathematics or in statistical mechanics means a statement about our degree of knowledge of the actual situation. In throwing dice we do not know in fine details of the motion of our hands which determine the fall of the dice and therefore we say that the probability for throwing a special number is just one in six.The probability wave of Bohr, Kramers, and Slater, however, meant more than that; it meant a tendency for something. It was a quantitative version of the old concept of 'potentia' in Aristotelian philosophy. It introduced something standing in the middle between the idea of an event and the actual event, a strange kind of physical reality just in the middle between

[1]M. Born, "On the quantum theory of collisions", loc. cit.
[2]ibid.
[3]See for instance M. Born (ed.), *The Born-Einstein letters,* op. cit. p. 186 f.
[4]Einstein expressed his hopes about a future complete theory on several occasions. See e.g. A. Einstein, B. Podolsky, and N. Rosen, "Can quantum-mechanical description of physical reality be considered complete?" loc. cit.; M. Born (ed.), *The Born-Einstein letters,* op. cit. p. 173. Born, instead, noticed: "(Einstein) calls my way of describing the world 'incomplete'; in his eyes this is a flaw which he hopes to see removed, while I am prepared to put up with it", ibid. p. 189
[5]W. Heisenberg, AHQP, interview on February 15, 1963, quoted by M. Jammer, *The conceptual development of quantum mechanics,* op. cit. p. 187

possibility and reality"[1]. The interference effects, as well as Von Neumann's so-called "no-hidden variable theorem" which appeared to preclude interpretation (i) in the strong version defended by Einstein, could not, according to Heisenberg, be accounted for by making "cheaper" assumptions.

(iii) The ψ-wave is plainly "real". This happened to be Schrödinger's position in the beginning of 1926 and in the 1950's. But of course, the content of the word "real" was not exactly identical in the two occurrences. In the first months of 1926, Schrödinger still considered that the ψ-wave is "real" in the strongest and most common sense, namely that it is both "effective" and "factual"; that it does not only "cause" events to occur but represents itself the network of the natural events; that therefore the ψ-function is a faithful *picture* of the events occurring in space-time. Several years later, from the end of the 1940's on, Schrödinger gave up this prescription of faithful picturing of the events occurring in space-time. Even though he still thought that the ψ-function can be considered as a picture "of something" *in abstracto,* he insisted that "the (observable facts) are not in one-one correspondence with (this wave picture)"[2]. And he thus explicitly precluded the idea that the ψ-wave is directly endowed with "factuality". However, this was according to him no obstacle to construing the ψ-waves as "real". Just as, in his early comment about the BKS paper, he was led to ascribe full "reality" to a theoretical entity in so far as it is "effective", he came to consider in the 1950's that "effectivity" was already a good reason (even if by no means the only one, as we shall see in chapter 4) to call the ψ-waves "real". Commenting on de Broglie's concept of "guiding wave", he noticed: "Something that influences the physical behaviour of something else must not in any respect be called less real than the something it influences - whatever meaning we give to the dangerous epithet 'real'"[3].

This is indeed a crucial sentence, which may help us to understand quite efficiently the basic metaphysical assumptions underlying the different conceptions about the "reality" of the ψ-waves, and the originality of Schrödinger's position. Its key word is *"less"*. Schrödinger just could not understand why one should call the "effectively guiding" ψ-wave *less* real than the supposedly "guided" particles. He willingly acknowledged that it is "useful to recall at times that all quantitative models or images conceived by the physicists are, epistemologically, only mathematical devices for computing observable events"[4]. He did recognize that "the wave functions are mental material (...)"[5] or that they are elements in a

[1]W. Heisenberg, *Physics and Philosophy,* op. cit. p. 40-41
[2]E. Schrödinger, *Science and Humanism,* op. cit. p. 41
[3]E. Schrödinger, "What is an elementary particle?", Endeavour, 9, 109-116, 1950
[4]ibid.
[5]E. Schrödinger, *Transformation and interpretation in quantum mechanics,* (Dublin seminar 1952, in: E. Schrödinger, *The interpretation of quantum mechanics (Dublin seminars 1949-1955 and other unpublished texts),* op. cit., p. 82)

"model of our thoughts"[1]. But such a remark could not lead one to put waves (of whatever kind) and particles on a different footing. The use of "mental material" to perform "thought experiments (...) is precisely the same pattern physics has always followed, also when one worked with atoms, molecules, electrons and light waves. The armoury has been enhanced and the analytical methods have changed, that is all"[2]. As we noticed in section 1-5, Schrödinger considered that atoms, molecules and the like are just as much theoretical constructs as ψ-waves. The macroscopic *things* are themselves, according to him, but constructs of the mind, made out of pure perceptual and intellectual Machian "elements"[3]. This being admitted, there seems to be no reason left to consider that ψ-waves are *less* real than particles or than anything else whose "reality" is never doubted in everyday life.

We here again bump into an apparently very paradoxical feature of Schrödinger's thought. His enduring "realism" about ψ-waves (or BKS radiation field) is associated with the most extreme metaphysical *anti-realism*. This is not to say either that Schrödinger wavered between realism and anti-realism, or that he adopted some intermediate position[4]. This plainly means that his realist vocabulary and attitudes are *rooted into* an uncompromising version of metaphysical anti-realism; that his being so eager to defend the "reality" of ψ-waves *arises from* a very acute critique of the constructive procedures which yield the entities of physics and of daily life. But how can this possibly happen? How can an apparently realist attitude be *grounded* on an underlying anti-realist doctrine?

S. Blackburn's recent work on what he calls "quasi-realism"[5] completely clarifies this issue. Basically, the quasi-realist is "someone who, starting from an anti-realist position finds himself progressively able to mimic the thoughts and practices supposedly definitive of realism"[6]. The quasi-realist, however, does not content himself with a half-convinced mimicry: he "becomes an embarrassingly enthusiastic mimic of traditional realist sentiments, and in his very zeal, we might expect him to differ from the real realist"[7]. Why is it so? Let us consider for instance two sentences (p_1) and (p_2), which state the principle of causality following respectively an anti-realist approach and a realist approach:

[1]ibid. p. 81

[2]E. Schrödinger, *Transformation and interpretation in quantum mechanics,* (in: E. Schrödinger, *The interpretation of quantum mechanics (Dublin seminars 1949-1955 and other unpublished texts),* op. cit. p. 82)

[3]See E. Schrödinger, *William James lectures (c. 1954), 3rd lecture* (in: E. Schrödinger, *The interpretation of quantum mechanics (Dublin seminars 1949-1955 and other unpublished texts),* p. 145-149; E. Schrödinger, *Nature and the Greeks,* op. cit. chapter 7. An analysis of Schrödinger's conception of the "thing" of everyday life is provided in chapter 5.

[4]The idea of an "intermediate position" is defended by Y. Ben-Menahem, "Struggling with realism: Schrödinger's case", in: M. Bitbol & O. Darrigol, (eds.), *Erwin Schrödinger, Philosophy and the birth of quantum mechanics,* op. cit.

[5]S. Blackburn, *Essays in Quasi-realism,* Oxford University Press, 1993

[6]ibid. p.15

[7]ibid. p. 28

(p_1) "Inquire into nature *as though* every event had a cause"
(p_2) "It is the case that every event has a cause"

It is clear that a "real realist" has strong reasons to prefer (p_2) over (p_1). But has an anti-realist any good reason to prefer (p_1) over (p_2)? S. Blackburn's answer is flatly: *no*. For after all, the very attempt at rejecting (p_2) after having accepted (p_1) can but have *one* sound motivation, namely the belief that "*in reality* it is *not* the case that every event has a cause". If one instead adopts a fully consistent anti-realist position, there is nothing more involved in (p_2) than in (p_1). The quasi-realist kind of anti-realist can thus accept (p_2) without any problem, *as* his realist colleague does; and he is even willing to accept propositions expressed by (p_2)-like sentences *in situations where the realists cannot*. His zeal in calling all sorts of entities "real" is then *just* what makes him so (paradoxically) different from the "real realist". Last but not least, a correlative characteristic of the quasi-realist is that, not without some good reasons, he is "often charged with 'scientism' at this point, or in other words with confining genuine reality to an ontology and a set of features delineated by some favoured fundamental science, such as physics"[1].

In view of such a description, I think Schrödinger can be categorized as one of the most typical quasi-realist thinker of the twentieth century[2]. His "realist" zeal, to begin with, is not to be insisted upon any longer. Schrödinger's absolute lack of precautions about the "unreal" or the "virtual" character of the ψ-waves (and of the BKS radiation field) was motivated by the remark that, in so far as we have to inquire into nature *as though* there were (3-dimensional *or* 3n-dimensional) interfering waves, in so far as "we must *think* in terms of spherical waves emitted by the source"[3], there is no good *a priori* reason for depriving them of the epithet "real"; except of course if we are "real realists" who seriously believe that tables, chairs (and particles) are *real* in a sense no mind-construct can be. Secondly, in the same way as Blackburn's archetypal quasi-realist, Schrödinger focused his notion of "reality" on scientific constructs. Even the ordinary "thing" was identified by him to a kind of scientific theory by which the infant begins to orient himself in the world[4]. The extreme difficulty most commentators have had in tackling Schrödinger's position about realism could thus serve as a perfect illustration of Blackburn's warning: "in the philosophy of these things, it is not what you end up saying but how you get to say it, that defines your 'ism'"[5]. It is not Schrödinger's extensive use of the concept of reality

[1]ibid. p. 8
[2]M. Bitbol, "Quasi-réalisme et pensée physique", Critique n°564, 340-361, 1994
[3]E. Schrödinger, *Science and Humanism*, op. cit. p. 47
[4]E. Schrödinger, *William James lectures (c. 1954)*, (in: E. Schrödinger, *The interpretation of quantum mechanics (Dublin seminars 1949-1955 and other unpublished texts)*, op. cit. p. 145-149); and chapter 5 for further developments and analysis.
[5]S. Blackburn, *Essays in Quasi-realism*, op. cit. p. 7

which defines his 'ism'; it is *how* he got to insist on it. Now, as we have repeatedly showed (especially in section 1-5), his insistence arose from anti-realist premises. Schrödinger therefore sides *unambiguously* with anti-realists, and even with the anti-realists of the most accomplished kind, namely *quasi-realists.*

Here we see most clearly, from the point of view of philosophical positions, what we had previously (§1-8) outlined in terms of an artistic metaphor. A quasi-realist thinker like Schrödinger is a post-modernist in philosophy. He opposes the dogmatism of the classicist and the naively representationalist attitudes of the realist, but he does not see why the anti-realist should confine us to the immanence of experimental manipulations and facts. Schrödinger considers that, in his opposing classical realism, the modern anti-realist tends to overreact and to become just as dogmatic as his opponent. After all, if we do not believe that reality is already given with all its structures *out there,* we become free to shape out representations and systems of references, at least in so far as they do not contradict the results of our experimental activity. Moreover, these representations and systems of references play a role, as regulative principles, in the process of scientific research. Such a regulative function of representations is so important that it cannot be dispensed with.

This was the basic idea which guided Schrödinger throughout his career, implicitly at first, and then more and more explicitly.

2-2 *Holism and wave-packets (1925)*

Holism was already part of Schrödinger's methodological project when he first attempted to formulate his own version of the theory of atomic phenomena in the mid-twenties. But this project could not reach its complete development as long as it was associated with the belief that a proper theory of microscopic phenomena must offer some exact counterpart to the notion of localized re-identifiable body. The holistic project could not be consistently worked out as long as its generic concepts had to coexist with at least some equivalent of the concept of permanent localized individuals. Schrödinger realized this difficulty progressively, and by the summer of 1926 he became fully aware of it. However, it was not before the end of the thirties and the beginning of the forties that he was able to accomodate a world-picture which was plainly deprived of any remnant of corpuscle-like entities. The major philosophical step which allowed this transformation is what we have called Schrödinger's post-modern turn. From then on, the theory was no longer supposed to be a reflection of natural phenomena, but only to provide one with a model connected to the phenomena through a set of correspondence rules. Accordingly, Schrödinger's priority was no longer to display a direct equivalent of the particles in the wave-mechanical formalism, but only to show that this formalism, together with its correspondence rules, is compatible both with certain sets of discontinuous observations and with the *appearance* of permanent body-like behaviour at the macroscopic scale. It is the need of this important shift that delayed Schrödinger's complete reappraisal of his own initial wave interpretation until the beginning of the fifties.

Let us then begin to scrutinize the conflict between Schrödinger's holistic framework of thought and his persistent corpuscular representations. The import of holism on Schrödinger's theoretical attitudes can already be detected during the prehistory of quantum mechanics, in his 1924 paper about the BKS theory. Here, after having suggested that the linear increase of the energy and momentum fluctuation with time could be compensated by transferring it to an ever-increasing number of molecules, he wrote: "a certain stability of world events *sub specie aeternitatis* can only exist through the connection of each individual system with the rest of the world. A separated individual system would be, from the point of view of the whole, chaos"[1]. But it was not until 1925 that he began to develop his own stringent version of wholeness, where the very concept of individual atomic system would progressively dissolve. That year, Schrödinger wrote a paper[2] wherein he criticized Planck's attempt at taking into account the indistinguishability of N molecules by dividing the Boltzmann-Gibbs statistics by the number N! of their permutations. He explained that a proper "statistical grounding"[3] of the numerical consequences of indistinguishability would rather lead one directly to the Bose-Einstein statistics. But the new method of counting which was involved in the Bose-Einstein statistics, namely equiprobability of the various distributions of the *numbers* of molecules in energy cells, did not fit with the idea of an *independent* distribution of the individual molecules. Einstein therefore introduced, in his second paper on the new statistics[4], the notion of a "mysterious" kind of mutual interaction between molecules (or light-quanta) which would explain the discrepancy of his formula with respect to the standard case of equiprobability of the distributions of *individual molecules* in cells. Schrödinger very soon realized the problem. In the absence of a detailed understanding of Einstein's "interactions", however, he suggested that it was more appropriate to consider the gas as a whole from the outset than to correct the molecular picture retrospectively: "one can also take the view that such a thorough-going independence of gas molecules from one another does not exist quantum mechanically (...), that certain quantum states with completely determined energy levels are to be attributed not to the single

[1]E. Schrödinger, "Bohrs neue Strahlungshypothese und der Energiesatz", loc. cit.; translated by O. Darrigol, "Schrödinger's statistical physics and some related themes", in: M. Bitbol & O. Darrigol (eds.), *Erwin Schrödinger, Philosophy and the birth of quantum mechanics*, op. cit.

[2]E. Schrödinger, "Bemerkungen über die statistische Entropiedefinition beim idealen Gas", Berlin Akademie der Wissenschaft Sitzungsberichte, 1925, 434-441

[3]ibid.

[4]A. Einstein, "Quantentheorie des einatomigen idealen Gases", Berlin Akademie der Wissenschaft Sitzungsberichte, 1925, 3-14, §7

molecules, but to the body of gas as a whole"[1]. Accordingly, the ordinary Boltzmann-Gibbs statistics, which had been dismissed for molecules, could be now applied to the individualized energy states of the whole gas.

But Schrödinger had not really abandoned the molecular picture at this stage. He believed that it would prove possible in the future to derive energy levels of each single gas molecule from the energy level distribution as a whole. And he still accepted the idea that the holistic treatment of the gas could prove to be a "temporary" trick aimed at compensating for our lack of knowledge on the detailed processes which take place at the microscopic level.

By the end of 1925, however, after having thought about de Broglie's matter wave concept[2], he started thinking that the methods which were so efficient to deal with the gas as a whole also provided him with a clue for understanding microscopic phenomena. As M.J. Klein[3] rightly noticed, it was his work on quantum statistics, rather than his scattered interest in atomic physics, that paved the way towards wave mechanics. As a first step, taking very seriously the idea of an inversion of the roles between the concepts "variety of the energy states" and "variety of the bearer of these states", Schrödinger proposed to replace the idea that n_s molecules are present in a state ε_s by the idea that the s-th degree of freedom of the whole gas is in its n_s-th level[4]. The whole system was then considered as an aggregate of harmonic oscillators in various quantized states of excitation rather than as a set of molecules. Despite this momentous advance, Schrödinger thought that the corpuscularian picture could still arise as, so to speak, a by-product of the quantized matter wave scheme. He did not content himself with *replacing* the plural molecular representation by a global *hohlraum* oscillator representation; he wanted the first one to appear as a consequence of the second one. Or, in more metaphoric terms imitated from the expressions he used in the introduction to his paper, he did not restrict his theory to a description of a system of standing waves; he wished to single out, near the crest of the waves, some "froth" which could be considered as a satisfactory "ersatz" of the molecules. The latter result could be obtained by considering the "signal" resulting from the superposition of a great number of waves with a quite narrow range of frequencies. Having realized this, Schrödinger formulated his celebrated idea of representing each particle by a wave packet, in the fifth paragraph of his paper about Einstein's gas theory. But at this point, he had to deal with three major difficulties.

1) The region of constructive interference of the waves composing the wave packet had to be made narrow enough (in the three dimensions of

[1]E. Schrödinger, "Bemerkungungen über die statistische Entropiedefinition beim idealen Gas", loc. cit.; translated by L. Wessels, *Schrödinger's interpretations of wave mechanics,* op. cit. p. 37

[2]see V.V. Raman and P. Forman, "Why was it Schrödinger who developed de Broglie's ideas?", HSPS, 1, 291-313, 1969, for a detailed study of Schrödinger's reception of de Broglie's work.

[3]M.J. Klein, "Einstein and the wave-particle duality", The natural philosopher, 3, 3-49, 1964

[4]E. Schrödinger, "Zur Einsteinschen Gastheorie", Phys. Zeits., 27, 95-101, 1926

space) to fit the actual size of the particle. This result could in principle be reached, according to Schrödinger, by a method due to Debye and Von Laue. The method consists in combining waves whose propagation vectors have slightly distinct directions.

2) The classical wave theories predicted that wave packets would undergo a fast dispersion process, thus departing from the requirement that corpuscles are permanently well localized. Schrödinger fully acknowledged the problem in the very paragraph where he had proposed the wave-packet concept. But he hoped that a forthcoming new propagation law would allow him to circumvent the obstacle.

3) The necessity of superposing waves with a (quasi-) continuous spectrum of frequency, in order to obtain a wave packet for each molecule present in the gas, is flatly inconsistent with Schrödinger's representation of the whole gas as a discrete system of standing waves. The latter flaw was pointed out and carefully discussed by L. Wessels: "For each gas molecule a large number of waves is required in the construction of a relatively small wave packet, while according to the restriction on the energy distribution (...), the greatest number of frequencies that could possibly appear in the gas at any one time is just the number of gas molecules itself" [1].

These problems reflected the extreme difficulty, and perhaps the intrinsic impossibility, of reconciling the corpuscularian representation with the holistic standing wave model of the gas. Far from being solved by the advent of wave mechanics and its propagation law (the Schrödinger equation), they reocurred in it and became increasingly acute. Schrödinger's major move during the last weeks of 1925 and the beginning of 1926 consisted in transposing the type of standing wave model which proved so fruitful in the case of a whole gas, to the description of the atom. The confinement in a box had just to be replaced by a confinement in a potential. But at the same time, some of the most striking features of the gas theory disappeared. The atomic theory was no longer concerned with describing the population of each vibrating mode in terms of an excitation number; it just aimed at dealing adequately with the proper vibrations themselves. In other terms, it reverted to a first quantization scheme, whereas the former gas theory was already much in the spirit of the forthcoming second quantization. Being a standing wave model, and having purposely departed from de Broglie's model of progressive waves on closed orbits, Schrödinger's atomic theory was nevertheless bound to inherit the major conundrum of the gas theory[2]. How could one represent the particles and their motion? Was it reasonable to hope that the old picture of electrons running on Keplerian orbits

[1]L. Wessels, *Schrödinger's interpretations of wave mechanics*, op. cit. p. 146-147

[2]According to L. Wessels (in: "Schrödinger's route to wave mechanics", Stud. Hist. Phil. Sci., 10, 311-340, 1977), one of the reasons Schrödinger did not use the language of matter waves but rather a formal condition on Hamilton's equation for particles in his first paper on wave mechanics, was that he did not know whether the difficulties linked with the concept of wave packets could be met by his newly formulated theory.

round the nucleus would be retained in any way? In the second paper of his series "Quantization as a problem of proper values"[1], Schrödinger still thought he could find a compromise between the fruitfulness of the standing wave model and the persistence of a corpuscularian representation. On the one hand, one had to acknowledge that for closed orbits "whose dimensions are of the order of the wave length", the notion of a "system path" is meaningless, for "the wave group not only fills the whole path domain all at once, but also stretches far beyond it in all directions". On the other hand, the concept of a particle travelling on an orbit still retained its relevance when the de Broglie's wave length is much smaller than the curvature radius of the orbit. In the latter case one could indeed rely on the fact that the wave packets obey "the same laws of motion as a single image point of the mechanical system", and that they therefore "give, so to speak, an *equivalent* of the image point". But even this compromise proved utterly unsatisfactory. The three difficulties met by the wave packet concept in the gas theory here combined to form an inextricable complex. As Lorentz pointed out in his very long letter to Schrödinger of May 27, 1926[2], the dispersion of the wave packet (and thus the loss of identity of the particle) occurs very soon, unless one assumes that its dimensions are large compared to the wave length. But then, the requirement that the dimensions of the wave packet be very small in comparison with the curvature radius of the orbit is not fulfilled any longer. In the same spirit, Schrödinger had to postulate that the wave packet extends "over a large number of wave lengths, if it is to be approximately monochromatic", for "the wave group must move about as a whole with a definite group velocity and correspond to a mechanical system of definite energy"[3]. Then, again, the risk was to increase the dimension of the wave packet until it exceeds the size of the orbit. Lorentz's objection thus showed that the dimension requirement (1) of the wave packet representing the particles is inconsistent with the non-dispersion requirement (2); and Schrödinger's own postulate indicated that the dimension requirement (1) is inconsistent with the discrete energy scheme of the model (3).

Schrödinger was fully aware of these problems, even before having received the letter of Lorentz[4]. In his answer to Lorentz, he thus enclosed a paper, entitled "The continuous transition from micro- to macro-mechanics"[5], wherein he demonstrated that, *at least in the case of a*

[1]E. Schrödinger, "Quantization and as a problem of proper values II", in: *Collected papers on wave mechanics*, op. cit. (see p. 19 f.)
[2]H.A. Lorentz to E. Schrödinger, May 27, 1926, in: K. Przibram (ed.), *Letters on wave mechanics*, op. cit. p. 47
[3]E. Schrödinger, "Quantization and as a problem of proper values II", in: *Collected papers on wave mechanics*, op. cit. p. 20
[4]E. Schrödinger to H.A. Lorentz, June 6, 1926, in: K. Przibram (ed.), *Letters on wave mechanics*, op. cit. p. 59: "You see from the text of the note, which was written before I received your letter, how much I too was concerned about the 'staying together' of these wave packets"
[5]E. Schrödinger, "The continuous transition from micro- to macro-mechanics", in: *Collected papers on wave mechanics*, op. cit. p. 41

harmonic oscillator, the proper vibrations add up to form a wave packet which "does not spread out into larger regions as time goes on". The two former inconsistencies appeared to be solved at the same time. No other frequencies than those of the finite set of proper vibrations were required, and moreover the dispersion phenomenon was completely eliminated even though the wave packet retained reasonable dimensions (a few wavelengths). But this remarkable result was rapidly challenged, due to its complete lack of generality. Lorentz showed that what had been demonstrated by Schrödinger in the case of the harmonic oscillator did not hold any more in the case of the hydrogen atom[1]. Several months later, Heisenberg made the same point in his celebrated "uncertainty relations" paper[2], which incorporates a section entitled "the transition from micro- to macro- mechanics". Heisenberg did not work out the complete calculation, as Lorentz did, for it was sufficient to notice that the stability of wave packets is incompatible with the fact that the radiation frequencies of the hydrogen atom are never integer multiples of the basic frequency. Then, one had to admit, with Heisenberg, that "Schrödinger's reasoning is only viable in the case of the harmonic oscillator treated by him; in all other cases a wave packet spreads out in the course of time over the whole immediate neighbourhood of the atom". Schrödinger's early compromise between holism and the concept of localized spatio-temporal continuant had thus failed. The only thing which survived, out of his work about the transition from micro to macro-mechanics, was the very fact of the permanence of wave packets for the case of harmonic oscillators, which proved extremely important in recent years for the study of "coherent states" (in lasers, for instance)[3].

But then, what was it which prevented Schrödinger from discarding immediately the concept of corpuscular spatio-temporal continuant and from pushing the holistic model to its ultimate consequences? We already know the answer, in contemporary terms: before Schrödinger could abandon the last remnants of the corpuscular representations, he had to solve the *measurement problem,* or at least to show that its solution was not urgent. In other terms, more appropriate to the historical situation, he had to deal with the experimental discontinuities which are traditionally associated with the corpuscularian picture. As Lorentz noticed, "if we decide to dissolve the electron completely, so to speak (...)", this has an important disadvantage. The disadvantage is this: "(...) how am I to understand the phenomena of photoelectricity and the emission of electrons from heated metals? The particles appear here quite clearly and without alteration; once dissolved, how could they condense again?"[4].

[1]H.A. Lorentz to E. Schrödinger, June 19, 1926, in: K. Przibram (ed.), *Letters on wave mechanics,* op. cit. p. 70

[2]W. Heisenberg, "The physical content of quantum kinematics and mechanics" in: J.A. Wheeler & W.H. Zurek, *Quantum theory and measurement,* op. cit. p. 72

[3]F. Steiner, "Schrödinger's discovery of coherent states", Physica B 151, 323-326, 1988

[4]H.A. Lorentz to E. Schrödinger, May 27, 1926, in: K. Przibram (ed.), *Letters on wave mechanics,* op. cit. p. 48

Schrödinger's attempts to deal with discontinuities without any corpuscularian representations, by transferring them from the *object* of physics to the *interaction* between this object and the measuring apparatus, date back to 1927[1]. I shall postpone (until chapter 4) the discussion of these early attempts at establishing a relation between the discontinuities and the measuring interactions. It is sufficient at this point to indicate that they by no means *solve* the measurement problem, but rather constitute its first *statement*. Having realized that, Schrödinger gloomily acknowledged, in his Nobel lecture of 1933, that "the wave theory cannot meet this case (i.e. the apparent particle tracks in a cloud chamber), except in a very unsatisfactory way"[2]. But later on, after he had completed his critical analysis of the current Copenhagen interpretation of quantum mechanics, and after he had carefully circumscribed the measurement problem[3], his appraisal of the situation began to reverse. He insisted more and more on the inconsistencies of the corpuscularian concept of localized and individualized spatio-temporal continuant[4], whose intervention was still considered indispensable by his Göttingen-Copenhagen colleagues (at least as a symbolic tool). Besides, he grew ever more convinced that the solution of the measurement problem was not at all crucial for a consistent treatment of concrete physical problems.

Feeling that he was thus delivered from the obligation to find a direct equivalent of permanent corpuscle-like entities in his theoretical pictures, Schrödinger reverted to the initial holistic themes and developed them ever more consistently. This new trend of thought, characteristic of the end of the thirties and the beginning of the forties is easily perceptible in two fields of investigation: statistical thermodynamics and unified field theories.

Let us begin with statistical thermodynamics. Here, Schrödinger basically redeveloped his 1925-26 idea of dealing with the gas as if it were a whole vibrating system, of exchanging the role of the states and the bearers of the states, and of considering the Bose-Einstein (and Fermi-Dirac) statistics as a quite tricky consequence of the fact that the only individuals of the vibrating system are its proper modes rather than the excitations of these modes (the "particles"). A major consequence of these remarks was that one had to abandon the Boltzmann counting method for the bearers of the states, and rather use the Gibbs method for the whole system. These ideas were first sketched in an unpublished manuscript of 1938: "The habit of speaking of a 'new statistics' has arisen from insisting upon the Boltzmann-Maxwell method, where it does not suit the purpose

[1] E. Schrödinger, "The exchange of energy according to wave mechanics" in: *Collected papers on wave mechanics*, op. cit. p. 140 f.

[2] E. Schrödinger, "The fundamental idea of wave mechanics", in: *Science and the Human temperament*, op. cit. p. 153

[3] Especially in the two 1935 papers: E. Schrödinger, "The present situation in quantum mechanics" loc. cit. and "Discussions of probability relations between separated systems", loc. cit.

[4] See Schrödinger's notes entitled "principium individuationis", 1939, AHQP, microfilm 42, section 9

and rarely does"[1]. They were then developed in his seminar lectures of the Dublin institute for advanced studies in January-March 1944, entitled "Statistical thermodynamics"[2]. There, Schrödinger first explained why he thought the Boltzmann method is not appropriate: "(The Boltzmann method) suffices only for dealing with a very restricted class of physical systems - virtually only with gases. (...) In a solid the interaction between neighbouring atoms is so strong that you cannot mentally divide up its total energy into the private energies of its atoms. And even a 'hohlraum' (an 'ether block' considered as the seat of electromagnetic-field events) can only be resolved into oscillators of many - infinitely many - different types, so that it would be necessary at least to deal with an assembly of an infinite number of different assemblies, composed of different constituents". In view of this weakness, one must revert to the Gibbs method, which "is applicable quite generally to every physical system" irrespective of its internal constitution, because its " N identical systems are mental copies of the one system under consideration"[3].

Two important distinctive features of the 1944 lectures, when compared to the 1925-26 papers, are:

(i) the complete relinquishment of the concept of wave packet, and,

(ii) the exclusive stress put on the field quantization formalism[4] which, for all statistical purposes, is equivalent to Schrödinger's initial quantized matter waves model.

True, the latter feature, namely the extreme confidence in the field quantization formalism, is somewhat surprising. How could Schrödinger be content with an abstract device which had apparently lost any contact with his former concrete picture of a discrete set of standing waves in a box? As I suggested at the end of section 1-3, the reason for his persistent adherence to the second quantization and the field quantization schemes is that he believed the gap between them and the wave picture is not as wide as it looks; he thought nothing prevented one from considering second quantization as an algebraic formulation of some concrete and continuous model, just in the same way matrix mechanics can be considered as the algebraic formulation of wave mechanics. By 1944, he had gained a very precise idea of what this concrete and continuous formulation could be. This breakthrough was related to his work in cosmology and unified field theory, which in turn had been initiated by his quite enthusiatic reception of Eddington's 1936 book[5]. The basic idea he borrowed from Eddington amounts to considering the "particles" as proper modes of vibration of the

[1]E. Schrödinger, "The so-called new forms of statistics", 1938, AHQP, microfilm 42, section 7

[2]E. Schrödinger, *Statistical thermodynamics*, Cambridge University Press (1st edition 1944; second edition 1952)

[3]ibid. p. 3

[4]ibid. p. 49

[5]A. Eddington, *Relativity theory of protons and electrons*, Cambridge University Press, 1936; After his first enthusiasm had died away, Schrödinger noticed several mistakes in Eddington's reasonings: "it is unfortunately not very hard to find major errors in this ingenious book", E. Schrödinger to A. Einstein, July 19, 1939, in: K. Przibram (ed.), *Letters on wave mechanics*, op. cit. p. 33

closed universe as a whole: "Einstein's finite universe is in itself the natural and wall-less box, which engenders atomicity by the necessary discreteness of its proper modes of vibration"[1]. Schrödinger's initial picture of quantized matter waves was not only retained and extended beyond the domain of a gas sample, but the boundary conditions it needed (the "box") was made completely "natural" by invoking the finiteness of the universe. Accordingly, holism acquired very strong explanatory virtues, for in Eddington's model "the conditions prevailing in our laboratory experiments are essentially determined by the state of the universe as a whole". The reference of quantum mechanics to the whole is made imperative, according to Schrödinger, by the fact it has "to abandon the idea of individuality in different particles of the same kind". Conversely the theory of the whole (namely general relativity) must also deal with the detailed structure of matter, for it "has quite unexpectedly provided the only sound means for explaining atomicity"[2]. This is by far the most radical move one can conceive towards the completion of a program which had already been outlined by Einstein and which was praised by Schrödinger, namely the "amalgamation between matter and space-time", or in other words the disappearance of any difference of nature between the stage (space-time) and the actors (matter)[3].

Schrödinger's work, from 1937 to the fifties, was therefore mostly devoted to these exciting developments of unified field theories. He studied the proper vibrations of the universe in the static case and in the expanding case[4], with the explicit twofold motivation that "wave mechanics imposes an *a priori* reason for assuming space to be closed; for then and only then are its proper modes discontinuous and provide an adequate description of the observed atomicity of matter and light. Einstein's theory of gravitation imposes an *a priori* reason for assuming space to be, if closed, expanding or contracting". He then started a thorough correlative study of spatio-temporal affine connections which turned out to be extremely close to Einstein's own investigations[5].

But, as A. Rüger[6] cogently pointed out, there was one fundamental difference between Einstein's and Schrödinger's perspectives. This difference bears on the description of matter. Einstein was not satisfied with his own way of dealing with the problem of matter (the corpuscular "sources of the field") in general relativity. According to him, the right-hand member of the gravitational field law, namely the momentum-energy tensor T_{ik}, was but a "provisional means of representing matter"[7],

[1]E. Schrödinger, "World structure", Nature, 140, 742-744, 1937

[2]ibid.; see also E. Schrödinger, "Sur la théorie du monde d'Eddington", Nuovo cimento, 15, 246-254, 1938

[3]ibid.; see a full exposition of this idea in: E. Schrödinger, *Nature and the Greeks,* op. cit. p. 14

[4]E. Schrödinger, "The proper vibrations of the expanding universe", Physica, 6, 899-912, 1939

[5]E. Schrödinger, "The final affine laws", Proc. Roy. Irish Acad. 51A, 163-171, 1947

[6]A. Rüger, "Atomism from cosmology: Erwin Schrödinger work on wave mechanics and space-time structure", HSPS, 18, 378-401, 1988

[7]A. Einstein, *The meaning of relativity,* (1921), Princeton University Press, 1974, p. 82

in so far as "in reality, matter consists of electrically charged particles". In order to go beyond this preliminary step, and in order to account for matter in geometrical terms, one would have to analyze first the regions of very intense fields which are the most natural counterpart of particles in a pure field theory. In 1945, Einstein thought he had virtually found the solution of this problem[1]. By contrast, Schrödinger was not very eager to look for particle-like solutions of the field equations. On the one hand he felt almost sure that this attempt would fail: "we know that the classical interaction of such dainty little (geometrical) toys is altogether not competent to describe the actual electromagnetic interaction of the ultimate constituents of matter"[2]. On the other hand, he did not even believe that this was necessary. The proper "theory of everything", whose program is defined in the introduction of his book *Space-time structure,* was not supposed to display any "ersatz" of the classical little individual re-identifiable bodies, but only to account for "atomicity" in the wider sense, namely to yield integers: "In so far as any progress in the more complex features of this interaction (...) has been made at all, it rests not on very complex classical solutions of the type alluded to above, but on much simpler ones, to wit plane sinusoidal waves which are just simple enough to be subjected to quantum mechanical considerations. (...) This way is not likely to lead over very complicated 'particle-like' solutions"[3]. Thus, in contrast with the 1925-26 gas theory which associated wave-packets with particles, the "cosmological gas theory" of 1938-50 was consistently holistic and plainly a-corpuscular. As Schrödinger explained most clearly to Einstein, his ultimate purpose was not to *substitute* something for the particles but to *replace* them by proper modes of vibration of the whole universe: "I believe that one has to introduce matter into the abstract general theory of relativity, which contains the T_{ik} only as 'asylum ignorantiae' (to use your own expression), not as mass points or something like that, but rather, shall we say, as quantized gravitational waves"[4].

2-3 *Holism and the three dimensions of space (1926)*

The conflict between Schrödinger's holistic views and his residual corpuscularian representations also manifested itself in another way. When a system of n particles is considered, ψ is a function in a 3n dimensional space. But then, what is the link between this ψ and the properties of *each* particle? Schrödinger had all the elements of this problem close at hand. In the *first* paper of his series "Quantization as a problem of proper values", he had already performed a general

[1]A. Einstein, "A generalization of the relativistic theory of gravitation", Ann. Math. Princeton, 46, 578; 47, 146,731, 1945

[2]E. Schrödinger, *Space-time structure,* Cambridge University Press, 1950, p. 116

[3]ibid.

[4]E. Schrödinger to A. Einstein, July 19, 1939, in: K. Przibram (ed.), *Letters on wave mechanics,* op. cit. p. 33

derivation of his equation for a many-particle system, and this involved a ψ-function in a 3n dimensional q-space. Then, in the second paper of the series, he made very clear that the wave-surfaces are surfaces in a q-space, and that he considered his theory as a mechanical equivalent of wave-optics in q-space[1]. The transition was not from 3-dimensional particle mechanics to 3-dimensional wave propagation, but from 3n-dimensional Hamiltonian mechanics to 3n-dimensional wave mechanics, or in other words from the motion of the image-point of a mechanical system in q-space to wave propagation in the same q-space. Now, there is a major difference between Hamiltonian mechanics and wave mechanics. From Hamilton's equations, one may calculate the (q,p) coordinates of the image point. These in turn can be separated in such a way that the i-th group (q_{1i}, q_{2i}, q_{3i}, p_{1i}, p_{2i}, p_{3i}) corresponds to the coordinates of the i-th particle of the system. In wave mechanics, the Schrödinger equation only enables one to calculate the *amplitude* of the 3n-dimensional ψ-function. And this 3n-dimensional amplitude is not unambiguously connected, in general, to the amplitude of n 3-dimensional wave-functions respectively associated to the n particles of the system. Schrödinger was also fully aware of this problem, to which he gave the first clear statement in June 1926: "(...)the difficulty of projecting the waves in q-space, when there are more than three coordinates, into ordinary three dimensional space and of interpreting them physically there"[2]. True, this sentence was written *after* Lorentz had raised the point in a letter of may 27, 1926. But in his answer to Lorentz, Schrödinger mentioned: "I have been very sensitive to this difficulty for a long time (...)"[3]. At this point, Schrödinger still thought that the problem could be solved in a straightforward manner. The end of the quoted sentence does not leave any doubt regarding this: "(I) believe that I have now overcome it".

The solution he proposed arose from his second (electrodynamic) interpretation of wave mechanics. Since the physical meaning had no longer to be sought in the amplitude of the ψ-function itself, but rather in the product $\psi\psi^*$, the problem no longer consisted in calculating n 3-dimensional wave-amplitudes from ψ, but rather in calculating n 3-dimensional densities of charge from $\psi\psi^*$. This result was obtained by a simple integration procedure:

$$\rho_i=-e\int_{r_1}\ldots\int_{r_{i-1}}\int_{r_{i+1}}\ldots\int_{r_n}\psi\psi^*d\mathbf{r}_1\ldots d\mathbf{r}_{i-1}d\mathbf{r}_{i+1}\ldots d\mathbf{r}_n$$

[1]E. Schrödinger "Quantization as a problem of proper values II", in: *Collected papers on wave mechanics,* op. cit. p. 18

[2]E. Schrödinger to H.A. Lorentz, June 6, 1926, in: K. Przibram (ed.), *Letters on wave mechanics,* op. cit. p. 55

[3]ibid.

In this formula, ρ_i represents the electric charge density of the i-th particle; the sum $\rho=\sum_i \rho_i$ then represents "the electric charge density in real space" for the system of n particles. However, as we shall point out in the next section, the electrodynamic interpretation making use of the product $\psi\psi^*$ could not *by itself* account for all the phenomena. The ψ-wave amplitude still retained a role at a certain stage, and Schrödinger soon recognized this new obstacle. Accordingly, when he worked out a more refined analysis of the "difficulty" of projecting the 3n-dimensional ψ-waves in ordinary space, he did not consider that the problem had just been solved in a satisfactory way by his integration method. Having demonstrated that the eigenfunctions of the wave equation for a system consisting of two particles are products $\phi_n\psi_m$ of the eigenfunctions ϕ_n for the first particle and of the eigenfunction ψ_m for the second particle, and that the general solution of the wave equation is a linear superposition of these products (with c-coefficients)[1], he complained in a letter to G. Joos that: "there are states of the combined system (i.e. c-distributions of this system) which just cannot be split into a c-distribution of the first system plus a c-distribution of the second. It is horribly trivial mathematically. An arbitrary linear combination of the *products* of eigenfunctions *cannot* in general even be represented as the product of two linear combinations"[2]. In his paper, he recognized he had to "(...) apply the *many*-dimensional form of 'wave mechanics' (...) instead of that *four*-(...) dimensional form which correspond more closely at the root of the matter, but which is meanwhile only prospective in character, because we do not yet understand how to formulate the problem for more than one electron by means of it"[3]. The strange point was that even though the necessity of introducing non-factorizable linear combinations of $\phi_n\psi_m$ products arises from the presence of coupling terms between the two parts of the system, this "entanglement" of the wave-functions persists *after* the interaction has vanished.

At the end of the hyper-creative years 1925-1928, Schrödinger was thus doomed to claim the irreducible holistic character of the wave-mechanical formalism, in connection with the 3n-dimentionality of the ψ-wave: "The ψ-function is in general (...) not a function of time and place, but it is a function of one, two, three ... places if the classical model of the system is made of one, two, three mass points. This is a very remarkable and deep-lying circumstance, which - I should mention - already makes the conception of the ψ-function as a collection of local states difficult"[4].

[1]This demonstration is given in: E. Schrödinger, "The exchange of energy according to wave mechanics" (1927), in: *Collected papers on wave mechanics*, op. cit. p. 140

[2]E. Schrödinger to G. Joos, November 17, 1926, AHQP, microfilm 41, translated by L. Wessels, *Schrödinger's interpretations of wave mechanics*, op. cit. p. 325.

[3]E. Schrödinger, "The exchange of energy according to wave mechanics" in: *Collected papers on wave mechanics*, op. cit. p. 137:

[4]E. Schrödinger, "Die Erfassung der Quantengesetze durch kontinuierliche Funktionen", Naturwissenschaften, 26, 486-489, 1929; translated by O. Darrigol, "Schrödinger's statistical physics and

In 1935, Schrödinger went one step further in his clear statement and analysis of the "entanglement" problem, in reaction to the Einstein-Podolsky-Rosen argument. In the very first sentences of his paper "Discussion of probability relations between separated systems", he noticed that "When two systems, of which we know the states by their respective representatives, enter into temporary interaction due to known forces between them, and when after a time of mutual influence the systems separate again, then they can no longer be described in the same way as before, viz. by endowing each of them with a representative of its own. I would not call that *one* but rather *the* characteristic trait of quantum mechanics, the one that enforces its entire departure from classical lines of thought"[1]. And he then summarized the situation in the following vivid manner: "the best possible knowledge of a whole does not necessarily include the best knowledge of all its parts, even though they may be entirely separated and therefore virtually capable of being 'best possibly known', i.e. of possessing, each of them, a representative of its own"[2]. Another "disconcerting" consequence of the holistic features of the wave mechanical formalism is what Schrödinger called the distant "steering" of a system, namely the fact that the ψ-function "arrived at for one system depends on the *programme* of observations to be taken with the other one"[3].

But although Schrödinger was certainly one of the most lucid analysts of the quantum mechanical holistic paradoxes in the mid-thirties[4], he appeared very skeptical concerning the possibility of incorporating them within a consistent world view. He still wondered whether the "entanglement" is not just a "convenient calculational trick", due to an unwarranted application of the non-relativistic formalism to situations which do not fall in its range of validity: "the conceptual joining of two or more systems into one encounters great difficulties as soon as one attempts to introduce the principle of special relativity into quantum mechanics"[5]. And he deemed that the reason for this postulated inadequacy of the non-relativistic formalism is that the "entanglement" seems to involve something like an "unretarded *actio in distans*" which can only take place if the whole system is small enough "to be able to neglect the time that light takes to travel across the system"[6].

some related themes", in: *Erwin Schrödinger, Philosophy and the birth of quantum mechanics,* op. cit. p. 259

[1]E. Schrödinger, "Discussion of probability relations between separated systems", loc. cit.; see a comment in: H.J. Treder and H.H. von Borzeszkowski, "Interference and interaction in Schrödinger's wave mechanics", Found. Phys., 18, 77-93, 1988

[2]ibid.

[3]E. Schrödinger, "Discussion of probability relations between separated systems", loc. cit.; see a comment in: M. Vujicic and F. Herbut, "Distant steering: Schrödinger's version of non-separability", J. Phys A (Math. Gen.) 21, 2931-2939, 1988.

[4]See especially B. Van Fraassen, "The Einstein-Podolsky-Rosen paradox", Synthese, 29, 291-309, 1974

[5]E. Schrödinger, "The present situation in quantum mechanics", loc. cit.

[6]E. Schrödinger, "Probability relations between separated systems", Proc. Camb. Phil. Soc., 32, 446-452, 1936

How are we to understand this Schrödinger's reluctance to accept all the consequences of wave mechanics? I think that, here again, we see the pre-1937 Schrödinger struggling in order to force a remnant of the corpuscularian representations into the holistic framework of wave mechanics. For, after all, the rebuttal of an instantaneous *"actio in distans"* (namely just the type of non-local effect that Einstein, Podolsky and Rosen[1] had ruled out from the onset) only makes sense if one considers that nature is made of localized parts of a system which may somehow communicate through their ψ-state. If one drops the localized interacting parts and retains exclusively the ψ-functions themselves, the very idea that the entanglement reflects some kind of permanent possibility of *action* of one part onto another becomes utterly irrelevant. Instead of mutual action, one should speak of "organicity" (like Bohm) or, more conventionally, of non-separability. As M. Lockwood pointed out, "it is crucial (...) to distinguish between non-local interactions and non-local entangled or correlated states"[2]. Accordingly, one of the available moves, in order to defuse the problems which are associated with the entanglement of ψ-functions, consists in *replacing completely* the semi-classical idea of interacting *parts* by the wave-mechanical description of a system as a *whole*. No plurality of objects, each of them having to be *in* a state: one overall state, full stop.

But this is *exactly* what Schrödinger himself did when he first formulated his own version of the gas theory. In his celebrated paper on the Einstein's gas theory[3], he already mentioned that the Bose-Einstein statistics *seems* to be underpinned by some unknown interaction between localized molecules, but that it would be much closer to "the true essence of the new theory" to treat the gas as a whole, and to describe it as a system of quantized standing waves in which the excitation number of a degree of freedom *replaces* the number of molecules in the corresponding state. This radical inversion of roles between the states and the bearer of the states, between the whole and the parts, had just to be transposed to the theory of atomic phenomena.

Now, the formalism of second quantization was perfectly suited for that purpose. It is thus not very surprising that, after he had completely integrated second quantization in his own system of thought in 1937-40, and after he had accordingly brought his criticism of the concept of individual corpuscle to its ultimate consequences, Schrödinger did not insist any longer on the possible disappearance of the entangled features of the quantum-mechanical description of phenomena in a future

[1]A. Einstein, B. Podolsky and N. Rosen, "Can quantum-mechanical description of physical reality be considered complete?", Phys. Rev., 47, 777-780, 1935; in: J.A. Wheeler and W.H. Zurek (eds.), *Quantum theory and measurement*, op. cit. p. 140: "(...)since at the time of measurement, the two systems no longer interact, no real change can take place in the second system in consequence of anything that may be done to the first system"

[2]M. Lockwood, "What Schrödinger should have learnt from his cat", in: M. Bitbol & O. Darrigol (eds.) *Erwin Schrödinger, Philosophy and the birth of quantum mechanics*, op. cit. p. 381

[3]E. Schrödinger, "Zur Einsteinschen Gastheorie", Phys. Z., 27, 95-101, 1926

relativistic theory. He rather pointed out that this entanglement (or the multi-dimensionality of the ψ-function) is just an appropriate expression of the fundamental lack of individuality of the so-called "particles"[1]; a perfectly appropriate formal counterpart of the priority given to the whole over the artificially isolated parts.

As for the apparent contradiction between the non-locality of a holistic wave-function and the locality of the macroscopic phenomena, it was virtually resolved by what I have called *Schrödinger's post-modern turn*. Indeed, according to Schrödinger (in the 1950's), the continuous wave-mechanical description is no longer supposed to provide a faithful reflection of "observable facts"[2] (or macroscopic phenomena), but only to be *connected* to them by means of a set of (probabilistic) correspondence rules. If the *non-local* wave-mechanical formalism eventually results in the successful probabilistic prediction of *local* macroscopic phenomena as well as of their (possibly EPR-like) correlations; and if, conversely, fulfilling the demand of *descriptive* locality, namely of locality of the entities which play a role in the formalism, yields inadequate predictions, then not only there is no true contradiction between the non-locality (or rather the a-locality) of the wave function and the locality of the observed macroscopic phenomena, but the latter is a predictive *consequence* of the former. In B. d'Espagnat's terms the underlying a-local quantum-mechanical representation yields the *"appearance* of a local world"[3], whereas no local formalism can yield an appropriate and exhaustive account of the appearances which are generated in our surrounding local world by a certain class of experimental manipulations; this is enough to ensure harmony in so far as "empirical reality" is concerned.

2-4 *Wave interpretation versus electrodynamic interpretation: a prehistory of the empirical correspondence rules*

Two interpretations of wave mechanics were put forward by Schrödinger during the first months of 1926: the pure wave interpretation and the electrodynamic interpretation. Both of them proved unable to account for the whole range of atomic phenomena. Both of them were successively relinquished by Schrödinger himself[4]. But in this section, I shall show that a certain *combination* of the two interpretations works; and that this combination is nothing else than a very sober version of contemporary quantum mechanics with its correspondence rules. The

[1]E. Schrödinger, *July 1952 colloquium*, (in: E. Schrödinger, *The interpretation of quantum mechanics (Dublin seminars 1949-1955 and other unpublished texts)*, op. cit. p; 32)
[2]E. Schrödinger, Science and Humanism, op. cit. p. 41
[3]B. d'Espagnat, "Appearance of a local world", Phys. lett., A 171, 17-20, 1992; *Veiled reality*, Addison-Wesley, 1995
[4]E. Schrödinger, "Quantization as a problem of proper values, IV", in: *Collected papers on wave mechanics*, op. cit. p. 120 (Schrödinger relinquishes the pure wave interpretation and adopts the electrodynamic interpretation); E. Schrödinger, *Four lectures on wave mechanics*, op. cit. p. 52 (here, Schrödinger aknowledges that the electrodynamic interpretation is "surely not quite satisfactory")

reason why Schrödinger did not realize this possibility will be discussed at the end of the section.

Schrödinger's first interpretation of wave mechanics consisted in taking the ψ-function at face value, and considering it as a faithful description of the processes which take place within the atoms: "we should try to connect the function ψ with some vibration process in the atom, which would more nearly approach reality than the electronic orbits, the real existence of which is being very much questioned today"[1]. This, at least, allowed one to account for Bohr's energy levels and quantization rules by a quite familiar standing wave model, and also to give correct predictions about the frequencies of the lines in certain situations where the old theory of quanta had failed. But several difficulties were still to be solved at this point. How is one to explain that the only observed frequencies are the *differences* between the oscillation frequencies of the proper modes rather than the latter frequencies themselves? How does the transition from some proper mode to another produce radiation? And how is it possible to calculate wave-mechanically the main characteristics of the lines, namely their intensity, their polarization, and their bandwidth? In order to answer the first question, Schrödinger proposed to consider that the proper frequencies have such large orders of magnitude that they are unable to "set the aether in motion"[2]. Radiation is thus produced not by the proper vibrations of the atoms, but by the *beats* resulting from the combination of two distinct proper vibrations: "one only needs to imagine that the light wave is causally related to the beats, which necessarily arise at each point of space during the transition"[3]. The characteristic frequency of these beats, namely "the number of times per second the intensity maximum of the beat-process repeats itself" is just identical to the difference of the proper frequencies. However, this circumstance is a purely mathematical one which does little to make us understand why one can *only* observe these differences and not neighbouring frequencies as well. As Lorentz pointed out, "no instrument (resonator, grating) whose operation is completely determined by linear equations would respond to these beats as it would to vibrations of frequency ν_1-ν_2"[4]. Moreover, no answer was given to the last question, namely the one bearing on the intensity, the polarization and the bandwidth of the lines.

A new perspective arose from Schrödinger's study of the equivalence between matrix mechanics and wave mechanics[5]. Heisenberg, Born and Jordan had succeeded in accounting for the intensities of the lines through

[1]E. Schrödinger, "Quantization as a problem of proper values, I", in: *Collected papers on wave mechanics,* op. cit. p. 9.

[2]E. Schrödinger, Letter to W. Wien, February 22, 1926, AHQP, Microfilm 41

[3]E. Schrödinger, "Quantization as a problem of proper values, I", in: *Collected papers on wave mechanics,* op. cit. p. 10

[4]H.A. Lorentz to E. Schrödinger, May 27, 1926, in: K. Przibram (ed.), *Letters on wave mechanics,* op. cit. p. 49

[5]E. Schrödinger, "On the relation between the quantum mechanics of Heisenberg, Born and Jordan, and that of Schrödinger" in: *Collected papers on wave mechanics,* op. cit. p. 45. See also "Quantization as a problem of proper values, III", in the same volume.

their matrix elements a_{kl}. Comparing the wave-mechanical expression for these matrix elements, and the classical expression of the z-component of the electric moment of a dipole which involves the electric charge density ρ, it was quite natural to assume that:

$$\rho=-eRe[\psi \frac{\partial \psi^*}{\partial t}],$$ and later on that:

$$\rho=-e\psi\psi^*,$$

where $\psi=\sum_k c_k \psi_k e^{2\pi i E_k t/h}$.

This being assumed, one had only to apply the laws of classical electrodynamics to the density $\rho=-e\psi\psi^*$ in order to calculate the main characteristics of the emitted radiation: "the intensity and polarization of the emitted light is thus intelligible on the basis of the Maxwell-Lorentz theory"[1].

Everything seemed to be clarified at this point:

(1) The appropriate frequencies $\nu_k-\nu_{k'}$ were the only ones which intervened in the expression of $\psi\psi^*$. This was enough to meet Lorentz' objection, in the very way Lorentz himself had suggested in his letter of may 27, 1926: replace the "beats" $\psi_k+\psi_{k'}$ by the "combination tones" $\psi_k\psi_{k'}$.

(2) The Schrödinger equation yielded a continuity equation for $-e\psi\psi^*$, which was perfectly isomorphic to the usual continuity equation for charge density.

(3) The Stark effect, the intensities, and the polarizations of the lines were accounted for by means of a classical electrodynamic calculation, using $\rho=-e\psi\psi^*$ as a source term.

(4) The stability and the lack of radiation of the atom in a proper state was explained. Indeed, "if only a single proper vibration is excited, the current component disappears and the distribution of electricity is constant in time"[2]. No accelerated charge could account for radiation in a proper state.

In spite of these startling successes, however, the pure electrodynamic interpretation soon showed its limits. The attempt at picturing the events which occur inside the atoms as classical electric charge density currents proved just as unsatisfactory as the pure wave representation:

(1') If one tries to take seriously the idea that the electron is but a cloud of electricity centered by the nucleus, the problem of the stability of this cloud must be addressed. True, the electromagnetic stability of the atom

[1]ibid. p. 47; see also p. 60: "(ψ) is perfectly capable of entering into the unchanged Maxwell-Lorentz equations between the electromagnetic field vectors as the 'source' of the latter"

[2]E. Schrödinger, "Quantization as a problem of proper values, IV", in: *Collected papers on wave mechanics,* op. cit. p. 123

can be explained by the stady state of the cloud, but one has then to explain how the cloud itself can be in a steady state. Unfortunately, this is not possible, as Schrödinger soon recognized himself. This impossibility was stated in a letter to Lorentz of June 1927, where Schrödinger wrote that if the cloud of electricity were ruled by classical electrodynamics, then the electron could not even "hang together"[1].

(2') The cloud picture is not consistent with the necessity of solving the Schrödinger equation with a potential corresponding to point-charge sources. As Lorentz pointed out, "if one alters the term $\frac{e^2}{r}$, one runs the risk of losing the correct eigenvalues for E"[2].

(3') As L. Wessels noticed, there is a striking asymmetry between the way Schrödinger dealt with the characteristics of the emission lines and his study of the effect of incident radiation: "in calculating radiation intensities, he applied classical electrodynamics to the charge cloud determined by wave mechanics (...). But in determining the effect of radiation falling from without, Schrödinger did not correspondingly calculate the effect of such radiation on the charge cloud already present"[3]. He first solved his equation for ψ, using the point charge potential, and "only then did he turn to his electrodynamic interpretation to guide calculation of the intensity and frequency distribution of the secondary radiation thereby produced"[4]. Similarly, when he attempted to describe the Compton effect[5], he could not avoid using extensively the wave formalism, thus confining his electrodynamic interpretation to some peculiar features of the behaviour of the electrons.

I think that these difficulties of the electrodynamic interpretation of quantum mechanics all merge into one. Let me describe it. As we have noticed previously, it is just as inappropriate to consider that the events occurring within the atom are described by an electric cloud $-e\psi\psi^*$ ruled by the laws of classical electrodynamics, as it is to consider that they are described by a semi-classical vibration process. All the calculations must be performed by using the wave formalism (namely the Schrödinger equation) with point-source potentials, whereas the square modulus of the wave function must only be calculated *at the very last stage,* in order to bridge the gap between the formalism and the "observed facts" (such as line intensities, polarization etc). Much attention has to be paid in order not to mix up the two stages of the process, and thus to avoid the series of misgivings which led to the final dismissal of Schrödinger's interpretation(s) of quantum mechanics from 1927 on. Schrödinger was

[1]E. Schrödinger to H.A. Lorentz, June 23, 1927, AHQP, microfilm 41

[2]H.A. Lorentz to E. Schrödinger, June 19, 1926, in: K. Przibram (ed.), *Letters on wave mechanics,* op. cit. p. 71

[3]L. Wessels, *Schrödinger's interpretations of wave mechanics,* op. cit. p. 265. This asymmetry can readily be detected by comparing §5 and §9 in: *Four lectures on wave mechanics,* op. cit.

[4]ibid.

[5]E. Schrödinger, "The Compton effect", Ann. der Phys., 82, 1927; in: *Collected papers on wave mechanics,* op. cit. p. 124

right to hold on to the principle of superposition throughout the calculation of the wave function, thus derivating the proper interference effects. But he was wrong to take it at face value during the last stage of the experiment, and to perform his final $-e\psi\psi^*$ assessment as if the terms of the superposition *still* retained actual "simultaneous existence", with the status of terms of a distribution of charge whose electromagnetic radiation is ruled by Maxwell-Lorentz equations. At any rate, this latter conception was clearly dismissed by an experiment performed by E. Gaviola in 1929[1]. Even though this experiment could still have been accomodated with a moderate version of Schrödinger's electrodynamic interpretation, namely one in which the proper boundaries were chosen between the intermediate ψ-calculation and the final $-e\psi\psi^*$ assessment, it was taken as a refutation of Schrödinger's conception, and accordingly as a proof of Born's interpretation (see below).

To summarize, the wave-like picture, by which the atomic processes were supposed to be described in Schrödinger's two first papers on wave mechanics, and which cannot be dispensed with in intermediary calculations, is not isomorphic to the "observed facts". Conversely, the electrodynamic picture, which is perfectly adapted, with some precautions, to yield many important features of the "observed facts", cannot be used for the calculations between them. Hence a new type of dissociation between the theoretical representation and the facts, and the correlative necessity to consider the electrodynamic picture as a set of empirical correspondence rules rather than a proper description of whatever fraction of the atomic processes.

It is widely accepted that it was Born, with his probabilistic interpretation of quantum mechanics, who first realized the necessity of ascribing to $\psi\psi^*$ the status of an empirical correspondence rule rather than that of a faithful (and even 'mimetic') description of microscopic phenomena. But actually, things are much more intricate. Interpreting a certain expression as a density of probability does not by itself solve all the problems.

Firstly, the difficulty of stating explicitly the boundary between the situations where one must use a ψ-calculation, and those where a $\psi\psi^*$-calculation is appropriate, proved to be just as considerable within Born's framework of thought as it was in Schrödinger's. In his paper about the collision theory, Born started to identify the probability with a certain component of ψ, and it was only in a short footnote added in proof that he mentioned the *squared* component of the ψ-function as a better candidate for the role of probability of a certain outcome of the collision[2]. Then, after Pauli's contribution[3], Born's probabilistic interpretation focused on

[1]L. Wessels, *Schrödinger's interpretations of wave mechanics,* op. cit.; also: E.M. Mac Kinnon, *Scientific explanation and atomic physics,* The University of Chicago Press, 1982, p. 260

[2]M. Born, "On the quantum mechanics of collisions", in: J.A. Wheeler and W. H. Zurek, *Quantum theory and measurement,* op. cit. p. 52

[3]W. Pauli to W. Heisenberg, October 19, 1926, quoted by O. Darrigol, *From c-numbers to q-numbers,* University of California Press, 1992, p. 335

$\psi\psi^*d\tau$. ψ itself was only an intermediate mathematical fiction allowing one to formulate the probability laws which rule the motion of the particles. This choice being made, the necessity of performing calculations on ψ itself rather than directly on $\psi\psi^*$, became somewhat mysterious (the only clue being the correspondence between the probability and the intensity of a classical wave). And it was even made much more obscure by the *subjective* interpretation of probabilities which was adopted more or less explicitly by Born from the outset. If $\psi\psi^*$ merely reflects our ignorance about the particle motion, about its going here (event e) or there (event e'), why is it that the probability p(e or e') is not equal to the sum of the probabilities p(e) + p(e'), but rather contains *interference* terms?

Secondly, a probabilistic interpretation must state most clearly what $\psi\psi^*d\tau$ is the probability *of.* As it appears from the previous discussion, Born was manipulating at least four definitions of the type of event which $\psi\psi^*d\tau$ is supposed to be the probability of, sometimes mixing them or leaving some ambiguity about the one he was referring to:

(i) *The outcome of an experiment,* namely the only acceptable definition of an event in an operationalistic framework of thought. In this case, the dividing line between the domain where ψ-calculations have to be performed and the situations where $\psi\psi^*$ probabilistic evaluations can be used, is very easy to draw. ψ-calculations apply *between* experiments (or between a preparation and an experiment), whereas $\psi\psi^*$ probabilistic evaluations are only relevant for experimental outcomes.

Even though Born sometimes comes very close to such an operationalistic definition of the type of event to which the $\psi\psi^*$ probabilistic evaluation apply, it is difficult to find it in isolation when going across his writings. For instance, in a sentence like "One gets no answer to the question 'what is the state after the collision', but only to the question 'how probable is the specified outcome of the collision'"[1], the word *outcome* could either mean *the intrinsic direction of particle scattering,* thus pointing to case (ii) below, or *a spot on a scintillation screen,* thus restricting the conception of what probabilities are about to a set of operationally defined experimental outcomes.

(ii) *The particle intrinsic scattering direction, or its intrinsic motion:* "$\Phi_{n\tau m}(\alpha,\beta,\gamma)$ gives the probability for the electron, arriving from the z-direction, to be thrown out into the direction designated by the angles

[1]M. Born, "On the quantum mechanics of collisions", in: J.A. Wheeler and W. H. Zurek, *Quantum theory and measurement,* op. cit. p. 54

α,β,γ (...)"[1]. And, more generally: "the motion of particles is ruled by the laws of probability (...)"[2].

(iii) *The particle intrinsic location* : " $|\psi|^2$dv is the probability that the electron (regarded as a corpuscle) *is* in the volume element dv"[3].

(iv) *The particle location (or motion) as it is* found *in an appropriate experiment* : " $|\psi|^2$dv is the probability that an electron will be *found* precisely in the volume element dv"[4]. The idea that an experimental outcome consists in *finding a particle* in such and such volume of the (p,q)-space is popular but it is quite ambiguous: it implies referring to particles, but it also restricts the scope of probabilities to what is *found* experimentally about them. It is somehow intermediate between the purely operational interpretation of definition (i), and the two quasi-classical corpuscularian interpretations ((ii) and (iii)).

Schrödinger had no *a priori* reluctance to accept the *first* (purely operational) definition of the set of events. As many physicists, he was perfectly aware of the fact that, when discrete experimental events are at issue, the concept of intensity (or more precisely of line intensity), is but a way to express the probability of these events. And since, in Schrödinger's electrodynamic interpretation, $-e\psi\psi^*$ is the appropriate quantity which enables one to calculate the line intensities, nothing really new is added if it is rather said: *$\psi\psi^*$ is the appropriate quantity which enables one to calculate the probability of the corresponding spots on a scintillation screen.* In the first paragraph of his paper of spring 1926 on the relations between Heisenberg's matrix mechanics and his wave mechanics, Schrödinger aknowledged explicitly a correspondence between: (a) the transition probabilities (which can be considered, according to the most prudent interpretation, as the projection in the matrix-mechanical model of the probabilities of the experimental events), (b) the intensities of the spectral lines, and (c) the amplitudes of oscillation of the electric moments of the atomic dipoles[5]. As we have mentioned in §1-6, the only true concern of Schrödinger was about the attempt at going one step further and interpreting $\psi\psi^*$ as the probability

[1]ibid. p. 54. Mara Beller ("Born's probabilistic interpretation: a case study of 'concepts in flux'", loc. cit.) has rightly pointed out the importance of Born's having focused his original probabilistic interpretation on the energy and momentum variables rather than on position variables.

[2]M. Born, "Quantenmechanik der Stossvorgänge", Zeitschrift für physik, 38, 803-827, 1926

[3]M. Born, *Atomic Physics,* op. cit., p. 139

[4]ibid. p. 140

[5]E. Schrödinger, "On the relation between the quantum mechanics of Heisenberg, Born and Jordan, and that of Schrödinger" in: *Collected papers on wave mechanics,* op. cit. Schrödinger's further investigations on probabilities are to be found in: E. Schrödinger, "Sur la theorie relativiste de l'électron et l'interprétation de la mécanique quantique", Ann. Inst. H. Poincaré, 2, 269-310, 1932, and E. Schrödinger, "The foundations of the theory of probability", Proc. R.I.A., 51A, 51-66, 141-146, 1947. A very interesting comment on the 1932 paper can be found in: J.C. Zambrini, "Probability in quantum mechanics according to Schrödinger", Physica B 151, 327-331, 1988.

of some stochastic microscopic process occurring between *the preparation and the detection stage of the experiment.* In May-June 1926, the electrodynamic interpretation was aimed at avoiding any recourse to such intermediate stochastic processes which were so strongly suggested by a (too) straightforward realist interpretation of the transition probabilities as the probabilities of some quantum jumps occurring by themselves in the atoms. At that time, the only way Schrödinger could challenge an ontological interpretation of the transition probabilities was to replace it by an ontological interpretation of the density $-e\psi\psi^*$: "(...)one obtains for an atom with many electrons exactly what Born-Heisenberg-Jordan designate as the transition probability, with the new and plausible meaning 'component of the electric moment'"[1]. Later on, after the formulation of Born's probabilistic interpretation of quantum mechanics, a proper analysis of Schrödinger's criticisms shows that they were once more directed against the spurious attempt at describing the intermediate microscopic events by means of a stochastic model, whereas they did *not* preclude the simple use of $\psi\psi^*$ as an algorithm to calculate the probability (or intensity) of the *final* discrete outcome of a spectroscopic study. Indeed, the major problem Schrödinger raised was that, if $\psi\psi^*$ only represents the probability of an underlying "definite" microscopic configuration, and if, accordingly, "it does not relate to a single system at all but to an assemblage of systems", then we are offered "no explanation whatever why the quantities a_{kl} yield all the information which they do yield"[2]. By contrast, the electrodynamic representation, which applies to each individual system, provided one with a satisfactory explanation of the empirical content of Heisenberg's matrix elements, namely their ability to yield the intensities (or equivalently the probabilities of the *final* discrete experimental events).

Schrödinger was thus confronted very early with one of the central dilemmas of quantum mechanics. Its best known expression was given by Bohr at the fall of 1927: the description of phenomena in space-time and the causal description are "complementary"[3]. As we shall see in paragraph 6-2, one of the plausible interpretations of this original version of Bohr's complementarity was that the description of phenomena in space-time means the attempt at reconstituting particle paths from sequences of measurements submitted to the uncertainty principle, whereas the causal description refers to the (deterministic) wave-mechanical law of evolution. In terms that Schrödinger could have used in 1926, one would formulate this idea as follows: we need not one but *two* representations of the atomic phenomena. The first one, namely the electrodynamic representation, is directly related to the "observed facts", but it does not

[1]E. Schrödinger to H.A. Lorentz, June 6, 1926, in: K. Przibram (ed.), *Letters on wave mechanics,* op. cit. p. 56

[2]E. Schrödinger, *Four lectures on wave mechanics,* op. cit. p. 52

[3]N. Bohr, *Atomic theory and the description of nature,* Cambridge University Press, 1934; see also a clear exposition in: W. Heisenberg, *The physical principles of the quantum theory,* The University of Chicago Press, 1930, p. 65.

provide the link *between* subsequent observed facts; in a word, it is "factual" but not "effective". The second one, namely the wave representation, is perfectly able to provide a link between the observed facts, but not to account for certain facts; it is "effective" but not "factual". As long as Schrödinger wanted to merge the "effective" and the "factual" into a single representation, according to the classical ideal, the persistent duality of the models had to be considered as a symptom of failure. But as soon as he accepted, according to the post-modern turn, to dissociate completely the representation from what it is usually supposed to be a representation of (namely the facts), the aim was no more to unify two representations but to define a relation between them. One of the two initial representations (the "effective" wave-representation) accordingly gained a privileged status, whereas the other one (the "factual" electrodynamic representation) was considered as nothing more than a formulation of the sought link with the observed facts. In other words, the expressions which were once considered typical of the electrodynamic interpretation were ascribed the status of an empirical correspondence rule.

The 1952 Dublin seminar, entitled *Transformation and interpretation in quantum mechanics* [1], is partly devoted to an analysis of this link (that we call the empirical correspondence rules, and that Schrödinger called "the interpretation" in a restricted sense of the word): "The wave-function, characterizing the state of the system, and the operator, characterizing the experimental device, together are supposed to give a certain information on the observed value. We shall discuss this connection - which we call the interpretation - in detail"[2]. This connection can be expressed in two ways: one could either (i) take (the expression of the expectation value *of the experimental results* + the Bohr correspondence principle about observables) as a compound axiom, and derive the probability algorithm in a diagonal frame as a theorem; or, alternatively, (ii) take the probability algorithm in a diagonal frame as an axiom and derive the expression of the expectation value as a theorem. For most contemporary authors, the two procedures are allowed, and they can indifferently be used according to the type of calculation one wishes to perform[3]. Schrödinger rather insisted on their epistemological differences and chose to focus on the first one for reasons which are deeply connected to his own interest in the quantum theory of measurement (see §4-3).

At any rate, the initial conflict between the wave-interpretation and the electrodynamic interpretation was solved at this point. It retrospectively appears as a (too) concrete projection of the methodological requirement concerning the link between the representation and the observed facts.

[1]in: E. Schrödinger, *The interpretation of quantum mechanics (Dublin seminars 1949-1955 and other unpublished texts)*, op. cit. p. 39
[2]ibid. p. 50
[3]B. d'Espagnat, *Conceptual foundations of quantum mechanics*, W.A. Benjamin, 1976, p. 30 f.

2-5 *The lack of pictures*

Schrödinger's desire to draw a clear picture of the atomic processes, that is to say his alleged incapacity to renounce the images in physics, are often ascribed to his attachment to the *bild*-conception tradition of german physics which prevailed during the second half of the nineteenth century[1]. His very insistence on the idea that the theoretical picture does not necessarily mimic the observed facts has a (somewhat weaker) counterpart in Hertz' and Boltzmann's[2] writings. To begin with, Hertz thought that the aim of physics was to reduce natural processes to mechanics. This implied ascription of a preeminent role to pictures in physics, which was clearly explained in the very first page of the introduction of Hertz' well-known treatise of mechanics: "The most direct and in a sense the most important problem which our conscious knowledge of nature should enable us to solve is the anticipation of future events, so that we may arrange our present affairs in accordance with such anticipations. (...) In endeavouring thus to draw inferences as to the future from the past, we always adopt the following process. We form for ourselves images or symbols of external objects; and the form we which we give them is such that the necessary consequents of the images in thought are always the images of the necessary consequents of the things pictured"[3]. In this sentence, however, one finds some interesting reservations. It is not said that pictures are required at any cost, but rather that *we always adopt*, as a rule, a procedure involving the use of pictures. Moreover, Hertz insisted that no other similarity had to be sought between the picture and the things it pictured than the set of relations (or chains of consequences) it enabled one to anticipate. As for Boltzmann, he insisted even more directly that: "It is precisely the unclarities of the principles of mechanics that seem to me to derive from not starting at once with hypothetical mental pictures but trying to link up with experience from the outset"[4]. A picture which has not been freed from the necessity of strict mimicry of the experimental facts is thus, according to Boltzmann, in great danger of becoming "unclear".

Schrödinger's life-long praise of the notion of cultural tradition in science[5], and his repeated references to Boltzmann's epistemology, makes the historical interpretation of his position quite compelling. But, here,

[1]see e.g. S. d'Agostino, "Continuity and completeness in physical theory: Schrödinger's return to the wave conception of quantum mechanics in the 1950's", in: M. Bitbol and O. Darrigol (eds.), *Erwin Schrödinger, philosophy and the birth of quantum mechanics*, op. cit. p. 339

[2]S. d'Agostino, "Boltzmann and Hertz on the Bild-conception of physical theory", History of Science, 28, 380-398, 1990.

[3]H. Hertz, *The principles of mechanics*, Mac Millan, 1899

[4]L. Boltzmann, *Theoretical physics and philosophical problems*, (B. Mac Guinness, ed.), Reidel, 1974, p. 225

[5]E. Schrödinger, "Ist Naturwissenschaft milieubedingt?", in: *Über Indeterminismus in der Physik*, Barth, Leipzig, 1932; Trad.: *Science, Theory and Man*, Dover, 1957; E. Schrödinger, "Are there quantum jumps?", loc. cit.

we cannot content ourselves with asserting the historical influences to which he was submitted. Several questions have to be addressed about the personal way Schrödinger was carrying his burden of tradition, in order to sketch a coherent view of his own position. Was this burden recognized by him through direct reference to history, or did it (also) take other, more cryptic, forms? Was he completely dominated by this tradition, or did he submit it to critical scrutiny? And, finally, what were the original views he arrived at, by combining historical influences in some unique proportions?

Let us begin with some bare statements of Schrödinger's need for pictures. His boldest expression of this need is contained in a paper of 1948 about the historical and cultural significance of science: "The picture is not only a permissible tool, but also a goal"[1]. Such an order of priorities clearly departs from Mach's conception of the theoretical picture as a tool for the purpose of economy of thought. It is rather strikingly reminiscent of the views Boltzmann expressed in his popular lectures: "I am of the opinion that the task of theory consists in constructing a picture of the external world. (...) (The guiding star of the physicists) was not practical gain but the picture of nature within the intellect"[2]. This stress put on pictures as a final aim played, at any rate, an important role in the history of quantum mechanics, by prompting Schrödinger to formulate wave mechanics, in spite of his being perfectly aware of matrix mechanics. For he was just looking for something matrix mechanics could not provide: "I naturally knew about this theory, but I was discouraged, if not repelled, by what appeared to me as a very difficult method of transcendental algebra, and by the want of *anschaulichkeit*" [3].

These assertions, which amount to considering the picture as an aim by itself, or which apparently recommend recovering *anschaulichkeit* at any cost, sound quite dogmatic. They sound as if Schrödinger had imitated for his own sake, some centuries later, the "fathers of modern science" who, by reviving Greek science and philosophy, took over "pre-conceived ideas and unwarranted assumptions"[4]. In his case, everything looks as if he took over the pre-conceived ideas and unwarranted assumptions about *anschaulichkeit* of the previous generation of classical physicists. And everything also looks as if these preconceived ideas were deeply rooted in the most affective layer of his view of the world.

The impression that his taste for pictures was at least partially grounded in a deep-lying and emotively loaded presupposition, is supported by an oft-quoted sentence from the second paper of his series

[1]E. Schrödinger, "Die Besonderheit des Weltbilds der Naturwissenschaft", Acta Physica Austriaca, 1, 201-245, 1948; English translation in: E. Schrödinger, "On the peculiarity of the scientific world-view", *What is life? and other essays,* Doubleday anchor, 1957

[2]L. Boltzmann, *Theoretical physics and philosophical problems,* op. cit. p. 33

[3]E. Schrödinger, "On the relation between the quantum mechanics of Heisenberg, Born and Jordan, and that of Schrödinger" in: *Collected papers on wave mechanics,* op. cit. p. 46 (footnote)

[4]E. Schrödinger, *Nature and the Greeks,* op. cit. p. 16

"Quantization as a problem of proper values": "it has even been doubted whether what goes on in the atom could ever be described within the scheme of space and time. From the philosophical standpoint, I would consider a conclusive decision in this sense as equivalent to a complete surrender. For we cannot alter our manner of thinking in space and time, and what we cannot comprehend within it, we cannot understand at all"[1]. This conservative statement is all the more striking since Schrödinger insisted repeatedly on the historical relativity of so many human forms of thought[2], and since he later insisted "that one is very easily deceived into regarding an acquired habit of thought as a peremptory postulate imposed by our mind on any theory of the physical world"[3]. What is then the nature of the deeply entrenched presupposition which led Schrödinger to make an exception and to reject any mobility of our manner of thinking *(Denkformen),* when our tendency to draw spatio-temporal pictures is at stake? There are two possibilities. Either this presupposition had historical roots, or it had metaphysical roots. Either Schrödinger thought this manner of thinking is made *almost* immutable because it represents one of our most venerable legacies (a legacy which was, moreover, part of his own *weltanschauung),* or he held it is *completely* immutable, due to a compelling metaphysical reason.

In favour of the first possibility, there are several texts where Schrödinger grounds his own conception of quantum mechanics in the necessity to unify the various parts of our cultural inheritance. In August 1926, he wrote a letter to Wien where he explained: "Bohr's standpoint, that a space-time description is impossible, I reject *a limine*. Physics does not consist only of atomic research, science does not consist only of physics, and life does not consist only of science"[4]. And much later, in 1952, he rejected the current (non-figurative) interpretation of quantum mechanics by arguing that it takes the risk of "getting severed from its historical background", just in the same way as abstract (non-figurative) painting has shaken off "the indebtedness to our predecessors"[5]. Following this trend of thought, the loss of pictures in physics is logically conceivable, however culturally disastrous. But accordingly, valuing pictures in physics appears just as metaphysically weak as it is historically sound.

One is thus led to wonder whether Schrödinger's resistance to the relinquishment of pictures could not have had a deeper, and metaphysically stronger, reason. A metaphysical reason which nothing prevents us however from considering as an hypostasized version of the

[1]E. Schrödinger, "Quantization as a problem of proper values (II)" in: *Collected papers on wave mechanics,* op. cit. p. 26-27

[2]E. Schrödinger, *Nature and the Greeks,* op. cit. p. 17; E. Schrödinger, "Ist Naturwissenschaft milieubedingt?", loc. cit.

[3]E. Schrödinger, *Science and Humanism,* op. cit. p. 48

[4]E. Schrödinger to W. Wien, August 26, 1926, quoted and translated by W. Moore, *Schrödinger, Life and thought,* Cambridge University Press, 1989, p. 226.

[5]E. Schrödinger, "Are there quantum jumps?", loc. cit. p. 109

historical reason, or in other terms as an unconscious burden of culturally relative tradition under metaphysically absolutized disguise, for after all it also derives from some wide-ranging philosophical inheritance. This reason is indeed related to the "doctrine of identity" that Schrödinger borrowed from the Indian *Vedanta,* that he also related to some trends of western mysticism, and that he shared with Schopenhauer. According to this monistic doctrine, everything happens on the surface of "this one thing - mind or world"[1]; there is nothing like a world-in-itself acting on our senses; there is no ontological duplication (even though there can be a methodological one), between a representation and what it is supposed to reflect: "the story is occurring only once, not twice"[2]. With such a strong sense of the identity between the world and the world-picture, it becomes clear that the latter is not something which could be suppressed without harm. One may easily conceive an alteration of some of the historically inherited "special features" of our scientific world-picture[3], but certainly not a breaking of its general (spatio-temporal) frame[4]. For to modify this frame is to modify the world itself. Here, the loss of pictures in physics is not only culturally dangerous; it appears to be metaphysical nonsense.

Having reached Schrödinger's highest point of resistance to the loss of pictures in physics, we must now realize that he by no means contented himself with proclaiming repeatedly his position from a sort of historical or metaphysical stronghold. However deeply he may have been persuaded of the metaphysical inevitability of picturing the world in general, he fully acknowledged the necessity of giving epistemological justifications to our need for pictures in each particular science and at each particular stage of history. For, after all, the observed discontinuous facts and the theory, which have a distinct methodological status, were both, according to his own metaphysics, only *part of* the overall picture of the world. It was thus not *a priori* obvious that the theory itself had to be a true picture, in the same sense as the overall one in which it was embedded. The only thing Schrödinger's metaphysics made definitely impossible to believe was that the discontinuity of the experimental events is a reflection of some discontinuous processes occurring "out there", and that it is therefore *in principle* vain to look for a continuous theoretical picture to which the experimental events be connected. All things considered, the special issue of pictures in physics was left quite open by Schrödinger. In spite of his personal and philosophical reluctance to do so, he then considered very seriously the possibility of dispensing with pictures at some stage of physics. Let us see how he proceeded.

[1]E. Schrödinger, *Mind and matter,* op. cit. p. 157; see also p. 138 for another striking version of this identity: "The reason why our sentient, percipient and thinking ego is met nowhere within our scientific world-picture can easily be indicated in seven words: because it is itself that world picture"

[2]ibid. p. 156

[3]E. Schrödinger, *Mind and matter,* op. cit. p. 88

[4]see M. Bitbol, *La Clôture de la représentation,* in: E. Schrödinger, *La nature et les grecs,* Seuil, 1992, p. 27-33 for a more detailed discussion.

Although he did not very much appreciate the light-heartedness with which positivist thinkers jettison the venerable attempt at understanding nature by moulding a picture, Schrödinger underlined that, nevertheless, the mere unification of the observed facts into a system of mathematical relations "(...) is so striking and interesting, that for our eventual grasping and registering them the term 'understanding' seems very appropriate"[1]. He even pointed out that there might exist, in some situations, very sound reasons for preferring a non-figurative theory over a figurative theory. In March 1926, he was trying to compare Heisenberg's matrix mechanics and his own wave mechanics. If he had been so dogmatic about the necessity of pictures in physics, he could have contented himself with proving that matrix mechanics can be derived from wave mechanics. Together with the fact that wave mechanics allowed a figurative interpretation, this would have prompted him to declare both the mathematical derivativeness of matrix mechanics and the epistemological superiority of wave mechanics. But he did not adopt this straightforward strategy. He rather pointed out that "(...) there might perhaps appear to be a superiority in the matrix representation because, through its stifling of intuition, it does not tempt us to form space-time pictures of atomic processes, which must perhaps remain uncontrollable"[2]. He thus felt obliged to perform the demonstration the other way round, i.e. to show the possibility of deriving wave mechanics from matrix mechanics, in order to make it clear that the wave-functions "(...) do not form, as it were, an *arbitrary* and *special* 'fleshy clothing' for the bare matrix skeleton, provided to pander to the need for intuitiveness"[3]. In so far as the (proper) wave-functions can be constructed from given matrices, just as the matrices can be constructed from the (proper) wave-functions, one cannot contend any longer that wave mechanics is an *ad hoc* device for creating pictures.

This very lucid requirement of reciprocal equivalence shows that, unlike the later proponents of hidden variable theories, Schrödinger did not consider it satisfactory to add an empirically void "clothing" to the structure of quantum mechanics just for the sake of recovering the classical ontology or for the sake of satisfying the desire of pictures. What he wished to demonstrate was rather that there exists an adequate picture and a (non-classical) ontology which arises quite naturally from unmodified quantum mechanics itself. He did not want to recover pictures at any cost, but only to show that the present situation of physics was not as averse to space-time picturing as the Göttingen-Copenhagen physicists had believed. According to him, the only reason why his Göttingen-Copenhagen colleagues thought they were bound to abandon the dream of framing a picture of the physical processes in space-time, was that the

[1]E. Schrödinger, *Nature and the Greeks*, op. cit. p. 90

[2]E. Schrödinger, "On the relation between the quantum mechanics of Heisenberg, Born and Jordan, and that of Schrödinger", in: *Collected papers on wave mechanics*, op. cit. p. 58

[3]ibid.

corpuscularian representation, with its well-defined trajectories of individual material points, had failed: "all these assertions systematically contribute to the relinquishing of the ideas of 'place of the electron' and 'path of the electron'"[1]. This failure had been incorporated in both matrix mechanics and wave mechanics, but it did not preclude *a priori* the possibility of outlining another, non-corpuscularian type of image. The wave-picture was, of course, considered by Schrödinger as the most likely candidate for replacing the particle-picture. But there was a persistent obstacle to this replacement, quite similar to the one which hindered a full development of Schrödinger's holistic views until the late thirties. This obstacle was the necessity, as long as the link between micro- and macro-mechanics had not been clarified, to retain something of the old corpuscularian representation (for instance the concept of wave-packet) in order to make sense of localized and discontinuous experimental phenomena.

The problem was thus not only one of replacement of an image by another, but one of accomodation of a picture with the other one: "(...)it seems to me that the only reason for the iconoclastic uproar is the following: the corpuscle concept has, it is true, become the unquestioned and inalienable possession of the physicist who continuously uses it as a mental construct (...) but (...) it leads to considerable embarassment because we have not yet succeeded in fusing it with the wave concept"[2]. However, when the concept of corpuscle came to be "questioned" systematically by Schrödinger, the difficulty of the "fusion" faded away, and the attention reverted to the perspective of truly *replacing* every previous pictures by the wave picture. The link between micro- and macro-mechanics accordingly took a new (and completely non-corpuscularian) aspect, which associated a clear statement of the empirical correspondence rules and an attempt at solving the measurement problem.

This association of a thorough criticism of the atomistic pictures with a promotion of the wave picture, provides us with a very striking example of Schrödinger's own way of associating Mach's positivistic influence with Boltzmann's Bild-conception of physical theories. Schrödinger considered that Mach's philosophy had in some way come closer than any other one to a proper metaphysical foundation of physics. For he accepted that the world is essentially a construct out of an enriched set of Machian "elements"[3]. And he also acknowledged that positivism may help to fight against the tendency manifested by many physicists to confuse a

[1]E. Schrödinger, "Quantization as a problem of proper values I", in: *Collected papers on wave mechanics,* op. cit. p. 26

[2]E. Schrödinger, "Die Besonderheit des Weltbilds der Naturwissenschaft", Acta Physica Austriaca, 1, 201-245, 1948; English translation in: E. Schrödinger, "On the peculiarity of the scientific world-view", *What is life? and other essays,* op. cit.

[3]E. Schrödinger, *Nature and the Greeks,* op. cit. p. 92; *Mind and matter,* op. cit. (chapter 1, first sentence). By saying so, Schrödinger was quite close from Boltzmann himself who claimed that "(...) not only matter, but also other people are for me mere mental symbols, just an expression of equations between complexes of sensations" (L. Boltzmann, *Theoretical physics and philosophical problems,* op. cit. p. 15)

descriptive achievement with a proper understanding of the phenomena[1]. But on the other hand, he believed that Mach's metaphysics and methodological criteria were but an analytical step before fulfilling Boltzmann's epistemological requirements. His mixture of positivist-like acuteness in discriminating the empirical and the theoretical sides of the scientific problems, and of determination to overcome this purely analytical stage of thought, was thus deservedly praised by Einstein[2] who had to use the same two-step strategy in order to clear up the field of concepts on which he was to build the theory of relativity.

We must now look more closely at how Schrödinger managed to articulate his positivistic trends with the high standards he wished to impose onto theories. Firstly, he thought that Boltzmann's urge for clear 'pictures' did not contradict Mach's warnings: "Boltzmann's idea consisted in forming absolutely clear, almost naïvely clear and detailed 'pictures' - mainly in order to be *quite* sure of avoiding contradictory assumptions. Mach's ideal was the cautious synthesis of observational facts that can, if desired, be traced back to the plain, crude sensual perception (pointer reading). (...) However, we decided for ourselves that these were just two methods of attack and that one was quite permitted to follow one or the other provided one did not lose sight of the important principles that were more strongly emphasized by the followers of the other one, respectively"[3]. The prescription to picture complemented the positivist-like cautiousness, by encompassing a proper recognition that the "elements" which are used in our construction of the world are not only bare sensations but also images and thought, and that "imagination and thought take an increasingly important part (...) as science, knowledge of nature, progresses"[4].

Secondly, Schrödinger had sound reasons to believe that Boltzmann's "clear pictures" are the most efficient instruments theoretical physicists can make use of. His major argument in favor of the latter statement was stated again and again in his writings: these pictures are the only "mental help", the only "tool of thought"[5], at our disposal to be "quite sure of avoiding contradictory assumptions"[6] when we synthesize observational facts. Boltzmann's tendency of being "childishly precise" about his models cannot be taken as a symptom of his inability to recognize that the experimental evidence is forever incomplete; it is rather an expression of his having understood that "(...) without an absolutely precise model,

[1]E. Schrödinger, *Nature and the Greeks,* op. cit. p. 89
[2]A. Einstein to E. Schrödinger, August 8, 1935. French translation in: A. Einstein, *Oeuvres choisies, I, Quanta,* Seuil, 1989, p. 238
[3]E. Schrödinger to A.S. Eddington, March 22, 1940 (in: E. Schrödinger, *The interpretation of quantum mechanics (Dublin seminars 1949-1955 and other unpublished texts),* op. cit. p. 121)
[4]E. Schrödinger, *Nature and the Greeks,* op. cit. p. 92
[5]E. Schrödinger, *Science and Humanism,* p. 22
[6]E. Schrödinger to A.S. Eddington, March 22, 1940 (in: *E. Schrödinger, The interpretation of quantum mechanics (Dublin seminars 1949-1955 and other unpublished texts),* op. cit. p. 121)

thinking itself becomes imprecise, and the consequences to be derived from the model become ambiguous"[1].

2-6 *The lack of continuity*

How can one obtain the required *absolute precision* in our theoretical models, and a perfect clarity in our pictures? According to Schrödinger, this result can but be reached if the model goes well beyond the previously observed facts: "(...)the desire for having a *clear* picture necessarily led one to encumber it with unwarranted details"[2]. One must integrate in the model not only the *actual* experimental results, but an infinity of *possible* results; one must perform a systematic "completion in thought"[3] of the recorded observations. And the only acceptable proof that this process of *completion in thought* has been completed is the disappearance of any *gap* in the picture, namely its *continuity*.

This continuity condition was already stated by Schrödinger in 1929[4], when the prospect of forming a satisfactory continuous picture of atomic phenomena seemed quite remote: "(...)we are bound to supplement our immediate observations, in order not to be left with a patchwork of individual facts instead of reaching some sort of 'weltbild'"[5]. It was then reformulated more assertively in 1950, after a post-modern distance between the observed facts and the represented content of the picture had been established: "(...) from an incomplete description - from a picture with gaps in space and time - one cannot draw clear and unambiguous conclusions; it leads to hazy, arbitrary, unclear thinking - and this is the thing we must avoid at all costs!"[6]. According to Schrödinger, there was thus no doubt that the gaps in our pictures *had* to be filled in. The only problem was that, at the birth of quantum mechanics, nobody could figure out how this aim was to be reached.

The most natural way of filling the gaps, when what is observed consists, for instance, of a series of dots in a cloud (Wilson) chamber, is bare interpolation[7]. One is thus tempted to insert more and more imaginary dots between the actual dots, and to make them smaller and smaller until they form something like a continuous trajectory. But, Schrödinger says, Heisenberg's uncertainty relations have demonstrated that this is just impossible[8]. The process of making the dots smaller and smaller, as well as closer and closer, would only result in their increasing dispersion. It would lead to a cloud of points which by no means looks

[1]E. Schrödinger, *Science and Humanism*, p. 25
[2]ibid. p. 24
[3]E. Schrödinger, "Might perhaps energy be a merely statistical concept?", loc. cit. p. 169
[4]E. Schrödinger, "Conceptual models in physics and their philosophical value" and "Indeterminism in physics", in: *Science and the human temperament*, op. cit. p. 62, 121, 124, 128, 131
[5]E. Schrödinger, "Indeterminism in physics", in: *Science and the human temperament*, op. cit. p.62
[6]E. Schrödinger, *Science and Humanism*, p. 40
[7]E. Schrödinger, "Indeterminism in physics", in: *Science and the human temperament*, op. cit. p. 60
[8]ibid.

like a corpuscular trajectory. In other terms, no deterministic link between the observed dots, in which the former infinitesimal fragment of trajectory (together with the local potential) fixes univocally the later infinitesimal fragment of trajectory, can be established. The only link between the observed dots is a probabilistic one.

The alternative solution is then to fill in the gaps by means of the very continuous theoretical entity which serves as a tool for calculating the probabilistic link between the observed dots, namely the ψ-wave. But this procedure sounds utterly unnatural, due to the obvious heterogeneity between the extended wave-like filling material and the point-like observed facts. As Heisenberg first pointed out in his celebrated 1927 paper, if it is to fit with each observed dot, the ψ-wave must be "reduced" to a wave packet whose size is of the order of magnitude of the precision with which the corresponding position measurements have been performed[1]. Thus, even if a kind of q-space continuity between the cloud chamber dots is established by means of the ψ-wave, one still has to cope with the *temporal* discontinuity implied by the successive "reductions" of the ψ-wave. As I shall explain at length in chapter 4, Schrödinger did not retain the idea of the "reduction" of the ψ-wave, and accordingly did not recognize the necessity of a temporal discontinuity in our theoretical picture either. His strategy consisted in doing much more than just filling in the gaps *between* the observed dots. It consisted in ascribing an *absolute priority* to the continuity of the ψ-wave picture over the discontinuity of the dots. In short, his quite bold prescription could be formulated thus: *if the observed facts do not fit with the continuity of the picture, then just eliminate the facts from the picture* (even if it means pushing the concept of fact to the edges of the scientific thought, and relating the facts indirectly to the picture through the correspondence rules ...).

But, at this point, an embarassing question arises: is this purely continuous theory, completely freed from the obligation of incorporating something of the experimental discontinuities in the course of the time-development of its entities, able to account for the experimental effects which prompted the introduction of the quantum of action by Planck?

Here, we are reaching one of the most surprising chapters of the history of quantum mechanics, a chapter where the emotionally rooted convictions and the sociological predominance of the Göttingen-Copenhagen group managed to mould the opinion of the majority of physicists during a full half-century, even against the clearest theoretical evidence.

During the first half of 1926, Heisenberg's reaction to Schrödinger's wave mechanics had been extremely negative. After having attended Schrödinger's conference in Munich, in July 1926, he wrote his impressions to Pauli: "Schrödinger throws overboard everything which is 'quantum theoretical': namely, the photoelectric effect, the Franck [-

[1]W. Heisenberg, "The physical content of quantum kinematics and dynamics", in: J.A. Wheeler and W.H. Zurek, *Quantum theory and measurement*, op. cit. p. 74

Hertz] collisions, the Stern-Gerlach effect, etc. It is not then difficult to establish a theory. However, it does not agree with experience"[1]. A few months later, the conviction that wave mechanics would prove unable to account for properly "quantum theoretical" effects was expressed directly by Bohr and Heisenberg to Schrödinger, during the latter's visit to Copenhagen[2]. According to Bohr, they managed to convince Schrödinger that "(...) a continuity theory in the form indicated in his last paper at a number of points leads to expectations fundamentally different from those of the usual discontinuity theory"[3]. Actually, Schrödinger felt both quite embarassed by Bohr's contentions and not fully convinced. In his letter to Bohr, written a few weeks after his stay in Copenhagen, he acknowledged that "(...) the psychological effect of these objections - in particular the numerous specific cases in which for the present my views apparently can hardly be reconciled with experience - is probably even greater for me than for you"[4]. However, he did not believe that this incompatibility was something which hindered the very possibility of using continuous pictures: "I do not consider it inconceivable to construct pictures that actually reproduce the above circumstances"[5]. He even suspected that the difficulties which Bohr had indicated were really no more than *apparent,* and he did not therefore renounce finding a clue to reconcile pure wave mechanics with the most striking discontinuous aspects of atomic processes. One of his priorities in the following years was then to formulate wave-mechanical accounts of all the known "quantum theoretical" effects. The task did not prove untractable, even though it was considerably delayed by the long maturation which eventually led Schrödinger to find an appropriate articulation between the "effective" ψ and the "factual" $\psi\psi^*$.

In 1927, he provided a wave-mechanical demonstration of Planck's radiation law[6]. He insisted at the end of paragraph 4 of his paper that this result was obtained *without* using the postulate of quanta properly speaking.

He also gave during the same year a wave-mechanical account of the Compton effect[7], by considering it as a phenomenon of diffraction of high frequency electromagnetic waves on a moving grating of electronic charge distribution. The resulting directions of propagation, and the Doppler effect, proved equivalent to the values Compton was able to

[1]W. Heisenberg to W. Pauli, July 28, 1926, quoted in: J. Mehra and H. Rechenberg, *The historical development of quantum mechanics,* 5-2, Springer-Verlag, 1987, p. 822

[2]W. Heisenberg, *Physics and beyond, encounters and conversations,* George Allen and Unwin, 1971, p. 75

[3]N. Bohr to R. Fowler, October 26, 1926, quoted in: J. Mehra and H. Rechenberg, *The historical development of quantum mechanics,* 5-2, op. cit. p. :26

[4]In: N. Bohr, *Collected works,* E. Rüdinger (gen. ed.), vol. 6, J. Kalckar (ed.), North-Holland, 1985, p. 12

[5]ibid. p. 13

[6]E. Schrödinger, "The exchange of energy according to wave mechanics", in: *Collected papers on wave mechanics,* op. cit. p. 143-145

[7]E. Schrödinger, "The Compton effect", in: *Collected papers on wave mechanics,* op. cit. p. 124

predict previously by using a corpuscular model. Actually, Schrödinger's calculation looks highly unsatisfactory by present days standards due to his reluctance to use 3n-dimensional ψ-functions for composite systems, and to his recurrent taste for 3-dimensional waves and charge density clouds. But at any rate, by providing a semi-classical wave alternative to Compton's original semi-classical calculation in terms of particles, Schrödinger had demontrated that one has no reason to consider the Compton effect as a convincing *proof* that corpuscularian representations cannot be dispensed with at one stage or another of the account of microscopic phenomena. As more recent work on that topic[1] has shown, the multiplicity of calculations which are able to yield satisfactory prediction of Compton's parameters (provided they incorporate conservation of energy and momentum) can rather be cited as a typical case of underdetermination of theories by experiments, than as a definitive evidence of the "photon nature of light".

One may add to these preliminary achievements of Schrödinger's at least two wave-mechanical calculations performed by other physicists. In the late 1926, G. Wentzel proposed a wave-mechanical treatment of the photo-electric effect[2]. And later on, in 1929, N.F. Mott published a deservedly well-known paper entitled "The wave mechanics of α-ray tracks"[3]. In this article, Mott did not limit himself to giving a remarkably sober wave-mechanical account of the particular problem of the tracks in cloud chambers; he defined a general method allowing one to account for *any* kind of discontinuous phenomena by the continuous formalism of wave mechanics. The principles he used are the following:

(1) define the degrees of freedom of a relevant composite system (possibly including part of the measurement device);

(2) solve the Schrödinger equation for the multi-dimensional wave-function of this composite system;

(3) postpone indefinitely (until the "act of observation", for all practical purposes) the reference to the discontinuities: "the wave mechanics unaided ought to be able to predict the possible results of any observation that we could make on a system, without invoking, *until the moment at which the observation is made,* the classical particle-like properties of the electrons or α-particles forming that system"[4].

The teaching of this series of papers, and especially of Mott's paper, which avoids some pitfall's of Schrödinger's initial interpretation of wave mechanics, is unambiguous: it is perfectly possible to account for typically "quantum theoretical" effects without introducing any intermediate temporal discontinuity. One just has to use extensively the multidimensional wave-mechanical formalism (with its eigenfunction

[1]R. Kidd, J. Ardini, & A. Anton, "Compton effect as a double Doppler shift", Am. J. Phys., 53, 641-644, 1985; J. Strnad, "The Compton effect: Schrödinger's treatment", Eur. J. Phys., 7, 217-221, 1986
[2]G. Wentzel, "Zur Theorie des photoelektrischen Effekts", Z. Phys., 40, 574-589, 1926
[3]N.F. Mott, "The wave mechanics of α-ray tracks", Proc. Roy. Soc. Lond., A126, 79-84, 1929; in: J.A. Wheeler and W.H. Zurek, *Quantum theory and measurement,* op. cit. p. 129
[4]ibid.

scheme) *for a sufficiently large system,* and to restrict the probabilistic scheme to the connection between the *final* outcome of the calculation and the relevant experimental events. At no intermediate point between the preparation of the experiment and the experimental events have discontinuities and probabilistic considerations to be introduced. Moreover, since the criteria which must be used to stop the time-development of the ψ-function are purely practical, nothing prevents one from prolonging it indefinitely, and from taking into account more and more degrees of freedom. In other terms, the temporal discontinuities are by no means an integral part of the predictive power of quantum mechanics; *they just have to be* related *to the formalism, whenever it is suitable in practice*[1].

Schrödinger did not work out this idea in its full generality before the late forties and the beginning of the fifties. But as soon as he came to realize its perfect coherence and soundness, he resumed very actively his early attempt at giving a wave-mechanical account of the "quantum theoretical" effects. He focused his attention once more on Planck's radiation law, which he managed to demonstrate for his Dublin seminar lectures of 1949, in a section characteristically entitled "Planck-black-body-radiation (without discontinuity!)"[2]. At the end of the sub-section about the Bose-Einstein statistics, he points out that his derivation does *not* rely either on the idea that each system is always in an eigenstate of some observable, or on the related idea that systems jump from one eigenstate to another: "(...) on (the ordinary) photon view one implicitly admits that not only the whole body of radiation but every simple "oscillator" (or proper mode) is always in a state of sharp energy. We have assumed nothing of the kind. The concept of eigenstates and of their degeneracy (or multiplicity) is given, and it is unavoidable in quantum mechanics. It takes the role of the permutation number in Boltzmann's original reasoning". *Atomicity,* namely discreteness of the level scheme of continuous wave processes, here again replaced *atomism,* namely discontinuity of the processes and entities themselves. Accordingly, in the 1952 edition of his book *Statistical Thermodynamics* (whose first edition dates back to 1944), he inserted an important "appendix" where he purported to demonstrate that "the thermodynamical functions depend on the quantum-mechanical level-scheme, not on the gratuitous allegation that these levels are the only allowed states", and that they do not depend either on the view that "(...) a physical process consists of continual jump-like transfers of energy parcels between microsystems"[3]. Finally, the same year 1952, Schrödinger wrote a paper wherein he gave an outline of a

[1]For a philosophical analysis of the pragmatic aspects of the interpretation of quantum mechanics, see: M. Bitbol, *Mécanique quantique: une introduction philosophique,* Flammarion, 1996; M. Bitbol, *De l'intérieur du monde* (in preparation).
[2]E. Schrödinger, *Notes for seminar 1949,* (E. Schrödinger, *The interpretation of quantum mechanics (Dublin seminars 1949-1955 and other unpublished texts),* op. cit. p. 115)
[3]E. Schrödinger, *Statistical thermodynamics,* Cambridge University Press, 1952 (introductory "note on second edition")

wave-mechanical account of the Franck-Hertz experiments[1]. This account is also contained in a series of lectures given at the Dublin institute for advanced studies[2].

With these results in mind, the repeated reminder, by the members of the former Göttingen-Copenhagen group, that Schrödinger's preference for a continuous picture in quantum mechanics did not allow him to yield the properly "quantum theoretical" effects, sounds utterly surprising. It sounds all the more surprising since these physicists speak as if they had not really understood the reason why the first Schrödinger's attempt had failed. Let us read for instance Rosenfeld's commentary on Born's probabilistic interpretation, written in 1971: "Would this formal equivalence (between matrix mechanics and wave mechanics) clinch the issue in favor of Schrödinger's contention that the proper quantal concepts can altogether dispensed with? Far from it, Heisenberg had at once seen that this contention was untenable: Schrödinger's way of treating the charge density as a classical source of radiation would even prevent him from obtaining Planck's law for the distribution of thermal radiation"[3]. It was indeed impossible to obtain Planck's law by a classical electrodynamic theory applied to the charge cloud, namely to the "factual" expression $\psi\psi^*$; but it was *not* impossible to obtain Planck's law by making calculations on the level scheme displayed by the "effective" *continuous* ψ-function, as Schrödinger had demonstrated in his 1927 paper and in his 1952 appendix. Thus, in a certain sense, the "proper quantal concepts" could be dispensed with. And in another sense, they were retained by wave mechanics in the form Schrödinger had given them in his first paper of 1926, namely in the form of a system of eigenfunctions (the "level scheme"). As for the mathematical equivalence between wave mechanics and matrix mechanics, it had only to be complemented with a full recognition of the status of $\psi\psi^*$ as an empirical correspondence rule in order to become a full *physical* equivalence.

There is then no conflict between wave mechanics and matrix mechanics concerning the status of Rosenfeld's "quantal concepts". Actually, the common framework which one may sketch out when these two initial versions of quantum mechanics have carefully been worked out, is likely to prove closer to Schrödinger's position than to Heisenberg's. As Mara Beller cogently pointed out, it was not only wave mechanics, but *also* matrix mechanics which "undermined the fundamental role of *a priori* stationary states and 'irreducible' quantum jumps"[4]. Matrix mechanics, as wave mechanics, incorporates a level

[1]E. Schrödinger, "Are there quantum jumps?" loc. cit.

[2]E. Schrödinger, *July 1952 colloquium,* (in: E. Schrödinger, *The interpretation of quantum mechanics (Dublin seminars 1949-1955 and other unpublished texts),* op. cit. p. 27)

[3]L. Rosenfeld, Commentary on Born's probabilistic interpretation, in: J.A. Wheeler and W.H. Zurek, *Quantum theory and measurement,* op. cit. p. 50

[4]M. Beller, "Schrödinger's dialogue with Göttingen-Copenhagen physicists", in: M. Bitbol and O. Darrigol, *Erwin Schrödinger, Philosophy and the birth of quantum mechanics,* op. cit. p. 286. One must notice that Bohr was much more prudent (and more consistent) in this respect than either Heisenberg or

scheme, not the necessity of considering that microscopic objects jump from one level to another. Quantum mechanics *in general,* not only wave mechanics, is alien to the concept of "quantum jump". If the "quantum jumps" were really indispensable in order to derive the Planck's radiation law, no *interpretation* of quantum mechanics could be sufficient *as such* to perform this derivation. One would not only have to *interpret* the formalism, but to *add* something to it, namely the idea that "quantum jumps" *really occur.* It is thus on this metaphysical issue of the "reality" of quantum jumps between two observations that the debate actually centered. According to Mara Beller, "(...) the whole controversy gains intelligibility only when we assume that not only Schrödinger but also Heisenberg (sincere or not) had some very strong opinions about the way unobservable processes *really* occur in nature"[1]. Now, on this ontological issue, I believe Schrödinger's position was much more consistent than that of his opponents. Schrödinger could argue that, since the "quantum jumps" are not necessary in order to predict any observable effect, and since they are not even an integral component of the quantum mechanical formalism, they can be dispensed with in virtue of the Ockham's razor rule. His own strategy of ontologizing the entities of the most economical (and at the same time adequate) physical theory, could by no means lead him to ontologize the "quantum jumps" (or to endow them with "reality"), for the said quantum jumps are just an additional "convenient metaphor"[2] serving to illustrate the level-scheme of quantum mechanics. Schrödinger's ontological elimination of "really occurring quantum jumps" reflects a sound version of the principle of economy of thought. His asserting the unreality of quantum jumps was not grounded on bare beliefs, but on a certain set of consciously manipulated criteria allowing one to endow theoretical entities with "reality" and to avoid postulating unnecessary levels of "reality" (see chapter 4 for more details about these criteria).

By contrast, Heisenberg's insistence in 1927 on the essential character of the discontinuities for the theoretical description of fluctuation phenomena[3] had much weaker justifications. Firstly, as we have already noticed in previous paragraphs, it proved quite easy for Schrödinger to demonstrate that these discontinuities can perfectly be dispensed with, even when one has to account for the fluctuations of energy between two interacting atoms[4]. Secondly, as Schrödinger pointed out somewhat ironically, Heisenberg's underlying idea that systems occupy one level and

Born. According to Bohr in 1929, "(...) it might be said that the concepts of stationary states and individual transition processes within their proper field of application possess just as much or *as little* 'reality' as the very idea of individual particles. In both cases we are concerned with a demand of causality complementary to the space-time description (...)".

[1]ibid.

[2]E. Schrödinger, *Statistical thermodynamics,* op. cit., p. 90

[3]W. Heisenberg, "Schwankungerscheinungen und Quantenmechanik", Z. Phys., 40, 501-506, 1927

[4]E. Schrödinger, "The exchange of energy according to wave mechanics", in: *Collected papers,* op. cit. p. 137-146

then jump to another level (or that they undergo temporal discontinuities, from one eigenstate of the Hamiltonian to another), is definitely inconsistent with both the structure of quantum mechanics and the epistemological decision to limit physics to the description of "observable facts". It is "irreconciliable with the very foundations of quantum mechanics"[1], for it jettisons the principle of superposition which indicates that there are available states which are *not* eigenstates. And it is inconsistent with the decision to limit physics to the description of "observable facts" because it surreptitiously tells something about what the systems "really" do when no observing subject and no measuring apparatus interfere with them: "(The assumption that each gas-molecule is always in one of its stationary states) is in violation of that precious principle that the same school of physicists is so anxious to put across, namely that we must never admit anything to *be* except what we have measured"[2].

The strategy sketched in the sentence just quoted, namely that which consisted in displaying internal contradictions within the position of the Göttingen-Copenhagen physicists, is a constant feature of Schrödinger's attitude, from 1927 until his death. Einstein adopted the same strategy to some extent (see the EPR paper), but he did not go as far as Schrödinger in the perfect assimilation of the basic positions of his opponents. Schrödinger was able to imitate these positions with such plausibility that many authors thought he had been sincerely converted to them. At the end of certain pages of careful and apparently convincing analysis of the epistemic interpretation of the ψ-function, even somebody well-informed might be quite surprised to read the following remark: "I actually *do not* think this is to be the appropriate way of looking at things. I have explained it here (...) because one can only with some clarity say one disagrees with a view, after one has explained it"[3].

The next chapter is then devoted to an evaluation of Schrödinger's original way of proving (or at least favouring) his interpretation of quantum mechanics by a *reductio ad absurdum.*

[1]E. Schrödinger, *Statistical thermodynamics,* op. cit., p. 89
[2]ibid. p. 90, 93; see also Notes for seminar 1955 (E. Schrödinger, *The interpretation of quantum mechanics (Dublin seminars 1949-1955 and other unpublished texts),* op. cit. p. 109)
[3]E. Schrödinger, *Transformation and interpretation in quantum mechanics,* (E. Schrödinger, *The interpretation of quantum mechanics (Dublin seminars 1949-1955 and other unpublished texts),* op. cit. p. 82)

CHAPTER 3

THE ANALYTICAL STANCE

The period going from 1928 to 1935 looks like a puzzle for the Schrödinger scholars. One usually considers that, after having realized the failure of his early interpretative Tattempts, Schrödinger accepted the broad lines of the Copenhagen interpretation and reverted more and more of his scientific attention to non-quantum mechanical branches of physics. However, the idea that Schrödinger agreed with the current views on quantum mechanics does not fit very well with his increasing tendency to depart from quantum mechanical studies. According to L. Wessels[1], the reason why he devoted his work to other theoretical fields was that, at bottom, he could not accept the loss of any physical picture. His behaviour could well prove, even against some of his explicit assertions, that his acceptance of the Copenhagen interpretation was, to say the least, very critical. But I think we can go beyond this opposition between texts and scientific behaviour by reading carefully the papers in which Schrödinger expressed his positions during the period 1928-1935, and by classifying them according to their expected audience.

One of the most striking expressions of Schrödinger's allegiance to the dominant views is contained in a lecture he delivered in Munich, in May 1930, at the *Deutsches Museum*. Here, developing on his favourite theme of interpolation (or completion in thought), he explained: "when we interpolate the actual measurement by the best possible means, they are embedded in continua (...) that do not represent the natural object in itself, but rather the relation between subject and object"[2]. This was meant as an acknowledgement of what he had rejected completely four years earlier, namely that the ψ-function represents a relation between subject and object, or alternatively "(...) the *knowledge* that we possess at any given time of the observations actually carried out". Now, what about pictures? Did Schrödinger accept the relinquishing of pictures during these years, just as he had renounced the ascription of a status of "reality" to the ψ-waves? Not so. According to him, it was only the *epistemological status* of the wave picture which had to be revised: "We feel it as a painful limitation of our right to truth and clarity, that our symbols and formulas and the pictures connected with them do not represent an object independent of the observer but only the relation of subject and object. But is this relation not basically the one true reality that we know?". The pictures have not been lost as such, according to Schrödinger, but our conception of what they represent has changed. They do not represent what they appear to represent, namely "real" waves in q-space, but only

[1]L. Wessels, "Erwin Schrödinger and the descriptive tradition", in: R. Aris, H.T. Davis, and R.H. Stuewer (eds.), *Springs of scientific creativity,* University of Minnesota Press, 1983, p. 265

[2]E. Schrödinger, "Die Wandlung des physicalischen Weltbegriffs", in: E. Schrödinger, *Gesammelte abhandlungen,* Verlag der österreichischen Akademie der Wissenschaften, Friedrich Wievweg & Sohn, 1984, vol. 4, p. 600-608; quoted and translated in: W. Moore, *Schrödinger, life and thought,* Cambridge University Press, 1989, p. 250

our *relation* to nature, or even our own cognitive status with respect to nature.

This proposition written in 1930 looks like an exact antithesis of his previous and later "realist" views. But things are much more intricate. In order to see this, we have to revert our attention to the last sentence of the previous quotation. In the light of Schrödinger's philosophical writings, this sentence expresses a personal thought, not one that was merely borrowed from his colleagues. The idea that we have no access to something "out there" but only to an extended set of neutral "elements" is part of his most deeply entrenched metaphysical views. Nothing therefore prevented him from going very far, and even farther than any other contemporary physicist, in the process of ontological deconstruction, until he had landed on the surface of the bare subject-object relatedness. But the problem is that he could not stop at this point. For him, the fact that the "only true reality that we know" is the pure Schopenhauerian object-for-a-subject relation rather than either subject or object, cannot be construed as an achievement of physics, but rather as its metaphysical ground: "Subject and object are only one. The barrier between them cannot be said to have broken down as a result of recent experience in the physical science, for this barrier does not exist"[1]. Schrödinger's lecture in Munich, in may 1930, is thus not to be taken as the manifestation of a first about-face which would be followed by another about-face in the fifties. It is rather the extreme point of an indispensable preliminary deconstructive process, before the ontological reconstruction might eventually take place. It represents the indispensable preliminary recognition of the Unity between subject and object, before a new step in the ever renewed objectivation process is taken.

As for the idea that the ψ-waves do not mimic a natural process but rather express a subject-object relation, it can very well appear retrospectively as a forerunner of what I have called the post-modern dissociation. In the 1930 text, Schrödinger opposed an intentional attitude (towards the transcendent object of physics) and a semi-reflective attitude (towards the immanence represented by the subject-object relation). In 1950, he rather dissociated *two intentional attitudes:* the first one directed towards the processes supposed to be contributing to the "observed facts", and the second one directed towards the objects which *appear* to be referred to by the theoretical entities (namely the 3-n dimensional waves in q-space or the states in Fock space).

True, the reflective (or semi-reflective) move was eventually discarded in the clearest way: "(The state vector) must not be regarded as 'hovering in empty space' between subject and object"[2]. But this early move also opened the way for a later recognition of the impossibility of merging the "factual" interpretation scheme and the "effective" theoretical entities.

[1]E. Schrödinger, *Mind and Matter*, op. cit. p. 137

[2]E. Schrödinger, "Might perhaps energy be merely a statistical concept?", Nuovo cimento, 9, 1958, p. 169; in: E. Schrödinger, *Gesammelte abhandlungen*, op. cit., vol. 1, p. 509.

The gap between transcendent nature and immanent subject-object relation had only to be replaced by a gap between the processes which express themselves in experiments and what ψ-waves appear to refer to; a gap between the hypothetical ground of the phenomena and the quasi-real (in Blackburn's sense) or internally real (in Putnam's sense) entities of modern physics.

That Schrödinger's reconstructive project was actually latent in the years 1928-1934, and not just abandoned, can be seen in many other texts of the same period. In his conference "Conceptual models in physics and their philosophical value"[1] of December 1928, he developed at length the concept of "completion in thought". And he argued that the possibility of ascribing a "real existence" to some theoretical entity depends on its ability to serve as a "scaffolding" for both actual and virtual observed facts. True, the very applicability of the concept of virtual observations (we would say "counterfactual experimental propositions") is doubted in certain cases. But, according to Schrödinger, the only conclusion to be derived from this is that the object referred to in quantum mechanics "is not a material point"[2]. By contrast, one may figure out a new, *non-corpuscular* representation such that the unavailable virtual observations are not conceived as *missing,* but just eliminated and irrelevant: "I definitely believe that the elimination ought to be possible without leading to the consequence that no visualizable scheme of the physical universe whatever will prove feasible"[3]. The project of restoring a figurative depth in physics, and of framing a new ontology, was thus still alive at the end of 1928. And it was again expressed in Schrödinger's Nobel lecture of 1933: "It is by no means a new demand to claim that, in principle, the ultimate aim of exact science must be restricted to the description of what is really observable. The question is only whether we must henceforth forego connecting the description, as we did hitherto, with a definite hypothesis as to the real structure of the Universe. To-day there is a widespread tendency to insist on this renunciation. But I think that this is taking the matter somewhat too lightly"[4]. This quotation, with its two steps, summarizes the whole situation. On the one hand, Schrödinger acknowledges the necessity of performing the Machian *tabula rasa* as completely as possible; and on the other hand he claims that he has not renounced the possibility of building something new on the ruins of the classical representations. One can only guess that, in such a difficult position, Schrödinger had to choose the aspect on which he would put emphasis according to his audience. Y. Ben-Menahem's suggestion[5], that

[1]E. Schrödinger, "Conceptual models in physics and their philosophical value", in: *Science and the human temperament,* op. cit.
[2]E. Schrödinger, "Indeterminism in physics", in: *Science and the human temperament,* op. cit. p. 58
[3]E. Schrödinger, "Conceptual models in physics and their philosophical value", in: *Science and the human temperament,* op. cit. p. 132
[4]E. Schrödinger, "The fundamental idea of wave mechanics" in: *Science and the human temperament,* op. cit. p. 153
[5]Y. Ben-Menahem, "Struggling with causality: Schrödinger's case", loc. cit.

he taught orthodoxy to his students because he had nothing better to propose, whereas he expressed more overtly his doubts and his projects when he was due to speak to more specialized audience, is quite convincing. More fundamentally, in the sort of low-water point he had reached, the only strategy which was available to him consisted in analyzing so carefully the phenomenal material ordered by the quantum description that, hopefully, the necessity for a reconstructive attempt would manifest itself from within. As we shall see, this methodologically phenomenalist attitude was not just a temporary mood. It was prolonged even after Schrödinger had re-elaborated his "realist" interpretation of quantum mechanics in the 1950's, and it has thus to be considered as the permanent foundation of his intentional directedness towards the entities of this theory.

3-1 *The ontological significance of the uncertainty relations*

Schrödinger never attempted, as Einstein occasionally did in his celebrated discussions with Bohr, to challenge the general applicability of the uncertainty relations. He had no reason to do so, for these relations are an integral part of his own wave mechanics. He would rather use the uncertainty relations as a weapon against two ontological elements which were still operating, although with lots of qualifications, in the current views on the interpretation of quantum mechanics. These two elements are quantum jumps and corpuscularian categories.

Let us then begin with quantum jumps. The basic argument was already formulated in a letter of May 5, 1928 to Bohr[1]. There, Schrödinger mentioned: "It seems to me that there is a very strange relation between Heisenberg's uncertainty relation and the claim of discrete quantum states. On account of the former, the latter can really not be experimentally tested". The argument was grounded on the uncertainty relation for action and angle variables (J and w):

$\Delta J . \Delta w = h$

In order to lose all knowledge of an angle variable whose period is 1, it is enough to put $\Delta w = 1$. But "then you have $\Delta J = h$, i.e. just equal to the difference in J-values of neighbouring quantum states". The traditional picture of steady states corresponding to some well-defined value of J is thus made irrelevant by the action-angle uncertainty relation. Accordingly, the idea of a quantum jump whereby an object undergoes a sudden change from one steady state to another becomes pointless. In his answer to Schrödinger, Bohr did not attack the very structure of the argument, but he noticed that one of its assumptions, namely the limitation of the angular uncertainty to one period, is unacceptable: "In the interpretation of experiments by means of the concept of stationary states, we are indeed always dealing with such properties of an atomic

[1]In: N. Bohr, *Collected works*, vol. 6, op. cit. p. 47

system as depend on phase relations over a large number of consecutive periods"[1]. The concept of steady state then retains its pertinence, provided one completely renounces any knowledge bearing on the value of the phase variables of the system. In his Frankfurt lecture of December 1928, Schrödinger acknowledged that Bohr's critical remark was perfectly sound. True, it was not correct to limit the phase uncertainty to one period. But this did not rescue in the least the idea that systems can be said to *occupy* steady states: "The physical meaning of having to admit an uncertainty of the angle variable, much greater than the period, is obviously that sharp quantization is not a property which the system can be said to possess at a *definite moment*" [2]. Here again, the *level scheme* is a permanent long-term feature of quantum mechanical descriptions, but certainly not the ascription of one particular level of this scheme to the system at every instant. The argument was made even more straightforward by using the Energy-time uncertainty relation:

$\Delta E.\Delta t = h$

Just as it was meaningless to enquire what is the "action" of the system in a definite *period*, it is meaningless to enquire "(...) what is the energy of a sytem at a definite *instant* (...)"[3]. And consequently, asking "(...)whether energy actually passed by jumps or in a steady flow from one atom to another, naturally becomes illusory". This line of thought was followed by Schrödinger throughout his subsequent career. In his very last paper of 1958, he repeated the argument in a virtually unchanged version: "(...) the levels are just so densely packed as to disallow one to distinguish unequivocally between neighbouring levels, on account of the uncertainty relations that hold between the pairs of conjugate variables"[4].

Now, what are we to think of this argument? Its validity has recently been challenged on account of its tacitly assuming a conception of the energy variable as c-number rather than as q-number (i.e. operator)[5]. But I think this remark essentially amounts to Bohr's objection, and it can thus be answered in the same way. In both cases, one insists on the discreteness of the energy levels, with the explicit *or implicit* assumption that Δt is arbitrarily large. Explicit in Bohr's case, and implicit when one insists on replacing c-numbers by q-numbers. Conceiving energy as an operator indeed yields a "level scheme", namely the set of all the eigenstates of the operator, which are to be construed as the *long-term* steady-states of the object; it can by no means have among its consequences that a system occupies one of these states at *every given instant*. Thus, if one keeps in

[1] N. Bohr to E. Schrödinger, May 23, 1928, in: N. Bohr, *Collected works*, vol. 6, op. cit. p. 49

[2] E. Schrödinger, "Conceptual models in physics and their philosophical value", in: *Science and the human temperament*, op. cit. p. 130

[3] ibid. p. 127

[4] E. Schrödinger, "Might perhaps energy be merely a statistical concept?", Nuovo cimento, 9, 1958, p. 164; in: E. Schrödinger, *Gesammelte abhandlungen*, op. cit., vol. 1, p. 504

[5] O. Darrigol, "Schrödinger's statistical physics and some related themes", in: M. Bitbol & O. Darrigol (eds.) *Erwin Schrödinger, Philosophy and the birth of quantum mechanics*, op. cit. p. 271

mind that Schrödinger's argument was not taken as purporting to show that quantum jumps do not exist, but only to bring out that nowhere in quantum mechanics can one find the slightest indication in favour of these jumps, it remains perfectly acceptable.

The second line of attack for the sake of which Schrödinger used uncertainty relations, was directed against the concept of particle. According to Heisenberg, "(...) the uncertainty relation specifies the limits within which the particle picture can be applied"[1]. By contrast, Schrödinger tended to demonstrate that, *even* within these limits, the particle picture is completely inappropriate.

In order to reach his target, Schrödinger made a quite original analysis of the *retrospective* significance of Heisenberg's uncertainty relations. Let us first recall what was Heisenberg's position about the retrospective value of his relations. According to him, the uncertainty relations state that any measurement of one variable V *alters* the value of the canonically conjugate variable V' of an undeterminable amount, such that after carrying out the measurement of V with a certain accuracy ΔV we cannot *predict* the value that we *shall* find for V' with an accuracy better than $\frac{h}{\Delta V}$. Then he concludes: "This formulation makes it clear that the uncertainty relation does not refer to the past"[2]. It is even very easy to demonstrate that the relations can be violated, for times previous to the first measurement. But now comes the difficult part of the argument, namely the evaluation of the epistemological significance of the retrospective violation of the uncertainty relations. According to the positivistically inclined young Heisenberg "(...) this knowledge of the past is of a purely speculative character, since it can never (...) be subjected to experimental verification"[3]. Such a methodological claim was obviously rejected by Einstein and the supporters of hidden variable theories, who rather tended to consider the asymmetry between prediction and retrodiction as a proof of the incompleteness of quantum mechanics. As for K. Popper, he criticized Heisenberg's insistence on *verification,* arguing that the retrospective ascription of values to a pair of canonically conjugate variables can perfectly make sense, provided one ascertains it is *not* experimentally *falsified* [4].

The three positions (namely Heisenberg's, Einstein's and Popper's) are distinct, but they share at least *one* common presupposition. This underlying presupposition is that the sequence of measurements, as well as the retrospective or prospective ascription of values, bear on a single well-defined object.

[1]W. Heisenberg, *The physical principles of the quantum theory,* The University of Chicago Press, 1930, p. 15

[2]ibid. p. 20

[3]ibid.

[4]K. Popper, *The logic of scientific discovery,* Hutchinson, 1968, chapter XI; see also K. Popper, *Quantum theory and the schism in physics,* Hutchinson, 1982, section 3, seventh these.

Let us now describe Schrödinger's approach, as it is outlined in his 1931 conference "Indeterminism in physics", and then developed at length in his Dublin seminars[1] of the years 1949 and 1955. Schrödinger's departure point is that Heisenberg's insistence on the "alteration" (or "disturbance") of a variable by measuring its canonically conjugate variable somehow paves the way to the supporters of hidden variable theories: "To this, one would say that it is all right, but if one accepts it, one grants to Einstein that quantum mechanical description is incomplete"[2]. If we say that a value is "disturbed", it looks as if there existed some unknown value, but our measurement devices are not refined enough to enable us to reach them without shuffling everything. The latent complicity between Heisenberg's positivist-like statements and Einstein's program of replacing quantum mechanics by a complete theory of the behaviour of individual objects, is made even more evident when one considers the discussion about retrospective variable ascriptions with arbitrary precision: "If it is possible to obtain simultaneous accurate values of location and velocity, albeit belatedly, then a description that does not allow one to express them is deficient"[3]. In other terms, the completeness of quantum mechanics could by no means be rescued by a positivist-like prohibition against a theory which would include non-verifiable value ascriptions. This was already pointed out by Schrödinger in 1928: "In the adequate conceptual scheme, it should no longer appear as if our possibilities of experience were limited through unfavourable circumstances"[4].

An edict of prohibition calls for its transgression. The only possibility left, in order to demonstrate that quantum mechanics is a satisfactory and exhaustive description, is to sketch out an interpretation of its symbolism such that the question of what is the velocity of a particle between two measurements *does not even have to arise:* "(W)hat is in principle unobservable should not at all be contained in our conceptual scheme, it should not be possible to represent it within the latter"[5]. But if the concept of a particle travelling between two small spatial domains K and K' is maintained, the question necessarily arises. On the one hand, a precise position measurement on the particle P in the domain K yields the prediction of a very wide range of velocities for P; and on the other hand, when the particle has been detected after a time Δt in the spatial domain K', one can retrospectively ascribe to P a quite sharp value of the velocity, namely $\frac{KK'}{\Delta t}$, and also a precise direction of motion. Isn't it

[1]E. Schrödinger, *Notes for seminar 1949; Notes for seminar 1955,* (E. Schrödinger, *The interpretation of quantum mechanics (Dublin seminars 1949-1955 and other unpublished texts),* op. cit. p. 104 and 110)
[2]E. Schrödinger, *Notes for seminar 1955,* (E. Schrödinger, *The interpretation of quantum mechanics (Dublin seminars 1949-1955 and other unpublished texts),* op. cit. p. 110)
[3]ibid.
[4] E. Schrödinger to N. Bohr, May 5, 1928, in: N. Bohr, *collected works,* op. cit. vol. 6, p. 47
[5] ibid.

tempting to assert that the particle P actually *had* the velocity $\frac{KK'}{\Delta t}$, but that quantum theory could not predict it? The only option which remains, in order to dismiss the hidden variable program without relying on positivist-like interdicts, is thus to deny that there is anything like a particle P travelling between K and K': "Before the second measurement, it is ubiquitous in the cloud (it is not a particle at all)"[1]. Or, in other more provocative words: "You have not *found* a particle at K', you have *produced* one there!"[2]. Indeed, if this is so, the location K is not relevant for the "particle" detected at K', and there is no reason left to ascribe it the velocity $\frac{KK'}{\Delta t}$. In this way, the predictive statement at K does not conflict any longer with a retrodictive statement at K', for the retrodictive statement has merely disappeared.

Schrödinger's recapitulation of what was at stake in the debate is worth quoting in its entirety: "Einstein was inclined to infer from these or similar considerations that quantum mechanical description is incomplete. I am inclined to avoid the incongruity by what I think is the only alternative, viz. there is no meaning in saying that I have observed at (K') the *same* particle (as at K). After an emission process, there is a probability of spotting *a* particle at (K') after time (Δt) or anywhere else at any other time. This probability (of what Margenau calls a firefly-event) is controlled by the wave-function"[3].

Here we have an excellent example of Schrödinger's most characteristic attitude in the philosophy of physics. Whereas the current debate was centered on *epistemological* issues (i.e. uncertainty versus indeterminacy, verification versus falsification, completeness versus incompleteness), he chose rather to orient it towards *ontological* issues. He did not think that the relevant question was about particles' having a well-defined position and velocity or not, in the case we cannot *in principle* predict them simultaneously, but about the very nature of the entity to which the two values (or the two domains of values) are ascribed. He believed that the issue was not about our ability to *know* something about pre-existent little bodies, but about our considering them or not as the *basic components of the phenomenal world.* Accordingly, he transformed the positivist-like prohibition which says 'do not ask anything about the past trajectory of the particle because it cannot be submitted to experimental verification', into an ontological description: 'there is no reason to inquire about the previous trajectory of the "particle" detected at K', for there *was* no such particle before its detection at K' (or even more radically, because there are no particles at all)'. More is to be said about this shift from

[1]E. Schrödinger, *Notes for seminar 1949* (E. Schrödinger, *The interpretation of quantum mechanics (Dublin seminars 1949-1955 and other unpublished texts)*, op. cit. p. 106)
[2]ibid.
[3]E. Schrödinger, *Notes for seminar 1955* (E. Schrödinger, *The interpretation of quantum mechanics (Dublin seminars 1949-1955 and other unpublished texts)*, op. cit. p. 114)

epistemology to ontology in the next section, and also about the general ontological issues in sections 4-1 and 4-2, as well as in chapter 5.

3-2 *The state vector as a catalog of information*

Schrödinger did not confine his simultaneous rejection of hidden variable theories and of the epistemological interpretation of quantum mechanics to a discussion about the significance of the uncertainty relations. He expressed it in all its generality, in connection with the status of the ψ-function.

Let us begin with Schrödinger's criticism of the hidden variable program, and with his correlative reluctance to consider the ψ-function as a mere expression of our (incomplete) knowledge of some underlying microscopic state of the individual system. A few weeks after having read the Einstein-Podolsky-Rosen paper, he wrote a long letter to Einstein, wherein he presented a personal version of Von Neumann's "no-hidden variable" theorem[1]. But Einstein replied that he felt unable to understand it at all. Schrödinger then sketched a simpler argument, that he explained to Einstein in a letter of October 4, 1935[2], and that he then published in section 4 of his "cat-paper"[3]. This argument is based on the incompatibility between the observed quantization of the angular momentum and the idea that each system actually possesses a certain value of this variable, out of all the values which are made possible by the classical definition of angular momentum. More generally, "(...) if I wish to ascribe to the model at each moment a definite (merely not exactly known to me) state (...), then there is no supposition as to these numerical values to be imagined that would not conflict with some portion of quantum theoretical assertion"[4]. Later on, in his Dublin seminars of 1952, Schrödinger became more and more caustic against the hidden variable program; he called the idea that a system actually *possesses* a well-defined value of a variable, even though our most appropriate theoretical description has no room for it, the "belief in predestination"[5] (namely the belief that each system is predestined to yield a certain value when the corresponding observable is measured on it). According to him, this idea is incompatible with the quantum predictions: "Is it still a more or less irrelevant question of philosophical attitude whether we choose to accept the 'belief' or reject it? No. If our previous assumptions as regards the actual behaviour of nature are adequate, the 'belief' is definitely

[1]E. Schrödinger to A. Einstein, August 8, 1935, in: A. Fine, *The shaky game*, The University of Chicago Press, 1986, p. 79

[2]ibid. p. 81

[3]E. Schrödinger, "The present situation in quantum mechanics", in: J.A. Wheeler and W.H. Zurek, *Quantum theory and measurement*, op. cit.

[4]ibid. p. 156

[5]E. Schrödinger, *Transformation and interpretation of quantum mechanics*, (E. Schrödinger, *The interpretation of quantum mechanics (Dublin seminars 1949-1955 and other unpublished texts)*, op. cit. p. 78)

inadequate; it may and must be rejected on physical grounds"[1]. The ψ-function was thus bound to represent something else, not to say something more, than our incomplete knowledge of the hidden microscopic individual states. Yet, Schrödinger knew that a return to his own original conception of 1926 according to which the ψ-function reflects faithfully a sort of "blurred" atomic reality was precluded. His celebrated cat-paradox was precisely aimed at demonstrating this impossibility: "That prevents us from so naively accepting as valid a 'blurred model' for representing reality"[2].

In such a difficult situation, Schrödinger was doomed to explore as thoroughly as possible the epistemological solution which was put forward by the "reigning doctrine"[3]. At first sight, this epistemological solution looks very much like the incompleteness interpretation, for it also considers the ψ-function as a catalog of information. But a crucial corrective is added, which embodies the difference between the *reigning (epistemological) doctrine* and the incompleteness interpretation: if I have obtained a *maximal* amount of information by my experiments, then "I can turn aside as meaningless any further questioning about the actual state"[4]. The positivist-like prohibition here again operates, and it is justified by a variety of metaphysical monism, whose basic prescription was stated by Schrodinger in the following terms: "(...) no distinction is to be made between the state of the natural object and what I know about it"[5].

In his reflections about the monistic-epistemic interpretation of the ψ-function, Schrödinger did not try to dismiss it from the outset. Instead, he went as far as possible with it and showed that when pushed to its ultimate consequences, this interpretation had eventually to free itself from the epistemological considerations which served as its departure point. As we shall see below, Schrödinger was here again suggesting an ontological conversion of epistemological requirements, namely a projection of the epistemic limitations onto an appropriate system of intentionally aimed at entities.

From a historical point of view, it is important to underline that in his Dublin lectures of 1952 he adopted the same attitude towards what I have called the monistic-epistemic interpretation of the ψ-function as in his 1935 cat-paper. In both cases, he first apparently adopted a blend of the current interpretation, then evaluated its consequences, and finally inserted it within an appropriate ontological frame. This suffices to rule out the dominant chronological account according to which his insistence on the epistemic interpretation of the ψ-function reflected a high level of allegiance to the Copenhagen views in 1935, and that his interpretation of

[1]ibid. p. 79
[2]E. Schrödinger, "The present situation in quantum mechanics", in: J.A. Wheeler and W.H. Zurek, *Quantum theory and measurement*, op. cit. p. 157
[3]ibid.
[4]ibid.
[5]ibid.

the 1950's consisted in reverting entirely to his former conception of the ψ-function as a faithful description of the atomic processes. His cat-paper of 1935 was *already* oriented towards a criticism of the monistic-epistemic interpretation of the ψ-function; and his Dublin lectures, which contain the very same analysis of the monistic-epistemic interpretation as the cat-paper, can by no means be considered as a return to his original "'blurred model' for representing reality". These lectures rather represent the highest point of a process leading on the one hand to retaining all the important features of the monistic-epistemic interpretation of the ψ-function, especially its explicitly stated distance between the theoretical model and the observed facts, and on the other hand to removing the epistemological tinge from it. Schrödinger's interpretation of the ψ-function in the 1950's is thus utterly different from his wave interpretation of 1926; it arose as what we have called an *ontological conversion* of the monistic-epistemic interpretation rather than as a return to some former pre-reflective ontology; it must be construed as an extreme achievement of the modern, non-figurative, anti-realist, Copenhagen conception, leading to its transformation into a post-modern quasi-realist interpretation, and not as a pre-modern straightforward realist view. The fact that this final result was surprisingly close to the original wave interpretation just indicates that in 1926 Schrödinger was formally separated by a hair's breadth from his mature conception; he would only have had to distinguish carefully between ψ as the "effective" entity, and ψψ* as the "factual" correspondence rule. But, conceptually and culturally, it was a very long way, and Schrödinger's accomplishment is all the more considerable.

At this point, we must give further precisions about what Schrödinger considered as his most decisive argument against the epistemological element which is an integral part of the Copenhagen interpretation of quantum mechanics. This argument is present in both the 1935 cat-paper and the 1952 Dublin lectures. It has exactly the logical structure of a *reductio ad absurdum*. Let us suppose that the ψ-function is but a catalog of information (or an expression of our knowledge). This catalog, according to the positivist-like prohibition rule, is *maximal*. But then, if the catalog is maximal, any change in it must combine deletions with additions. "In the catalog not just new entries, but also deletions must be made. Now knowledge can well be *gained*, but not *lost*. So, the deletions mean that the previously correct statements have now become incorrect. A correct statement can become incorrect only if the *object* to which it applies changes"[1]. As a consequence, Schrödinger considers that the ψ-function must represent the state *of* some object. This is not to say that the ψ-function can no longer be considered as knowledge of some sort. But one must not forget that this knowledge bears the mark of its being knowledge *of* something. "Undeniably our knowledge is different in both

[1]ibid. p. 159

cases and undeniably this our knowledge, and nothing else, is laid down in the wave function. But this does not mean that nothing is different. For in both cases our knowledge is in accordance with the actual behaviour of the system, which is different. No 'metaphysics' is dragged in if we call it a *real* difference"[1]. And also: "(...) knowledge is only knowledge in virtue of its agreeing with some reality. If my knowledge changes while the corresponding reality remains the same, it either was a mistake or it becomes a mistake. There cannot be two 'knowledges' referring to the same reality"[2].

In other terms, the claim that the ψ-function represents *knowledge,* and has therefore a somewhat *epistemic status,* is not false but irrelevant. What defines knowledge is its *aboutness,* its being part of an intentional attitude. One must take seriously the thing (or the state of affairs) it appears to be directed at. As the argument of maximality has shown, there is no better reason for not considering seriously the intentional directedness in the case of a ψ-function, than there is in the case of a classical state or of a proposition of ordinary language. It is this semantical circumstance (not any metaphysical creed) which forces us to say not only that a certain ψ-function bears information on experimental facts, full stop, but also that it is "(...) the marvellous tool that is supposed to embody all facts *concerning the behaviour of the system,* not only the values of those observables that have sharp values, but also the statistics of those that have not"[3]. Accordingly, "Quantum mechanics must regard (the ψ-function) as the full counterpart of the complete classical description of the system"[4].

We can perceive once more, at this stage, the sound foundations of Schrödinger's return to a "realist" interpretation of the ψ-function in the 1950's. This doctrinal outcome was not at all motivated by a bare rejection of the anti-realist account of quantum mechanics, and by the naive claim that there exists something "out there" which the physical theories must take as their natural object. It was rather obtained as the end-product of a thorough internal analysis of the anti-realist *tabula rasa,* leading one to disclose the intentional component which pervades its end-product.

[1]E. Schrödinger, *Transformation and interpretation of quantum mechanics,* (E. Schrödinger, *The interpretation of quantum mechanics (Dublin seminars 1949-1955 and other unpublished texts),* op. cit. p. 80)
[2]ibid. p. 83
[3]ibid. p. 80
[4]ibid.

CHAPTER 4

TOWARDS A NEW ONTOLOGY

Schrödinger has often been reproached for having formulated (or reformulated) a wave-interpretation of quantum mechanics in the 1950's, without even trying to ground it in some calculations. People usually regret that he limited himself to generalities or philosophical considerations. L. Wessels notices that "(...)unlike de Broglie, who at about that same time returned to the task of constructing a precise mathematical theory based on his own pilot wave interpretation, Schrödinger did not attempt to work out his new wave picture in detail. Where in 1925 such an idea had been the starting point the creation of a new physical theory, it now gave rise only to philosophical polemic"[1]. However, there is an obvious difference between de Broglie's and Schrödinger's positions, which should not have escaped the commentators. De Broglie was bound to formulate new mathematical laws, in order to rule the classical-like entities of the "sub-quantal" realm which were supposed by him to be incompletely (only statistically) described by quantum mechanics. By contrast, Schrödinger strongly rejected the idea that quantum mechanics is incomplete in the sense advocated by Einstein and de Broglie. He had no reason at all to "work out his new wave picture in detail", for this picture was mathematically identical to a second quantized version of his original wave mechanics, and it had thus *already* been worked out. The only loophole of the theory, according to him, was its lack of integration within the relativist framework. Accordingly, from the early thirties to the late forties, he thoroughly investigated the relations between quantum theory, special relativity and general relativity. But, in spite of some suggestions in this direction during the early thirties, he progressively aknowledged that the actual problems of interpretation of quantum mechanics had little to do with this theory's being coherently integrated in a relativistic framework or not. Schrödinger therefore had only to cope with two questions which are bound to be discussed from a purely philosophical viewpoint:

(1) Can one frame a new ontology in strict correspondence with the symbolic system of standard quantum mechanics?

(2) How is it possible to connect the entities of the new ontology with the "things" and the "experimental facts" of our familiar environment?

The first question can only be solved after a proper philosophical analysis of the ascription of an ontological status has been performed. As for the second question, no doubt it is greatly clarified by some preparatory calculations belonging to a quantum theory of measurement; but Schrödinger suspected that its ultimate answer, if any, might well prove to be, once again, of a purely philosophical nature.

[1]L. Wessels, "Schrödinger and the descriptive tradition", in: R. Aris, H.T. Davis, and R.H. Stuewer (eds.), *Springs of scientific creativity*, op. cit. p. 265-269

Let us then begin with shaping the new ontology. The preliminary step towards a new ontology consists in dismantling more carefully than ever the traditional ontology of localized bodies, and in seeing whether its elements can be transferred as they stand to a novel system of entities. The next section is therefore aimed at completing our study of Schrödinger's thorough criticism of the corpuscularian categories.

4-1 *The fading of the concept of particle*

What is a particle? It is a small localized body whose constitutive features are the following:

(i) it can be ascribed permanent *properties* which embody *virtual* observations expressed by *counterfactual* empirical propositions,

(ii) it has individuality,

(iii) it can be re-identified through time.

These three basic features are usually distinguished as a result of their expressing functionally distinct elements of speech. Ascription of properties corresponds to predication; individuality corresponds to indexical reference by demonstrative pronouns like "this"; and reidentifiability corresponds to reference by names, for, as Kripke rightly pointed out, naming involves implicit reliance on an initial act of baptism and on the possibility of monitoring the trajectory of the body from this act of baptism on. Yet, the three basic features of bodies are not necessarily independent from one another. True, one may consider, as Duns Scot did, that individuality goes beyond any ascription of universal properties, thus leading to separate (i) and (ii). But it has also been proposed in the history of philosophy that a body is individualized by the complete set of its properties. The latter conception is logically linked to Leibniz' principle of the identity of indiscernibles, for if two bodies have exactly the same properties (namely if they are strictly "indiscernible"), and if the only feature which individualizes them is their properties, then they must be considered "identical". Conversely, the very idea that one can ascribe a property *to* a particular something (i) appears to presuppose the individuality of this something (ii). And asserting the permanence of a property (*modulo* some possible disturbances) (i) must assume the reidentifiability of its bearer (iii). Indeed, the expression "permanent property of something" is likely to be grounded in the following procedure: the *same* outcome is obtained again and again when a given experimental procedure is applied repeatedly on the *same* thing. Finally, it may perfectly be argued that the individuality of something reflects at least partly its *history,* through some kind of mark of the past on it, thus making individuality (ii) depend somehow on temporal identity (iii).

Our strategy will thus consist in studying the three listed ontological features in sequence while mentioning, whenever necessary, their deep-lying inter-relations.

The concept of virtuality was very soon recognized as a corner-stone in the debate on the interpretation of quantum mechanics. As early as 1926, Einstein challenged Heisenberg's positivist-like strict adherence to the effectively performed experiments. He believed that one could not dispense with introducing some version of the modal category of the *possible* in the reasonings, and only retain the *actual* [1], lest one loses the very content of the notion of a real object *on which* experiments are performed. In a conference of 1928, Schrödinger went even farther than Einstein in his stressing the decisive importance of virtualities as a basic ontological constituent. Whereas Einstein considered the "virtual" or the "foreseeable" as a *component* of reality, Schrödinger *defined* reality as a construct made out of a proper combination of actual and virtual material: "That is the *reality* which surrounds us: some actual perceptions and sensations become automatically supplemented by a number of virtual perceptions and appear connected in independent complexes, which we call existing *objects*" [2].

This sentence, together with other similar ones (see a systematic study in chapter 5), defines Schrödinger's peculiar use of modalities in these circumstances. Firstly, the "virtual" perceptions, observations, or experimental results which constitute a real object are not *exclusive* of one another. They are associated in "complexes"; they are construed as co-existent; in short, they are listed like a *conjunction* rather than like a *disjunction*. Secondly, the justification of their being linked in such a way is that they are experimentally *accessible* at any moment. The virtualities are conceived by Schrödinger as the modal expression of expectations[3]. A virtual observation is not only an observation which could have been made, but an observation which can be made in the future provided the appropriate experimental conditions are fulfilled.

Of course, one has to qualify this condition of permanent accessibility to the virtual observations, in order to make it applicable to the most familiar situations of daily life. An ideal accessibility presupposes that no change whatsoever happens between the instant when the actual observation is made and the instant when the conditions of the expected observation are fulfilled. However, immutability usually does not obtain. Some disturbances may occur, or the system may be submitted to an evolution law which modifies its state in the interval. It is thus indispensable to *modulate* the condition of accessibility. An observation O_v is considered as virtually coexistent with the actual observation O_a, if there exists a certain observation $O_{v'}$ which can be made in the future and which is connected with O_v by some operator of evolution involving relevant disturbance factors. In other terms, an observation is considered as virtual and coexistent with the actual observation, if it can be

[1] W. Heisenberg, *Physics and beyond*, op. cit. chapter 5
[2] E. Schrödinger, "Conceptual models in physics and their philosophical value", in: *Science, theory and man*, op. cit.
[3] ibid. p. 120

performed in the future *modulo* a certain evolution factor. Let us consider, for instance, a classical material point. Its position $q(t_0)$ having been measured at time t_0, one has the right to say that the momentum value $p(t_0)$ is virtually coexistent with $q(t_0)$ provided a measurement of the momentum can be performed at any time $t>t_0$, and the result $p(t)$ is connected to $(q(t_0), p(t_0))$ through an appropriate operator of evolution.

From this condition, it becomes clear that the condition of accessibility of virtual observations is deeply connected with the possibility of *interpolating* between actual observations. The two modes of "filling with thought", namely in coexistence and in succession, depend on each other. In view of this connection, the fact that any interpolation of the trajectory of microscopic bodies is submitted to Heisenberg's uncertainty relations gave Schrödinger a good reason to be pessimistic about "(...)whether in this case, in principle, virtual observations are at all conceivable, on which the real existence of these objects can be based"[1]. True, the idea that particles have some underlying properties of the usual sort, even though they are disturbed[2] in an *uncontrollable* way by the measurement, could still be sustained at this early stage of the debate about the meaning of quantum mechanics. And such a possibility would have been sufficient to maintain, at least formally, the concept of virtuality in spite of the uncertainty relations: the value of any observable could have been considered as virtually coexistent with the effectively measured value of another incompatible observable, *modulo* an evolution factor involving appropriate (but uncontrollable) disturbance terms. But Schrödinger found it increasingly difficult to accept this very artificial conception. As we mentioned in section 3-2, he had formulated his own version of no-hidden-variable theorems in 1935, and he claimed in the 1950's that the "belief" according to which the particles possess virtual values of every observables, is not justified.

Of course, we know nowadays that such Von-Neumann-like no-hidden-variable theorems do not rule out *any* hidden variable theory, but only a very restricted class of such theories. We also know that further theorems about hidden variables, such as Bell's or Kochen's and Specker's, only rule out certain classes of theories: namely the local and non-contextual hidden variable theories. Some hidden variable theories, such as Bohm's, belong to the class of those theories which are ruled out by *none* of the listed theorems. Could then Schrödinger have changed his mind about the possibility of ascribing simultaneously values to all observables to particles, in view of Bohm's theory? I guess he would not have been convinced. For on the one hand, even though he was perfectly aware of the ancestor of Bohm's theory, i.e. de Broglie's pilot wave theory, he showed no sign of attraction towards it. On the other hand one may guess that Schrödinger would have shown very little enthusiasm for Bohm's

[1]ibid. p. 121
[2]For a criticism of the disturbance conception of measurements, see E. Schrödinger, "What is an elementary particle", loc. cit. p. 111

type of hidden variable theories, due to the fact that this theory stands quite far from the epistemological standards he was eager to maintain. Completion in thought could not mean for him completion by something which is *in principle* out of reach of any kind of experimental assessment. But the contextuality of Bohm's theory prevents one *in principle* from testing experimentally the claim it makes, namely that the value of all the relevant properties of a particle are simultaneously determined at any time. There is thus no reason to think that Schrödinger would not have applied the same kind of Ockham's razor to this theory as the one he was ready to apply to his own wave mechanics in 1926: do not add empirically empty "clothing" to the structure of quantum mechanics just for the sake of satisfying the desire of pictures.

When specifically directed to the observables **q** and **p**, Schrödinger's remarks about uncertainty relations and his rejection of hidden variable theories led him to the conclusion that the particles cannot even be ascribed anything like a continuous trajectory: "Observations are to be regarded as discrete, disconnected events. Between them there are gaps which we cannot fill in"[1]. More precisely, we cannot fill them in *according to a trajectory pattern.*

At this point, the over-revolutionary attitude of Schrödinger arises. Is it coherent to keep on speaking of "particles" if they have nothing like a trajectory? Schrödinger's answer is a definite *no*. When he asked "what is a particle which has no trajectory or no path?"[2], it was just a somewhat ironical way of emphasizing that "(...)the particles, in the naive sense of the old days, *do not exist*" [3]. Some years later, he confirmed most clearly this equivalence between *no trajectory* and *no particle at all* in a letter to Henry Margenau: "To me, giving up the path seems giving up the particle"[4]. The reason for this strict implication is to be found in Schrödinger's combined meditation about individuality and trans-temporal identity. The "individual sameness" of the macroscopic bodies which surround us is ascertained, according to him, by their "*form* or *shape* (German: *Gestalt)*" [5], including some imperceptible details which distinguish them permanently from other bodies of the same kind. The elementary particles can also be ascribed a form, even though it is likely to be non-sense in their case to say that this is the form *of some material substratum.* But the said form can but define their species; it does not help to single out each one of them and to identify it through time. Instead, one must revert to another criterion in order to ascertain the individuality and identity of the particles. The alternative criterion, in classical mechanics, is merely their having distinct positions at a given instant, these positions being connected to distinct past histories through different continuous

[1]E. Schrödinger, *Science and Humanism,* op. cit. p. 27

[2]E. Schrödinger, "L'image actuelle de la matière" in: *Gesammelte abhandlungen,* op. cit., vol. 4, p. 507

[3]ibid. p. 506

[4]E. Schrödinger to H. Margenau, April 12, 1955, AHQP, microfilm 37, section 9

[5]E. Schrödinger, *Science and Humanism,* op. cit. p. 19

trajectories. As Schrödinger himself noticed in his letter to Margenau, this criterion was already proposed by Boltzmann in his *Vorlesungen über die Principe der Mechanik* of 1897[1]: "The discontinuity removes the univocal identification. Would you believe it, that Boltzmann, in his *Principe der Mechanik,* right in the beginning, underlines this point in what he calls his *Ertes kinematisches Grundgesetz.* This was a few years before Planck's great discovery, I think about in 1897"[2].

In quantum mechanics, however, we already know that the particles, if any, cannot be ascribed a trajectory. The ultimate criterion of permanent individuality thus collapses. True, there is still a possibility to rescue something of the old concept of individual and trans-temporally reidentifiable body. It is to say, as most contemporary physicists do, that two groups of circumstances are to be distinguished: the circumstances where the range of uncertainty of two trajectories overlap, and the circumstances where they do not overlap. In the first case, the particles have no definite individual identity, whereas in the second case, they have one[3]. But Schrödinger rejected this expedient from the outset. According to him, "Even if you observe a similar particle a very short time later at a spot very near to the first, and even if you have every reason to assume a causal *connection* between the first and the second observation, there is no true, unambiguous meaning in the assertion that it is the same particle you have observed in the two cases. The circumstances may be such that they render it highly convenient and desirable to express oneself so, but it is only an abbreviation of speech; for there are other cases where the 'sameness' becomes entirely meaningless; and there is no sharp boundary, no clear-cut distinction between (the two types of circumstances), there is a gradual transition over intermediate cases"[4]. Even if two "particles" are experimentally located very far away from each other, even if their Δx do not overlap, there is still a small probability that an "exchange" has occurred between them. The distinction can thus be performed *in practice,* but its possibility is ruled out in *principle:* "I beg to emphasize this and I beg to believe it: It is not a question of our being able to ascertain the identity in some instances and not being able to do so in others. It is beyond doubt that the question of 'sameness', of identity, really and truly has no meaning"[5]. In principle, there is nothing like two *distinct particles.* There is thus nothing like an individual and trans-temporally reidentifiable particle; and, Schrödinger concludes, there is thus nothing like a particle: "(...) I must warn of a misconception which the preceding sentences may suggest, viz. that crowding only prevents us from registering the identity of a particle, and that we mistake one for the

[1]See: L. Boltzmann, *Theoretical physics and philosophical problems,* (B. Mac Guinness, ed.), op. cit. p. 230-231.

[2]E. Schrödinger to H. Margenau, April 12, 1955, AHQP, microfilm 37, section 9

[3] See e.g. M. Born, "Physical reality", loc. cit.

[4] E. Schrödinger, *Science and Humanism,* op. cit. p. 17

[5] ibid. p. 18

other. The point is that there are not individuals which could be confused or mistaken one for another. Such statements are meaningless"[1].

The final claim that the difficulties in identifying a given particle and distinguishing particles from one another makes the very concept of individual particle *meaningless* is not explicitly justified, but it is not difficult to figure out a sound reason for it. Indeed, if one cannot ascribe with certainty a given droplet in a cloud chamber to a given particle, then, one cannot in general ascribe the droplet to *another* given particle either. The absence of a criterion for ascertaining the sameness of *one* "particle" is all-pervasive and challenges the very possibility of making sense of the concept of an individual particle. Each observation must eventually be considered as an isolated event, not to be related to any kind of spatio-temporal continuant; the particle itself accordingly dissolves in one or several scattered events: "When you observe a particle of a certain type, say an electron, now and here, this is to be regarded an isolated event"[2]. It is only the superficial linear appearance of some gatherings of events (i.e. tracks in Wilson cloud chamber) which tend to remind one of the trajectory of a particle. But, according to Schrödinger this must be considered as an illusion: "(...) it is better to regard a particle not as a permanent entity but as an instantaneous event. Sometimes these events form chains that give the illusion of permanent beings"[3]. Just the same type of illusion as the one which is widely know in psychology under the name "phi-effect", where *two static* spots of light being successively (and very quickly) switched on, they are seen as a *single moving* spot.

4-2 *An ontology of state vectors*

In so far as our previous analysis has left us with but scattered events, or long strings of trajectory-like separated events, the question of their lawlike connection arises. We know that the pure corpuscularian representation can by no means afford the sought connection. This is so because, even if the concept of trajectory could be maintained, it would only provide us with a *longitudinal* linkage between the events, whereas quantum phenomena also display a *transversal* linkage which manifests itself through the interference patterns[4]. Are we then compelled to adopt something like Bohrian complementarity, between a symbolic particle picture expressing the longitudinal linkage of events, and a symbolic wave picture expressing their transversal linkage? Schrödinger did not think so. Waves can do both jobs at once. Indeed, while the concept of particle path only bears longitudinal linkage, the concept of (possibly multi-dimensional) wave synthetizes the *two* types of linkages: "In a wave

[1]E. Schrödinger, "What is an elementary particle", loc. cit. p. 116

[2] E. Schrödinger, *Science and Humanism*, op. cit. p. 17

[3]ibid. p. 27-28; also in "What is an elementary particle?", loc. cit. p. 115: "(...) in favourable circumstances, long strings of successively occupied states may be produced (...). Such a string gives the impression of an identifiable individual, just as in the case of any object in our daily surrounding"

[4]E. Schrödinger, "L'image actuelle de la matière" in: *Gesammelte abhandlungen*, op. cit., vol. 4, p. 506

phenomenon you have - not always, but in many cases - the two complementary features of wave - (or phase-) surfaces and of wave-normals or rays"[1]. It is not even useful to reconcile the remnants of the particle picture with the wave picture, for one is only left with two elements which are definitely non-corpuscular: a set of scattered experimental events and a wave-like structural linkage (longitudinal and transversal) between them. The longitudinal connection "(...) is in a time-like direction and is thus of the ordinary causal type. The other one, the transversal, is a relation between simultaneous world-points or at any rate such at a space-like interval, so that there can be no causal relationship between them. It is a relationship of common structure pointing to a common origin"[2].

Up to this point, however, the multi-dimensional wave-functions remained abstract entities, embodying the twofold quantum mechanical law-like connection between otherwise isolated experimental events. In some texts, Schrödinger even wrote that "The wave functions are mental material for building analytical pictures of real objects in one's mind and performing *thought experiments* on them"[3]. This is apparently tantamount to making a sharp distinction between the real objects on one side and the wave functions treated as pure "mental material" on the other. But, as we already know, this kind of epistemological dualism was nothing more, for Schrödinger, than a convenient way of speaking. The status he ascribed to pictures and especially to "analytical pictures" went much beyond that of a mere characteristic of the mind as opposed to some reality lying *out there*. For there is no way we can speak of "reality", except by means of our analytical pictures.

Schrödinger thus found that ψ-waves also have many characteristics which fully support their being ascribed *reality*. One could say, in other terms, that he endowed them with an *ontological* significance. Of course, we must be very careful about the connotations of these words "reality" and "ontology". We already know in what essentially *immanent* sense ascribing "reality" to a set of entities was acceptable to Schrödinger. As for the word "ontology", it was usually given by Schrödinger a *transcendent* sense which made it inappropriate, according to him, in any discourse about theoretical entities. This can be seen in a text of 1958 where he rejected explicitly the project of "framing ontologically" the elements of our physical pictures, after having implicitly ascribed a metaphysical meaning to the adverb "ontologically"[4]. But, as we shall see,

[1]E. Schrödinger, July 1952 colloquium 1952, (E. Schrödinger, *The interpretation of quantum mechanics (Dublin seminars 1949-1955 and other unpublished texts)*, op. cit. p. 20). See chapter 6 for more details about Bohr's and Schrödinger's views on "complementarity".

[2]ibid.

[3]E. Schrödinger, *Transformation and interpretation in quantum mechanics*, (E. Schrödinger, *The interpretation of quantum mechanics (Dublin seminars 1949-1955 and other unpublished texts)*, op. cit. p. 82)

[4]Schrödinger explicitly rejected the project of "framing ontologically" the elements of our physical pictures, if "ontologically" is taken *in the metaphysical sense*, see e.g. "Might perhaps energy be a merely statistical concept", in: *Gesammelte abhandlungen*, op. cit., vol. 1, p. 508

Schrödinger could also easily have accepted, in a Quinean spirit[1], that an ontology corresponds to a system of objective references able to organize the experimental domain without having recourse repeatedly to statements about our sense organs or our instruments. He would only have added, as a scientist, a strong commitment (expressed by his frequent use of the word 'real', see §5-9) to the system of objective references which is suggested by the structure of the most advanced theory of modern physics.

Let us then list Schrödinger's non-metaphysical criteria for ascribing "reality" to ψ-waves (and/or for considering them as elements of a Quinean "ontology"):

(1) Wave-functions are so defined that they do not share the major defect of the corpuscularian representation, namely that the latter "(...) constantly drives our mind to ask for information which has obviously no significance"[2]. For instance, the corpuscularian representation drives our mind to ask for the precise value of the momentum of a particle at the very instant when the position observable has been measured; and the fact that one cannot answer this question is ascribed to the measuring devices' being mutually incompatible. By contrast, a wave-function is perfectly defined when only *one* of the two canonically conjugated observables (position or momentum) has been ascribed a precise value. No mention of our instruments is required in order to prevent one from asking meaningless questions. In his oft-quoted letter to Bohr of May 5, 1928, Schrödinger already emphasized that "in the adequate conceptual scheme it should no longer appear as if our possibilities of experience were limited through unfavourable circumstances"[3]. Of course, Schrödinger did not ignore that the very definition of a wave function is relative to a certain measuring device. At any rate, he did not ignore this in the 1950's, for in his paper "What is an elementary particle?" he forcefully emphasized that the states "(...) are not absolutely defined (...)"[4]. But in the framework of Schrödinger's idealistic monism, associated with a quasi-realist attitude, this does not at all prevent one from ascribing an "ontological" status to the wave function. For defining an ontology here must be taken in a restricted (Quinean) sense of choosing an appropriate system of references, not in the sense of an act of picking out some set of intrinsically defined objects. In this context, defining an ontology simply means adopting an intentional attitude (towards what appears to be referred to by the selected entity) rather than insisting on a reflective attitude (towards the experimental means of attestation). In order to avoid being *compelled* to adopt the latter reflective or epistemological attitude, one only has to make sure that the newly defined ontology does not leave

[1]W.V. Quine, *The roots of reference,* Open Court, 1974, p. 88
[2]E. Schrödinger, "What is an elementary particle?" loc. cit. p. 111
[3]E. Schrödinger, letter to N. Bohr, May 5, 1928, in: N. Bohr, *Collected works,* vol. 6, op. cit., p. 47
[4]E. Schrödinger, "What is an elementary particle?" loc. cit. p. 115

(4) *Relativity.* "Relativity of states" gave its name to Everett's interpretation of quantum mechanics. Its principle was stated thus: "All statements about the subsystems (...) become *relative* statements, i.e. statements about the subsystem relative to a prescribed state for the remainder (since this is generally the only way a subsystem even possesses a unique state); and all laws are correlation laws"[1]. Under closer examination, it appears that this insistence on the relativity of states and on the correlations provides the only acceptable link between the idea that many branches coexist and the fact that our lives and our social agreement about the actual surrounding macro-world are only concerned with *one* branch. As S. Saunders[2] cogently pointed out, Everett's interpretation relies on an indexical conception of actuality. This indexical character of actuality can easily be displayed through an analogy with the more familiar characteristics of time. In Everett's interpretation, the specific connection between actuality and the variously possible outcomes exhibited by the branches of a holistic superposition follows just the same pattern as the connection between *now* and the various possible tensed propositions referring to a given event. At each step of Mc Taggart's well-known regress, the contradiction between two tensed propositions such as *e is past* and *e is future* can be removed provided one says explicitly relative to *which* events f and f' each proposition is true: *e is past relative to f* and *e is future relative to f'*. Likewise, in a quantum superposition, the contradiction between the factual propositions corresponding to each term can be removed provided one says relative to which *other* factual proposition each given proposition is true: "'Observable X has value r; observable X has value s' are inconsistent. But introducing a new observable Y, we may say instead 'X has r relative to u of Y; X has s relative to v of Y' and there is no longer a contradiction"[3].

This strategy is very close to Schrödinger's in his 1935 cat-paper. In paragraph 10 of this paper, Schrödinger returns to the cat paradox that he first explained in paragraph 5. In paragraph 10, he explains the relevance of the paradox for the measurement problem, whereas in paragraph 5 he just took it as a *reductio ad absurdum* of the most straightforward realist reading of the ψ-function. He then mentions that after the entanglement between the object, the apparatus, and the cat wave functions have taken place, the "catalogue of information" of the apparatus (and of the cat) is very incomplete, for it does not even indicate which result has been recorded by the apparatus; nor does it indicate the biological state of the cat. But the global "catalogue" at least affords a list of *conditional statements* of the following form: *if* the pointer observable of the

[1]H. Everett, "Theory of the universal wave function" in: B.S. De Witt and N. Graham, *The many-worlds interpretation of quantum mechanics,* op. cit. p. 118

[2]S. Saunders, "Time and quantum mechanics", in: M. Bitbol & E. Ruhnau (eds.), *Now, time and quantum mechanics,* Editions Frontières, 1994

[3]ibid.

prediction? Wouldn't it be more prudent to look for entities which are directly connected with sharp values of the observables? And if this is found to be impossible, wouldn't it be wiser to forego any ontological reconstruction? Schrödinger was not impressed at all by this argument, for he did not think that there is a crucial difference between a statistical prediction and a sharp prediction: "the *statistics* of (the observable) λ must be regarded as a well-defined characteristic of the first state, just as the value of λ naturally is in the second state (i.e. in the case when the state is an eigenstate of λ)"[1]. He might even have argued, as he did in his 1922 Zürich conference[2] and in a paper of 1948[3], that sharp predictions, far from being necessarily primitive, can perfectly reflect regularities arising from a stochastic background. Giving them an ontological priority over statistical predictions is not justified, at least not any more than the other way round.

When analysed in the light of such remarks about statistics, Schrödinger's attempt at ontologizing wave-functions becomes much more understandable.

As a preliminary step, it has to be realized that one of the basic difficulties which is met in quantum mechanics is objectivation of the properties and entities under investigation. The major criteria of objectivity are stability, repeatability, independence of each type of perception or experimental results from the idiosyncratic situation of the scientist who perform the experiment. These results must in particular be such that they can be given a reasonable amount of independence with respect to the spatial location of the experiment, and to the history of the measurement chain. But, usually, in the quantum domain, these criteria are not fulfilled. Measuring a given variable yields an outcome which cannot be reproduced if an intermediate measurement of some other incompatible variable is performed. It is only in special cases, when measurements of the *same* observable or of *commuting* observables are performed, that exact reproducibility of the results is to be expected. In every other situation, *each discrete experimental outcome* has to be regarded as an isolated occurrence depending on a set of uncontrollable factors pertaining to singular and irreproducible instrumental circumstances. This is in particular true of position and momentum measurements, which, when measured in alternation, play a crucial role in the definition of the individuality and temporal reidentifiability of classical particles. Thus, unless one invokes either disturbance theory or contextuality of determinations in Bohm's sense, which are both tantamount to deflecting the reproducibility criteria in such a way that they become only applicable *in abstracto* to *in principle inaccessible* processes, one has to recognize that the process of objectivation has failed to a certain extent in the quantum domain.

[1]ibid. p. 80

[2]E. Schrödinger, "What is a law of nature?", in: *Science and the human temperament*, op. cit.

[3]E. Schrödinger, "Die Besonderheit des Weltbilds der Naturwissenschaft", loc. cit.

The consequence of the previous remarks is the following. If one sticks to the domain of experimentally accessible processes and events (rather than having recourse to hidden processes), objectivation of the type of entities which may be thought of as *directly producing* each discrete experimental event and of their properties is precluded. Now, in the standard interpretation of quantum mechanics, the role of producing discrete experimental event is ascribed to the entities called *particles*. Therefore, under the no-hidden variable hypothesis, *particles* and their properties *cannot* be construed as objective features of the world; except, for all practical purposes, in very special cases of large spatial separation and/or of iteration of measurements of a single variable.

By contrast, once a preparation has been defined, the *statistical distribution* of each observable is a reproducible characteristic, irrespective of the order of measurement of several (possibly conjugate) other observables, in addition to the first. As Schrödinger insisted, "The statistics of λ is repeatable in any case"[1]. The statistics of *any* observable measured after a preparation associated with a certain wave-function has been defined, is stable, repeatable, and reasonably independent from contingent intermediate conditions. In other terms, the statistics of observables, or even better the *generator of these statistics,* viz. the wave-function associated with the preparation, can be considered as an *objective* feature of the world. First-level entities, namely localized entities such as particles which may be considered as directly producing localized events, cannot be objectivized; but second-level entities, namely entities embodying statistical (or, exceptionnally, sharp-valued) regularities of these events, can perfectly be objectivized. ψ-waves are objective, and this is certainly a good basis for their being ontologically construed, as Schrödinger tended to think[2].

(4) However, the project of ontologization of ψ-waves has further obstacles to overcome, once their objectivity has been recognized. According to Heisenberg "One may call the waves in configuration space 'objective' when one wants to say that these waves do not depend on any observer; but one can scarcely call them 'real' unless one is willing to change the meaning of the word"[3]. In order for the ψ-waves to become elements of a new ontology, construed in a reasonably strong sense and

[1]E. Schrödinger, *Transformation and interpretation in quantum mechanics,* (E. Schrödinger, *The interpretation of quantum mechanics (Dublin seminars 1949-1955 and other unpublished texts),* op. cit. p. 80)

[2]Unexpectedly, Born was not very far from this position either. A few sentences in his Waynflete lectures of 1948 could almost have been written by Schrödinger: "I personally like to regard a probability wave, even in 3N-dimensional space, as a real thing, certainly as more than a tool for mathematical calculations. For it has the character of an invariant of observation; that means it predicts the results of counting experiments, and we expect to find the same average numbers, the same mean deviations, etc., if we actually perform the experiment many times under the same experimental condition. Quite generally, how could we rely on probability predictions if by this notion we do not refer to something real and objective?". M. Born, *Natural philosophy of cause and chance,* Oxford University Press, 1949, chap. IX

[3]W. Heisenberg, *Physics and philosophy,* op. cit. p. 130

not only in the weak Meinongian sense of system of (possibly non-existent) objects, they would not have only to be 'objective', but also 'real'. But what does one mean by 'real', if any recourse to a transcendent world of things-in-themselves is precluded? Heisenberg insisted that the etymology of 'reality' is related to the latin word 'res', namely to the *'things'* of everyday life which are located in ordinary three-dimensional space; in order for a ψ-wave to be 'real', it would then have to be represented in ordinary three-dimensional space, which is not the case in general since ψ-waves are rather extended in configuration space. One could at this stage invoke second-quantized formalisms where it is no longer necessary to use ψ-functions in configuration space, since second quantization is performed on a ψ-function in ordinary 3-dimensional space. But this is quite secondary. For, in spite of the importance Heisenberg gives to it, the latin etymology of 'reality' only covers a small part of the semantic domain of this word. After all, something which can be represented in three-dimensional space (say a unicorn) is not necessarily real; and it is only by convention that one could restrict domain of real entities to the set of those which are *contained* in ordinary three-dimensional space, rather than to the larger set of those which *manifest* themselves in this space. Another component of the meaning of the concept of 'reality', already suggested by the latter idea of *manifestation* in ordinary space, has yet to be analyzed.

The reason why the latin etymology of 'realität' was so much insisted upon by Heisenberg is that the German language he uses has another, non-latin, word for some aspects of what native English speakers would subsume under the concept of 'reality'. This other word is 'wirklichkeit'. At the end of chapter VIII of his "Physics and philosophy", Heisenberg uses extensively the predicate 'wirklich', which is translated by 'actual' in the English version, in good agreement with the German etymology of the word which derives from the verb 'wirken' ('to do work, to have effect'). He specifically applies this predicate 'wirklich' to the 'real' *things* and processes of the macroscopic scale which are describable in terms of classical concepts. Thus, being 'real' in the sense in which a thing of everyday life is, does not *only* mean being located in ordinary space, but also being 'wirklich', or 'actual', by opposition to 'having no effect', 'virtual' or 'potential'. This supports P. Heelan's interpretation of Heisenberg's statement about the reality of ψ-function: "A wave function, (Heisenberg) says, is 'objective but not real', for 'real' or 'actual' implies an empirical content while 'objective' does not"[1]. According to this view, the strongest argument of Heisenberg against the 'reality' of ψ-functions is therefore that they have no empirical content of their own, namely that they only express a set of *potentia* for empirical appearances which are discrete, spot-like, and thus definitely closer to the expected empirical

[1]P. Heelan, *Quantum mechanics and objectivity*, Martinus Nijhoff, 1965, p. 150

content of a particle than of a wave (be it in ordinary space or in configuration space).

This challenge can be answered in two distinct ways. Firstly, one may try to insist that the above-mentioned set of *potentia* is behaving in such a way that its virtualities manifest themselves not only through their being actualized in such-and-such an experimental event exclusive of any other event, but also by modulating *as a whole* the characteristics or distribution of events. In this case, one could argue that the ψ-functions have at least an *indirect* distinctive empirical content: the global modulation effect which typically takes the form of an interference-like distribution of spots on a screen. Secondly, one may perfectly conceive new kinds of experiments in such a way that the *whole* distribution involved by ψ-functions becomes available at once. In this case, ψ-functions could be said to have *direct* empirical content, thus putting Heisenberg's contention of 'irreality', in the sense of a missing (empirical) 'actuality', under strong pressure.

Let us begin with the first line of argument, which was explicitly developed and almost completely worked out by Schrödinger. According to him, the wave-functions embody virtualities which are not *exclusive* of one another, and which must be construed as co-existent. Quantum mechanics directly indicates "simultaneous happenings" on a wave-surface, rather than "alternatives"[1]; it is this circumstance which gives rise to *interference patterns.* Consequently, the said (virtual) "happenings" are to be listed in the form of a *conjunction* rather than in the form of a *disjunction.* Schrödinger noticed very early that the use of conjunctions of coexistent (virtual) happenings was the crucial feature which distinguishes wave mechanics from particle mechanics: "We are confronted with the profound logical antithesis between

Either this or that (particle mechanics)
(aut-aut)
and
This as well as that (wave mechanics)
(et-et) "[2].

Later on, in 1952, he emphasized the significance of these conjunctions, quite consistently with his former definition of reality as a construct made of simultaneous occurrences: "Here 'real' is not a controversial philosophical term. It means that the wave acts simultaneously throughout the whole region it covers, not either here or there. (...) So the epithet 'real' means the momentous difference between 'both-and' *(et-et)*

[1]E. Schrödinger, *July 1952 colloquium,* (E. Schrödinger, *The interpretation of quantum mechanics (Dublin seminars 1949-1955 and other unpublished texts),* op. cit. p. 19)
[2]E. Schrödinger, "The fundamental idea of wave mechanics" (Nobel lecture, 1933), in: *Science and the human temperament,* op. cit.

and 'either or' *(aut-aut)"* [1]. A few years later, in a letter to B. Bertotti, Schrödinger made clear that it was exactly his wish to priviledge the idea of a wave considered as the object of experimental investigation, rather than the idea of a wave considered as the state of (a more or less corpuscular object, which made him so eager to retain the concept of frequency rather than the concept of energy as fundamental. He also mentioned that this shift from energy to frequency must be associated with a shift from disjunctions to conjunctions of 'happenings': "(...) in one case (sharp frequencies) one means the physical *nature* of the object in question, in the other (sharp energies) one means the *state* of the physical object. (...) why not keep the original meaning (or word) *frequency,* when the superposition or *simultaneity* gives us no trouble at all, while the word *energy,* by old habit, seems to demand an 'either-or' and thus entails the probability language"[2].

There was a very important difficulty, however, which Schrödinger fully recognized. Whereas his conception of reality involved an aggregate of simultaneously occurring *virtual* and *actual* happenings, the *actual* happenings of quantum mechanics could not be treated on the same footing as the *virtual* ones. The wave formalism articulated conjunctions of (virtual) happenings, but on the other hand, whenever actual facts are concerned, there is no way by which one can avoid making use of disjunctions: "The expectation-catalog of the object has split into a conditional disjunction of expectation-catalogs"[3]. This difficulty is intricate indeed, and it is obviously related to the measurement problem. Schrödinger therefore treated it as he treated the measurement problem itself (see section 4-3); essentially by postponing its solution and by proclaiming the priority of general laws over particular facts. But here, he had an excellent justification to give for his agnostic attitude about particular facts. Questions about particular facts are not only less interesting than questions about general laws; the standard quasi-corpuscular answer to the former questions make the latter more puzzling. Indeed, "explanation" of scattered spots on a screen by the representation of particles hitting the screen *either* here *or* there, is obtained at the cost of blurring completely what makes the specificity of the quantum law of evolution, namely *interference effects:* "if you accept the current probability views *(aut-aut)* in quantum mechanics, the single event observation becomes comparatively easy to tackle, but all the rest of physics (...) is lost to sight"[4].

Let us then come to the second line of argument about the 'reality' of the ψ-function. Is it possible to find experiments in which central features of the statistical distribution associated with ψ-functions become *directly*

[1]E. Schrödinger, "Are there quantum jumps?" loc. cit. p. 242
[2]E. Schrödinger to B. Bertotti, July 30, 1958, in: B. Bertotti & U. Curi (eds.), *Erwin Schrödinger scienziato e filosofo,* op. cit. p. 156
[3]E. Schrödinger, "The present situation in quantum mechanics", in: J.A. Wheeler & W.H. Zurek (eds.), *Quantum theory and measurement,* op. cit. p. 162
[4]E. Schrödinger, "Are there quantum jumps?" loc. cit. p. 242

available? This possibility exists, but it had not been fully realized until very recently. In a paper entitled "Meaning of the wave-function", Aharonov et al.[1] develop the concept of what they call "protective measurements". They start applying the standard quantum theory of interactions between a system characterised by the state vector $|\psi\rangle$ and a measuring apparatus. But, in the case on which they focus, the interaction Hamiltonian is not assumed to act during a very short time and with a strong intensity, as in the usual Von Neumann account. Rather, it is supposed to act during a very long time (less than one second, in practice) and with a very low intensity. This procedure gives rise to what one may call an "adiabatic" measurement. The most interesting feature of adiabatic measurements is that they *can* give *direct access,* through appropriate pointer observables, to the expectation value $\langle A\rangle=\langle\psi|A|\psi\rangle$ of any observable A. Moreover, they do so *without leading to any entanglement* of the state vector of the system with the state vector of the apparatus. After an adiabatic measurement has taken place, the state vector of the system can again be factorized, although the state vector of the apparatus has been modified due to the interaction hamiltonian. As for the state vector of the system, it is left unchanged, apart from the normal unitary evolution connected with the *system* hamiltonian alone. In other terms, the interaction hamiltonian has *no additional* effect on state vector of the system, and the evolution of this state vector proceeds exactly as if no measurement had taken place at all. As a consequence, it is perfectly possible to perform adiabatic measurements of expectation values on other observables B which *do not* commute with A, *without in any way changing* the result one would obtain if one repeated the adiabatic measurement of the expectation value of A. Measurements of expectation values of observables are *compatible,* even when measurements of the observables themselves are not.

Many other distributional features of $|\psi\rangle$ can be assessed by adiabatic measurements, such as for instance the standard deviation ΔA :

$$(\Delta A)^2=\langle\psi|(A - \langle A\rangle I)^2|\psi\rangle \ ,$$

And, if one chooses $|x\rangle\langle x|$ as the observable, one may also measure directly:

$$\langle\psi|x\rangle\langle x|\psi\rangle=|\psi|^2,$$

namely the square of the modulus of the wave function.

To summarize, adiabatic (or "protective") measurements are able to provide direct access to those distributional features of the ψ-function which are usually construed as arising from the statistics of many

[1]Y. Aharonov, J. Anandan, & L. Vaidman, "Meaning of the wave function", *Phys. Rev.* A47, 4616-4626, 1993; see also M. Dickson, "An empirical reply to empiricism: protective measurement opens the door for quantum realism", *Philosophy of science,* 62, 122-140, 1995

individual events. Moreover, each value obtained by means of an adiabatic measurement bearing on one observable is *reproducible* irrespective of any other adiabatic measurements which could be performed in the meantime. Thus, the distributional features of the ψ-function simultaneously fulfill the criteria of objectivity and of *direct empirical accessibility*. ψ-functions are both 'objective' and experimentally 'actual'. These are further non-metaphysical arguments for ascribing an ontological status to ψ-functions.

Such considerations on protective measurements are quite recent, and they were thus inaccessible to Schrödinger. However, his insistance on basing the empirical correspondance rules on expectation values rather than on individual probabilities (see paragraph 2-4 and 4-3 for more details) is not unrelated to what has just been said. Expectation values are distributional characteristics of the ψ-functions themselves, whereas individual probabilities refer to what ψ-functions enable one to predict about isolated experimental events. Focusing on expectation values means directing attention towards ψ-functions in configuration space, whereas focusing on individual probabilities means directing attention towards scattered (experimental) events in the spatio-temporal framework of the laboratory, and towards the kind of spatio-temporal continuants which are supposed to produce these events, namely particles.

From a phenomenological standpoint, one would express this as follows: the intentionality structure which is fulfilled[1] by providing distributional characteristics such as expectation values or standard deviations identifies itself with ψ-function; whereas the intentionality structure which is fulfilled by providing values of the observables themselves (especially position and momentum observables) is likely to be corpuscle-like. By choosing a formulation of the empirical correspondance rules that is based on the expectation values, Schrödinger clearly inclined towards the choice of ψ-functions as intentionality structures, leaving to future generations the task of showing how those intentionality structures could be fulfilled *directly* in appropriate experiments. Aharonov's protective measurements provides the proper tool for this direct experimental filling-out of ψ-function intentionality structure.

Actually, another set of considerations which were already available during Schrödinger's lifetime could have provided him with a good argument. As P. Heelan mentions[2], Einstein once developed a calculation about the probability distribution of a particle's position in a box. He

[1]See e.g. E. Husserl (1913), *Ideas (general introduction to pure phenomenology)*, Engl. Tr. G. Allen & Unwin, 1931, §136: "We have yet to note that the expression *'fulfilment' (Erfüllung)* has still another ambiguity which lies in a quite other direction: at one time it is 'fulfilment of intention', as a character which the actual thesis takes on through the special mode of meaning; at another it is precisely the peculiarity of this mode itself or the peculiar property of the meaning in question, to conceal 'rich resources' which motivate in accordance with reason". Sometimes 'Erfüllung' is also translated 'filling-out'. See also chapter 5.

[2]See P. Heelan, *Quantum mechanics and objectivity*, op. cit. p. 118.

assumed that this distribution is initially uniform. In this case, Einstein noticed that even if one lets the Planck constant tend to 0, the probability distribution of a particle in a box does *not* tend towards the extremely localized peak which would be associated with a classical trajectory; it remains perfectly uniform throughout. But is this result a shocking feature of quantum mechanics? Does it prove that quantum mechanics does not automatically enable one to recover the classical picture of the world when h tends to 0? Not exactly so. This result just proves that when h tends to 0, the limit of a quantum distribution of probabilities (with interference effects) is not a deterministic evolution but a classical distribution of probabilities (with no interference effects). It proves that the classical limit of quantum statistics is not classical mechanics but classical statistics. In order to recover the deterministic evolution of the position of a classical particle, one should not have considered the probability that such and such sharp value of the position is found in a measurement, but the *expectation value* of the measured position, which is related, *via* the Ehrenfest theorem, to the classical trajectory of a particle. Thus, the proper connection between quantum and classical mechanics is not to be sought in probability ascriptions for each value of (possibly incompatible) observables, but in the (systematically compatible) *expectation values* of these observables. When h tends to 0, the expectation values and their law of evolution remain unchanged, whereas the standard deviations are kept within an interval whose rate of expansion is smaller and smaller. After all, continuous monitoring of conjugate variables of a macroscopic object by (proportionally) low-energy interaction, is definitely more akin to adiabatic measurement of expectation values than to instant measurement of sharp values.

Let us push these considerations a little further. As we have just seen, from an experimentalist standpoint, the most natural extrapolation of macroscopic value-ascriptions is not microscopic value ascription, but expectation value ascription (and more generally ascription of values to distributional characteristics). Therefore, the most natural extrapolation of the macroscopic ontology in the microscopic realm, is not an ontology of bearers of sharp values but an ontology of bearers of distributional characteristics. In a word, the most natural extrapolation of the macroscopic ontology in the microscopic realm, is not an ontology of particles but an ontology of ψ-waves.

(5) The ψ-waves are *individuals.* They are individuals by virtue of their having a *form,* namely a wave-length and a (frequency or amplitude) *modulation*[1]: "(...) waves can easily be marked, by their shape or modulation. If you hear a good friend speaking on the wireless at New-York, you can tell with dead certainty that the wave which hits your receiver is the same which his voice has modulated many 1000 miles

[1]E. Schrödinger, "L'image actuelle de la matière" in: *Gesammelte abhandlungen,* op. cit., vol. 4

away. (...) These are trivial macroscopic examples. But the waves of quantum mechanics exhibit the same feature. They have to be treated as individuals"[1]. The major difference between these quantum mechanical individuals and the particles of classical mechanics thus bears on *location*. The particles have, by definition, a well-defined location in ordinary space at any instant, and it is this location, combined with their past locations (namely their trajectory), which classical physicists took as the criterion of their permanent individualisation. By contrast, the quantum mechanical individuals are entities extended in 3n-dimensional q-space. An interesting particular case is that of steady states (or standing waves), which are completely ubiquitous in the volume they occupy but which are individualized by their form. It is this case which was insisted upon by Schrödinger: "the proper modes have to be regarded as distinguished from one another, they have to be treated as true individuals"[2], even though nothing like location or trajectory could serve as a criterion of permanent individualisation.

Schrödinger's motivation for focusing on proper modes (or eigenstates) is to be found in quantum statistics, and in the fact that only eigenstates, not particles, obey the Maxwell-Boltzmann method of counting. In his paper "What is an elementary particle?", Schrödinger gave a very simple and very clear illustration of how the *new* (Bose-Einstein and Fermi-Dirac) statistics could be obtained[3]. Let us first suppose that we distribute a certain amount of money between several persons. Provided this amount of money is divided in finite quantities, the number of different distributions is given by the Bose-Einstein formula. Let us then suppose that we distribute "vacancies in a football team" between several persons. Once it has been noticed that one person cannot be offered more than *one* vacancy, it becomes clear that the number of different distributions is given by the Fermi-Dirac formula. The surprise comes when the metaphor is translated in terms of the relevant physical entities. The persons (individuals) stand for the *states*, not the particles; and the amounts of money or football club vacancies (non-individuals) stand for the particles. "The example may seem odd and inverted. One might think, 'why cannot the people be the electrons and various clubs their states? That would be so much more natural.' The physicist regrets, but he cannot oblige. And this is just the salient point: the actual statistical behaviour of electrons cannot be represented by any simile that represents them by identifiable things"[4]. With this illustration, one understands that quantum mechanics strongly suggests a kind of ontological inversion. In the classical paradigm, the particles were ascribed the grammatical status

[1]E. Schrödinger, *July 1952 colloquium* (E. Schrödinger, *The interpretation of quantum mechanics (Dublin seminars 1949-1955 and other unpublished texts)*, op. cit. p. 32)

[2]ibid. p. 32

[3]E. Schrödinger, "What is an elementary particle?" loc. cit.; see also "The nature of the elementary particles", in: *Notes for seminar 1949*, (E. Schrödinger, *The interpretation of quantum mechanics (Dublin seminars 1949-1955 and other unpublished texts)*, op. cit. p. 102-103)

[4]E. Schrödinger, "What is an elementary particle?" loc. cit.

of subjects of propositions and the states acted as predicates of the particles; but in the quantum paradigm, it is much more natural to consider states as subjects and the numbers of each variety of quanta in these states (or their statistical distributions) as predicates, in good agreement with the Fock-space second-quantized representation.

Let us summarize what has been said so far. ψ-waves remove the need to have *explicit* recourse to epistemological considerations in the formulation of quantum mechanics; they are ruled by the law of evolution of this theory; they bear coexistent virtualities; they are characterized by mutually compatible distributional characteristics which are natural extrapolations of macroscopic compatible determinations; and they are reidentifiable individuals. These circumstances support their being ascribed the status of entities of a new ontology. As we mentioned previously, the only weakness of this approach is related to the measurement problem: the actual sharp-value ascriptions cannot be united with the virtual ones in a single conjunction of coexistent occurrences; accordingly, it remains quite difficult to say that they inhere in a single "real" entity. A strategy which proved efficient was to shift attention from sharp values to distributional characteristics. But it is also true that this strategy can be perceived as a way of getting round the obstacle rather than facing it.

4-3 *The "blind spot" of quantum mechanics*

We must now confront the issue which is the key to any comprehensive interpretation of quantum mechanics: the measurement problem. We have seen throughout the present essay that this was *the* one crucial difficulty over which almost every interpretative option considered by Schrödinger stumbled.

Among the premises of Schrödinger's treatment of the measurement problem, there is the repeated rejection of any descriptive discontinuity. According to him, the idea of the "reduction of the wave packet" (or "wave packet collapse") initially suggested by Heisenberg in 1927[1], could *not* prove an acceptable account of what occurs during a measurement. But is there any alternative left? Schrödinger's arguments against the concept of "wave packet collapse" may at least help us to outline, by contrast, the most likely features of this sought alternative.

In the 1950's, Schrödinger stated most clearly his reluctance to include the *reduction (or collapse) of the wave packet* among the elements of the physical description: "Another disconcerting feature of the probability interpretation was and is that the *wave function* is deemed to change in two entirely distinct fashions; it is thought to be governed by the wave

[1]W. Heisenberg, "The physical content of quantum kinematics and dynamics" (1927), in: J.A. Wheeler and W.H. Zurek (eds.), *Quantum theory and measurement*, op. cit. p. 74: "Thus, every position determination reduces the wavepacket back to its original extension λ"

equation as long as no observer interferes with the system, but whenever an observer makes a measurement, it is deemed to change into an eigenfunction of *that* eigenvalue of the associated operator that he has measured"[1]. His strong no-collapse commitment relied on a requirement of internal coherence of the theory, and of uniformity of its law of evolution: "If one accepts this law - and it is universally accepted as a general law - one must stick to it. It must not be occasionally infringed upon by a man making a measurement"[2], for "To my mind it is patently absurd to let the wave function be controlled in two entirely different ways, at times by the wave equation, but occasionally by direct interference of the observer, not controlled by the wave equation"[3]. Another reason Schrödinger had to be so suspicious of the concept of collapse of the wave packet is that he could not see how this single effect could be produced by so many distinct experimental devices: "(...) there is usually more than one method for measuring the same thing, there is often a long list of different methods. It is highly improbable that they should all have precisely the same effect on the physical object in question"[4].

Accordingly, Schrödinger insisted upon formulating the probabilistic correspondence rules[5] in such a way that they automatically rule out discontinuous transitions from one eigenstate of an observable to another eigenstate, either between two measurements or during measurement processes.

As we already noticed in section 2-4 and section 4-2, Schrödinger knew one can choose between two (formally) equivalent formulations of the correspondence rules[6].

The *first formulation* requires two elements:

(i) an expression for the expectation value of an observable A, namely $\langle\psi | A | \psi\rangle$, and

(ii) the "axiom of correspondence", according to which if we associate an operator A to the class of experimental apparatuses able to measure the variable a, then an operator $f(A)$ has to be associated to the class of experimental apparatuses able to measure the variable f(a).

[1]E. Schrödinger, "The meaning of wave mechanics", in: A. George (ed.), *Louis de Broglie physicien et penseur*, op. cit., p. 18

[2]E. Schrödinger, *Transformation and interpretation in quantum mechanics*, (E. Schrödinger, *The interpretation of quantum mechanics (Dublin seminars 1949-1955 and other unpublished texts)*, op. cit. p. 83)

[3]ibid.

[4]ibid. p. 82

[5]Or the "interpretation", in the restricted sense Schrödinger ascribed to this word.

[6]E. Schrödinger, *Transformation and interpretation in quantum mechanics*, (E. Schrödinger, *The interpretation of quantum mechanics (Dublin seminars 1949-1955 and other unpublished texts)*, op. cit. p. 53)

As for the *second formulation* of the correspondence rules, it is founded on a "statistical axiom in the diagonal frame" of A. It amounts to focusing attention on the basis of eigenvectors $|a_i\rangle$ of A, and giving directly Born's probability: $|\langle a_i|\psi\rangle|^2$.

Schrödinger was aware that the two formulations are equivalent[1]. For in the first formulation, it is always possible to choose a variable f(a) such that, when the value of the original variable a is equal to a_i, the corresponding operator f(A) is such that: $f(A)|a_i\rangle=|a_i\rangle$, and $f(A)|a_j\rangle=0$ for any $j\neq i$. The operator f(A) is called the *projector* on $|a_i\rangle$: $f(A)=|a_i\rangle\langle a_i|$. This being done, one can obtain Born's probability by calculating the expectation value of f(A):

$$\langle\psi|f(A)|\psi\rangle=\langle\psi|a_i\rangle\langle a_i|\psi\rangle=|\langle a_i|\psi\rangle|^2$$

But Schrödinger then emphasized that, although the two formulations are *formally* equivalent, they are not *physically* equivalent: "the second is shorter, but decidedly more artificial. You have to swallow a greater lump at a time. You have to assume explicitly that the system can never be found in a non-eigenstate, when this quantity is measured!"[2]. The first formulation does not share this defect, for it does not incorporate any mention of the eigenstates.

Now, Schrödinger's last reference to (and criticism of) "finding a system in an eigenstate when a quantity is measured" is quite ambiguous in its context. Does this expression mean finding that the system *was* in an eigenstate of A *before* the measurement, or rather finding that it has been projected into an eigenstate of A *by* the measurement? If the first meaning were retained, the quoted sentence would belong to the long list of Schrödinger's criticisms of the concept of quantum jump (which presupposes that a system *is* always in some eigenstate of the relevant observable, and that it jumps from one eigenstate to another either spontaneously or under stimulation by a radiation field). But if the second meaning were retained instead, Schrödinger's attack on the formulation of the correspondence rules founded on a statistical axiom in the diagonal frame of observables would appear to have a much wider (and much more controversial) scope. After all, asserting that the system has been projected into an eigenstate of A *by* the measurement embodies nothing less than the basic requirement of any measurement theory, namely the *repeatability* of any given outcome[3]. Indeed, if the state of the system has been projected into the eigenstate which corresponds to the measured value a_i (or, equivalently, if the wave-packet has been reduced), a further

[1]About this equivalence see B. d'Espagnat, *Conceptual foundations of quantum mechanics*, op. cit. p. 30; B. d'Espagnat, *Veiled reality*, Addison-Wesley, 1995
[2]ibid.
[3]E. Schrödinger, "The present situation in quantum mechanics" in: J.A. Wheeler and W.H. Zurek (eds.), *Quantum theory and measurement*, op. cit. p. 158

measurement performed on the same system gives a_i with probability 1. In this case, Schrödinger's reluctance to accept the second formulation of the correspondence rules *seems* to threaten repeatability.

Actually, it *would* threaten it, if the wave packet collapse was not only a *sufficient* condition, but also a *necessary* condition for repeatability. But, as Van Fraassen has convincingly demonstrated[1], the wave packet collapse is *not* a necessary condition for experimental repeatability to be accounted for by quantum mechanics (see §4-5). Schrödinger was therefore fully entitled to dissociate the condition of repeatability, which he recognized as an integral part of scientific methodology[2], from the questionable concept of wave packet collapse.

Having found acceptable the dissociation between wave packet collapse and repeatability, we must now enquire about its significance. Isn't there a philosophical viewpoint from which this dissociation appears merely pointless? And in this case, can't we use the wave packet collapse as a convenient procedure allowing one to express repeatability? Let us first state the above-mentioned philosophical viewpoint. Then, we shall try to understand why Schrödinger thought that *even* in this case the dissociation between wave packet collapse and repeatability is not as pointless as it appeared to be at first sight.

If one retains the epistemic interpretation of ψ, there seems to be no reason to reject the concept of "wave packet collapse" as a straightforward way to impose the methodological requirement of repeatability upon the formalism. If the wave-function expresses *nothing else than our knowledge*, if an initial measurement of the observable *A* on a given system has given the result a_i, if we *know* that further measurements of *A* on this system can but yield the outcome a_i again and again, and if we realize that probability calculations about the outcome of further measurements of other observables *B* are to be performed by using the corresponding eigenstate of observable *A*, rather than the inital wave function, then *why* not claim that the wave "of knowledge" has been reduced by the first measurement? Schrödinger's rather negative answer to this question is related to his attitude towards the quantum theory of measurement, which we shall describe more carefully in subsequent paragraphs, but which can easily be outlined at this stage.

In his 1935 cat-paper, Schrödinger tried to work out as completely as possible all the consequences of the conception of the wave-function as a "wave of knowledge", or rather, to use his own vocabulary, as a "catalogue of information". From this viewpoint as from any other, the difficulty which has to be discussed is that the pure wave-mechanical account of the interaction between the system and the first measuring

[1]B. C. Van Fraassen, *Quantum mechanics, an empiricist view*, Oxford University Press, 1991, p. 252

[2]E. Schrödinger, E. Schrödinger, "The present situation in quantum mechanics"in: J.A. Wheeler and W.H. Zurek (eds.), *Quantum theory and measurement*, op. cit. §8; and also *Transformation and interpretation in quantum mechanics*, (E. Schrödinger, *The interpretation of quantum mechanics (Dublin seminars 1949-1955 and other unpublished texts)*, op. cit.)

apparatus does *not* yield a "collapse" of the system's wave function, but rather an *entanglement* of this wave function with the wave function of the apparatus. Following the indications of the wave-mechanical formalism, the wave function of the system *disappears* in the melting pot of the wave function of the whole during the measuring interaction. The basic effect of a measurement, in that respect, is to lead one to a *holistic* catalogue of information for the compound system (system+apparatus), and *not* to a sudden change of the symbols used to account *separately* for the catalogue of information of the system and for the catalogue of information of the apparatus. True, one can perfectly well use the information provided by the *actual outcome* of the measurement (which is a definitely non-theoretical element) in order to extract a *new* wave function *for the system alone* out of the combined wave function. However, this is by no means *a change* of the initial wave function of the system; this is a *redefinition* of it; this is a renewed decision to *separate* the elements of information which had been entangled by the measuring process: "(...)it would not be quite right to say that the ψ-function of the object which changes *otherwise* according to a partial differential equation, independent of the observer, should *now* change leap-fashion because of a mental act. For it had disappeared; it was no more. Whatever is not, no more can it change. It is born anew, is reconstituted, is separated out from the entangled knowledge that one has (...)"[1].

One can perfectly well *reconstitute* a system's wave-function by a "mental act", in order to predict as economically as possible the outcomes of subsequent measurements performed *on this system*, but it would be a category mistake (in G. Ryle's sense[2]) to *mix* this choice (or "mental act") with the objective description of what occurs to the catalogue of information of the composite system (system+apparatus). Even if one holds on to the epistemic interpretation of the ψ-function throughout, even if ψ-functions are construed as merely catalogues of knowledge, one has to distinguish carefully between *objective* knowledge and the contingent (possibly subjective) choice which consists in retaining *part of* this knowledge for further practical purposes. From the point of view of objective knowledge, one only needs *one* kind of "change": the one which is ruled by the Schrödinger equation, and which usually leads to entangled wave functions for composite systems. Indeed, it is perfectly possible to derive *directly* from such entangled wave functions, by general application of the empirical correspondence rules, and without any further "collapsing" manipulation, all the knowledge we have about the measured system; namely a list of probabilities that such and such result is obtained at the end of the measurement interaction. Moreover, it is also possible to derive joint probabilities for *sequences* of measurements, by application of the empirical correspondence rules on entangled wave

[1]E. Schrödinger, "The present situation in quantum mechanics" in: J.A. Wheeler and W.H. Zurek (eds.), *Quantum theory and measurement*, op. cit. p. 162

[2]G. Ryle, *The concept of Mind*, Hutchinson, 1949

functions corresponding to composite systems involving apparatuses for measurement of (possibly conjugate) observables A, B, C, ... Nothing else is required than holistic wave functions of ever increasing composite systems in order to make predictions. By contrast, the "wave function collapse" cannot be ascribed the status of objective "change": it is the outcome of a deliberate *choice*. The choice to pick out part of the composite system, to assign a wave-function to this part, and to describe exclusively the evolution of the latter, in order to avoid carrying the burden of the complex overall wave-function. What makes it obvious that this choice cannot partake in the objective description is that it is made at the cost of cutting off part of the information made available by the wave function of the composite systems; i.e. the residual (possibly negligible but never absent) interference terms.

In his 1952 seminar entitled "Transformation and interpretation in quantum mechanics", Schrödinger resumed his reflections on the epistemic interpretation of the wave function, trying once more to take it quite seriously and to analyze its consequences. The question about the status of the so-called collapse of the wave function was raised from the start of the seminar. There, Schrödinger noted that one would like to know "(...) to what extent (these changes) mean physical changes in the object or only changes in our knowledge about the physical object"[1]. However one has to be careful about what one means by "our knowledge" in this case. Even though Schrödinger recognized that "(...)undeniably (...) our knowledge, and nothing else, is laid down in the wave function"[2], he also warned that this formulation may be misleading. For "knowledge" is usually contrasted with "what is to be known"; and, as a rule, *what is to be known* is taken as much more extended than the *knowledge* we have of it. Insisting that the wave function *only* represents our knowledge thus prompts one to ask questions about what is beyond this (possibly *incomplete)* knowledge. And this in turn means becoming committed to the hidden-variable-like "belief" according to which systems already possess values of every variable, even though quantum mechanics implies that there is no experimentally available (simultaneous) knowledge of them.

If this consequence is to be avoided, knowledge has to be ascribed intentional directedness towards a representation which automatically embodies the *maximal* information which is available in experiments, and nothing more: namely, towards the representation which is just provided by wave mechanics (or pure unitary quantum mechanics). Once this is done, making a distinction between knowledge and what is known cannot mean opposing a wave function and a speculative set of intrinsic properties; it just means opposing the partial knowledge possessed by a

[1]E. Schrödinger, *Transformation and interpretation in quantum mechanics,* (E. Schrödinger, *The interpretation of quantum mechanics (Dublin seminars 1949-1955 and other unpublished texts),* op. cit. p. 39)

[2]ibid. p. 80

particular subject or a particular group, and the *maximal* knowledge embodied by a sufficiently inclusive wave function; or in other terms, it means opposing subjective knowledge to "reality" in the minimal, internalist, sense favoured by Schrödinger. Thus, unless one wants to. refer to cases where we happen to know *less than the maximal information,* namely to cases where the appropriate representation is provided by proper mixtures rather than pure states, there is no point insisting that quantum mechanical symbols *only* deal with our knowledge. And, similarly, unless one wants to refer to an additional voluntary act, performed in order to pick a certain restrictive amount of information out of the available one, there is no point insisting that the collapse of the wave function occurs but that it is *only* a change of our knowledge. For the collapse does not occur by itself, as a result of a change of our knowledge; *we* make it occur in our calculations in order to restrict our attention to the part of our knowledge we consider relevant in a given situation. "(They say) one must not call it a physical change, it is only a change in our knowledge. I consider this an unfair subterfuge - or plainly: non-sense"[1].

From a modern point of view, it is possible to see the previous remarks in a slightly different light. Traditional discussions on the collapse of the wave packet usually mix two steps; namely transition from the global wave-function of the composite system to a statistical mixture, and transition of the statistical mixture to a single event. The transition from a global pure state to a mixture can be accounted for, at least approximately (and using additional hypothesis which are to be discussed in paragraph 4-4), by pure wave-mechanical decoherence formalisms. Such a transition was implicitly assumed in Schrödinger's 1935 analysis of the measurement process when he pointed out that, after the measuring interaction, the catalogue of predictions represented by the global Ψ-function has been broken up into a conditional *disjunction* of predictions, whereas it had previously to be considered as a *conjunction*. Once the decoherence has taken place, the ignorance interpretation of the wave-function becomes (approximately) acceptable. And it then becomes just as unproblematic as in the classical theory of probability to modify this statement of ignorance by taking into account the newly acquired knowledge; i.e. to impose a transition from the overall mixture to a partial pure state corresponding to the observed pointer position on the apparatus, and then to factorize this pure state into a new state for the system and a new (pointer) state for the apparatus. But here again it would be a category mistake to confuse the reduction with the transition from an entangled pure state to a mixture. The reduction expresses the

[1]ibid. A similar remark was made recently by S. Y. Auyang, *How is quantum field theory possible?*, Oxford University Press, 1995, p. 115: "Consider Heisenberg's suggestion that the wavefunction represents not a microscopic system but our knowledge of it. The proposition sounds both indisputable and absurd. (...) All sciences are our knowledge, but the content of the sciences are features of the objective world".

intervention of a *particular* state of knowledge, whereas the transition to a mixture (namely the transition from a conjunction to an approximate disjunction) expresses a *generic* precondition of *any possible* knowledge. For any possible knowledge just consists in *determining which term of a disjunction obtains*.

Let us now conclude this discussion about the collapse of wave function by adopting a wider philosophical standpoint. True, *projecting* some epistemic elements onto the screen of an "objective world" in order to form new "properties", or new entities bearing these properties, is an old and well-tested procedure[1]. And we know that Schrödinger made an extensive use of this procedure when he tended to call 'real' an entity (namely the ψ-wave) whose dependence on the definition of the experimental context was fully recognized by him. So, why did Schrödinger find himself so eagerly opposed to projecting the sudden changes of our knowledge onto the quantum mechanical description? The reason is simple. After having projected *this type* of epistemic element onto the description, there is no way in which it can be forgotten and gradually absorbed into an all-comprehensive objective picture. Indeed the sudden change of knowledge about the value of a quantum observable is not uniformly reproduced, in general, under similar experimental circumstances, and it is thus not predictable. The decision about what to project must be taken *each time,* with no hope, in general, that it can be fixed in advance by such and such preliminary observation. The choice which consists in picking out part of the composite system can by no means be incorporated within a law-like sequence of events. It retains throughout its volitional status of choice, as well as the mark of its empirical motivation. The projection is a failure, because it is too obvious that it is *just a projection;* and also because, as we have pointed out, it is the projection of an isolated element of knowledge rather than of generic (or at least reproducible) features of knowledge. To recapitulate, one must be careful, when an ontological reconstruction is undertaken, to *project* only those epistemic elements which are both shared by any knowing subject, and which can be attached permanently (*modulo* an evolution law and some disturbance processes) to the entity referred to. Since, in the situation dealt with by quantum mechanics, the singular act of becoming aware of an experimental outcome does not fulfill the second condition, it must not be construed as a constitutive element of the newly ontologized entity (namely the ψ-wave). The "experimental fact" accordingly remains an outsider, something whose *irreducibly* epistemic character prevents one from assigning it a counterpart within the system of entities of the new ontology. An experimental fact can of course be probabilistically *related* to the theoretical description by means of the correspondence rules; but the theory can by no means mimic the experimental fact or take charge of its Humean *projection*.

[1]D. Hume, *A Treatise of Human Nature,* Book I, part III, section XIV, (ed. L.A. Selby-Bigge), Oxford University Press, 1960; see a discussion in: S. Blackburn, *Essays in Quasi-Realism,* op. cit. p. 55.

Hence Schrödinger's dominant attitude towards quantum mechanics: push that description of the entities which can be construed both objectively and repeatably to its limit; don't bother about its connection with experimental outcomes until the *very last* stage of the description; postpone the necessity of making this connection explicit for as long as you can. "(O)ne must, to repeat this, hold on to the wave aspect throughout"[1]. At most, one may rely on some loose stopping criteria which are sufficient *for all practical purposes:* "quantum mechanics stops as soon as anything reaches your senses (that has been said by Schopenhauer long ago)"[2], or "(...) *not until this inspection,* (...) does anything discontinuous, or leaping, take place. One is inclined to call this a mental action (...)"[3].

As a consequence Schrödinger appeared as one of the very few quantum physicists who felt motivated to formulate a genuine quantum theory of measurement. He began his undertaking very early in the history of quantum mechanics. In 1926, he was already invoking the peculiarities of the interactions between the radiation and the *receiver* in order to account wave-mechanically for the selective observation of the *differences* between the eigenfrequencies[4]. But the first step towards the modern quantum theory of measurement was made by Schrödinger in 1927[5], and this paper served as a paradigm of what he was intending to do throughout his career. Even though it contains no explicit reference to the measurement problem, Schrödinger quoted it repeatedly in subsequent articles (of 1936 and 1952) which were devoted to the quantum theory of measurement[6].

The immediate purpose of the 1927 paper was to account for the fact that "physical systems" (possibly an emitter and a receiver or, in view of later developments, an object and a measuring apparatus) "influence each other only when they agree in respect of a 'difference of level'"[7]. The

[1]E. Schrödinger, "The meaning of wave mechanics", in: A. George (ed.), *Louis de Broglie physicien et penseur,* op. cit. p. 26

[2]E. Schrödinger, *Short notes for Dublin seminar,* May 4, 1949; quoted in: E. Schrödinger, *The interpretation of quantum mechanics (Dublin seminars 1949-1955 and other unpublished texts),* op. cit. p. 98

[3]E. Schrödinger, "The present situation in quantum mechanics"in: J.A. Wheeler and W.H. Zurek (eds.), *Quantum theory and measurement,* op. cit. p. 162

[4]E. Schrödinger to W.Wien, February 22, 1926, quoted and translated by L. Wessels, *Schrödinger's interpretations of wave mechanics,* op. cit. p. 167

[5]E. Schrödinger "The exchange of energy according to wave mechanics", in: *Collected papers on wave mechanics,* op. cit.

[6]E. Schrödinger, "Probability relations between separated systems", Proc. Camb. Phil. Soc., 32, 1936, p. 451; E. Schrödinger, "Are there quantum jumps?"; loc. cit.; *July 1952 colloquium,* (E. Schrödinger, *The interpretation of quantum mechanics (Dublin seminars 1949-1955 and other unpublished texts),* op. cit.). In the two last cases, and as in the original 1927 paper, the idea that a quantum jump of energy $E_{kl}=h\nu_k-h\nu_l$ happens in the receiver (possibly belonging to a measurement apparatus) in order to compensate for a quantum jump of equal energy in the emitter was replaced by a concept of resonance between two oscillators O and O'. The condition of resonance is:

$\nu_k-\nu_l=\nu_{l'}-\nu_{k'}$.

[7]E. Schrödinger "The exchange of energy according to wave mechanics", in: *Collected papers on wave mechanics,* op. cit. p. 140

paper was also intended to show that "without quantum postulates" one can arrive "at an effect which is exactly the same as if the quantum postulates were in force"[1]. In order to obtain this result, Schrödinger had to solve his equation for the two interacting systems together, "united into *one* system" (let us call this interaction the "first-order measurement of energy" on one of the two systems by the other one). Then, the general solution of the compound equation was not a simple product of the proper wave functions ψ_k and ϕ_l for each system, but the linear superposition: $\Psi=c_1\psi_k\phi_{l'}+c_2\psi_{k'}\phi_l$. In modern terms, one would say that this calculation enables one to bring out a *correlation* between the two systems. And, from an experimental standpoint, "correlation" means that whenever a (second-order) energy measurement performed on the ψ-system gives E_k (resp. $E_{k'}$), then a (second order) energy measurement performed on the ϕ-system gives $E_{l'}$ (resp. E_l).

Accordingly, Schrödinger's calculation shows a definite relation between the results of two second-order measurement but it gives no indication whatsoever about their *actual result*. The description of the composite system is continuous and purely wave-mechanical, but it by no means clarifies the relation between the *superposition* Ψ and the fact that only *one* of the two pairs of values $(E_k,E_{l'})$ or $(E_{k'},E_l)$ obtains whenever a (second-order) energy measurement is performed. In short, wave mechanics provides us with a discrete *scheme* of levels and processes, but definitely has nothing to say about the singularity of an observed discrete phenomenon.

According to Bohr, this was due to a major methodological flaw in the calculation in Schrödinger's 1927 paper: "In the resonance problem mentioned, we are concerned with a closed system"[2]. In other terms, Bohr here emphasized that reference to the *second-order* measurements was unavoidable; that these measurements could themselves be described quantum mechanically, but that this would then call for reference to third-order measurements, etc.; and that the notion of a discrete experimental event would arise only by considering an *open* quantum system undergoing an unanalyzable interaction with an apparatus described in *classical* terms for the sake of unambiguous communication. It is thus clear that, even in retrospect, one cannot regard Schrödinger's paper of 1927 as having provided a *solution* of the measurement problem. However, this paper outlined some of the major themes of the measurement problem by means of some elegant statements. This would enable Schrödinger to consider it as his first contribution to the subject.

By discussing Schrödinger's earlier step towards a quantum theory of measurement, we have already identified two of the major themes which enter into the measurement problem: holism and infinite regress.

[1]ibid. p. 141

[2]N. Bohr, "The quantum postulate and the recent development of atomic theory", Nature, 121, 580-590, 1928, in: J.A. Wheeler and W.H. Zurek (eds.), *Quantum theory and measurement,* op. cit.

Schrödinger's later work shows that he was fully aware of the significance of these themes. Let us begin with holism. In the late twenties and the early thirties, the debate about quantum mechanics concentrated on Heisenberg's uncertainty relations. The dominant "explanation" of these relations arose from the Heisenberg microscope *gedankenexperiment*. According to Heisenberg[1], as well as Bohr in his early papers[2], the uncertainty relations are the expression of an unknown and (in principle) undeterminable disturbing influence exerted on the object by the very act of measurement. But in his 1931 paper on indeterminism, Schrödinger was certainly reflecting quite skeptically about this disturbance theory of the uncertainty relations when he wrote: "The question at issue is this: given any physical system, is it possible, at any rate in the theory, to make an exact prediction of its future behaviour, provided that its nature and condition at one given point of time are exactly known? It is assumed of course that no external and unforeseeable influences act upon the system from without; but such influences can always be eliminated, at least theoretically, if all bodies, fields of forces and the like capable of acting upon the system are included within it. (...) in order to do so the system under consideration has to be extended to comprehend the entire universe"[3]. In principle, therefore, a holistic move should enable one to dissolve a disturbance theory of measurement. But is it possible to do so in every case? Is this possible within the framework of quantum mechanics? In 1931, it was too early for Schrödinger to provide a definite answer; but at least the quoted paragraph shows that the general form of the solution was clearly understood by him.

Similar holistic ideas were developed by Bohr in 1935[4], in reaction to the EPR criticism of the disturbance concepts; and in the same year they were also given a precise quantum mechanical formulation by Schrödinger, in his theory of entangled wave-functions[5].

But on the other hand, the holistic stance had to be reconciled with the methodological requirement that *parts* of the world be isolated in order to define a "something" *on which* the measuring procedure is exerted and repeated. In Schrödinger's words, "(...) it is possible to *imagine* a finite, self-contained system, and in practice this abstraction is invariably made use of whenever a law of physics is enunciated"[6]. Thus, after a proper description of the entanglement has been worked out, it is necessary to think about the procedures of *disentanglement*. However, a procedure of disentanglement can but be grounded on the outcome of a measurement;

[1]W. Heisenberg, *The physical principles of the quantum theory*, op. cit. p. 20
[2]N. Bohr, "The quantum postulate and the recent development of atomic theory", loc. cit.
[3]E. Schrödinger, "Indeterminism in physics", in: *Science and the human temperament*, op. cit. p. 43-44
[4]N. Bohr, "Can quantum-mechanical description of physical reality be considered complete?", Phys. Rev. 48, 696-702, 1935, in: J.A. Wheeler and W.H. Zurek (eds.), *Quantum theory and measurement*, op. cit.; see an interesting comment in: J. Faye, *Niels Bohr, his heritage and legacy*, Kluwer, 1991, p. 205
[5]E. Schrödinger, "Discussion of probability relations between separated systems", (1935) loc. cit.
[6]E. Schrödinger, "Indeterminism in physics", in: *Science and the human temperament*, op. cit. p. 44

as we mentioned previously, it amounts to selecting part of the composite system and to ascribing it a wave function in good agreement with the result of an actually performed experiment. Unfortunately, if we are to remain consistent, this measurement must itself be described as a quantum-mechanical interaction governed by the Schrödinger equation, and it thus leads to further entanglement between the wave function of the second-order measurement apparatus and the wave function of the previous composite system. Hence the spectre of an infinite regress, whose seed was already present in the paper of 1927, but which was explicitly stated in 1935: "(...)this procedure will be called the *disentanglement*. Its sinister importance is due to its being involved in every measuring process and therefore forming the basis of the quantum theory of measurement, threatening us thereby with at least a *regressus at infinitum,* since it will be noticed that the procedure itself involves measurement"[1]. Schrödinger's study, in 1936[2], of the concept of improper mixture (in the sense of d'Espagnat[3]), could but confirm the difficulty. In 1935-1936, the conclusion of these studies was therefore essentially negative: the measurement version of the cat paradox, namely the fact that "our knowledge (about the biological state of the cat) has evaporated into conditional statements"[4], had received no solution within the framework of the quantum theory of measurement.

In the 1950's, however, Schrödinger decided to resume his studies of the quantum theory of measurement. He had renewed reasons to do so. The correspondence rules, which Schrödinger was analyzing more seriously than ever before, required a rationale. Accordingly, his preference for the expectation value version of the correspondence rules over the statistical algorithm in the diagonal frame was not just a matter of taste. It was motivated *inter alia* by the fact that the expectation value version of the correspondence rules fitted very well with the idea that each observable operates "(...) as a possible perturbing addition to the hamiltonian"[5], thus referring indirectly to a quantum theory of measurement. But, since 1935, the quantum theory of measurement had not progressed very much. In 1963, when Margenau listed the major advances in this field[6], he could but quote Von Neumann (1932), Schrödinger (1935), London and Bauer (1939), and himself in 1936-1938. Schrödinger thus felt very lonely in his late undertaking: "(...) quantum physicists bother very little about accounting, according to the accepted law, for the supposed change of the wave-function by

[1]E. Schrödinger, "Discussion of probability relations between separated systems", (1935) loc. cit.
[2]E. Schrödinger, "Probability relations between separated systems", (1936) loc. cit.
[3]B. d'Espagnat, *Conceptual foundations of quantum mechanics,* op. cit., p. 58
[4]E. Schrödinger, "The present situation in quantum mechanics" in: J.A. Wheeler and W.H. Zurek (eds.), *Quantum theory and measurement,* op. cit. p. 161
[5]E. Schrödinger, *Transformation and interpretation in quantum mechanics,* (E. Schrödinger, *The interpretation of quantum mechanics (Dublin seminars 1949-1955 and other unpublished texts),* op. cit. p. 53)
[6]H. Margenau, "Measurements in quantum mechanics", Annals of physics, 23, 469-485, 1963

measurement. I know of only one attempt in this direction (...) You find it in John Von Neumann's well-known book"[1]. Schrödinger made some isolated developments in this direction, especially in his 1952 lectures entitled *Transformation and interpretation in quantum mechanics;* but he very soon gave them up.

The reason for this ambivalent attitude towards the quantum theory of measurement, made of felt necessity and of renunciations, was that Schrödinger had not formulated a very clear idea of what was to be expected from it.

On the one hand, he had still hoped that the quantum theory of measurement could show how a perturbing operator turns "(...) the wave function as time goes on into an eigenfunction of the observable which is measured"[2], according to his reading of Von Neumann's 1932 attempt. And on the other hand, he realized the difficulties of this program: "I do not believe any real measuring device is of this kind"[3], namely of the kind that would force any wave function to transform into an eigenfunction of the observable that is being measured. He therefore suggested more and more insistently that such a program was not really worth pursuing, for the principle of superposition and the law of (wave-like) evolution are overwhelmingly more important than the mention of isolated facts. In this latter perspective, the aim of a quantum theory of measurement would not consist in displaying an equivalent of the observed discontinuities, but rather in showing that one can dispense with giving these discontinuities any descriptive counterpart.

To make these points more precise, we may consider that Schrödinger adopted two distinct (and somehow contradictory) attitudes toward the measurement problem:

(1) He sometimes suggested (but quite discreetly and with a strong note of skepticism) that the measurement problem would not prove intractable, provided a proper handling of quantum mechanical descriptions had been achieved.

According to this trend of thought, he looked in 1952 for a kind of compromise, based on the second quantization scheme, between (i) the current conception that discontinuous experimental events are produced by particles and that wave packet collapses occur, and (ii) his own no-particle and no-collapse position: "If, in the present case, you wish to

[1]E. Schrödinger, *Transformation and interpretation in quantum mechanics,* (E. Schrödinger, *The interpretation of quantum mechanics (Dublin seminars 1949-1955 and other unpublished texts),* op. cit. p. 83); see also E. Schrödinger, "The meaning of wave mechanics", in: A. George (ed.), *Louis de Broglie physicien et penseur,* op. cit., p. 18: "I know only of one timid attempt (J. Von Neumann, in his well-known book) to put this 'change by measurement' to the door of a perturbing operator introduced by the measurement, and thus to have it also controlled solely by the wave equation"

[2]E. Schrödinger, *Transformation and interpretation in quantum mechanics,* (E. Schrödinger, *The interpretation of quantum mechanics (Dublin seminars 1949-1955 and other unpublished texts),* op. cit. p. 83)

[3]ibid.

avoid the paradox while keeping as closely as possible to the 'particle-language' you need only accept that the number N of particles is not sharp but has a spread of $\sqrt{N}$. This means that the wave contains proper modes not only for N=1, but also for N=2,3,..., the lower one being very strong. Then, according to recognized rules, the 'finding of particles at B' leaves the conditions for finding one somewhere else *unchanged:* the wave does not collapse, the paradox is avoided"[1].

In another text of the same period[2], Schrödinger attempted to give a hint of a method which would allow one to bridge directly the continuous theoretical entities and the discontinuous experimental events without any concession to the "natural" ontology of localized bodies. However, the picture arrived at was by no means satisfactory. It consisted in ruling out the representation of an interaction between particles and apparatuses construed as systems of particles, but to retain the representation of an interaction between the (wave-like) system and the (wave-like) apparatus instead: "One must regard the 'observation of an electron' as an *event* that occurs within a train of de Broglie's waves when a contraption is interposed in it which by its very nature cannot but answer by discrete responses". It is quite obvious that this sentence was by no means intended as a solution of the difficulty, but as a metaphor of the kind of result which should *eventually* be reached. The problem is that this metaphor proved misleading, because it favoured the image of colliding de Broglie's waves in ordinary space rather than focusing on the appropriate holistic ψ-function in 3n-dimensional configuration space. It was also misleading because it retained the causal scheme of an interaction between the particle and the apparatus, merely projecting this scheme onto the interaction between the wave-object and the wave-apparatus. Schrödinger could not go much farther than replacing sentences like "the particle 'causes' a macroscopically observable change in the apparatus" by sentences which sounded like "the wave-object 'causes' a macroscopically observable discrete change in the wave-apparatus". But what was really needed was a full acceptance of the *parallelism* between the time-development of the holistic wave-function (object+apparatus) and the sequence of macroscopic events, rather than a new blend of the old idea of a causal interaction which takes place between objects and apparatuses in order to *produce* the events. This concept of *parallelism* is systematically developed in paragraphs 4-4 and 4-5, and then in chapter 6. It is shown that many aspects of Schrödinger's reflection on quantum mechanics point towards this direction.

(2) Much more frequently (and in his most lucid writings), Schrödinger behaved as if he thought, rather, that indefinitely postponing

[1]E. Schrödinger, *July 1952 colloquium,* (E. Schrödinger, *The interpretation of quantum mechanics (Dublin seminars 1949-1955 and other unpublished texts),* op. cit. p. 35)
[2]E. Schrödinger, "The meaning of wave mechanics", in: A. George (ed.), *Louis de Broglie physicien et penseur,* Albin Michel, 1953, p. 26

the solution of the measurement problem within a perfectly self-consistent quantum theory of measurement including the apparatus in the wave-mechanical description, was tantamount to ascribing it some (intrinsically elusive) kind of solution[1].

The latter attitude fits perfectly with Schrödinger's critical attitude towards science, and towards objective knowledge in general. Schrödinger was one of the few physicists who overtly recognized the impossibility of any scientific description's encompassing all of its own presuppositions. He often emphasized that sciences are ultimately embedded in human culture as well as in human practices, and that they rely on certain tacitly admitted assumptions which pre-exist to them[2]. Accordingly, he had no difficulty in aknowledging that there may remain some fundamental lacunae in physical theories, which just *reveal* the unavoidable dependence of science on its pragmatic background. He considered that any attempt at filling the lacunae in a systematic way, with the hope that none would remain, is more akin to dogmatic theology (or to scientism, which is a modern form of dogmatism) than to genuine science[3].

In quantum mechanics, one of the lacunae is quite obvious, according to Schrödinger: it is the fact that linear operators are associated with measuring devices by means of Bohr's correspondence principle. One has to take this association for granted "(...) though it remains the most delicate point, not to say the blind spot of the theory, which cannot be filled by pure mathematics"[4]. It would be an illusion to think that quantum mechanics, which incorporates such a blind spot in its foundations, as regards the structure of the macroscopic instruments which allow its experimental confirmation, could then conversely *account for* the said structure. Quantum mechanics can at most display its asymptotic *compatibility* with the *appearance* of a macro-world made of localized bodies in motion. Decoherence theories would just have to be conceived, according to that perspective, as a way of demonstrating this compatibility; and not, as some modern authors have claimed and as the Schrödinger of 1926 still hoped, as a way of displaying emergence of the macro-world out of a continuous wave background.

[1]This attitude evokes the idea of a Gödelian incompleteness of quantum theory, as H. Primas tends to think: "The proposition 'the cat is in a definite biological state' is endophysically undecidable". H. Primas, "A propos de la mécanique quantique des systèmes macroscopiques", in: M. Bitbol and O. Darrigol (eds.), *Erwin Schrödinger, Philosophy and the birth of quantum mechanics,* Editions Frontières, 1993, p. 401

[2]E. Schrödinger, "Quelques remarques au sujet des bases de la connaissance scientifique", Scientia, 57, 181-191, 1935; see a comment in: F. Nef, "A propos d'une controverse entre Carnap et Schrödinger", in: M. Bitbol and O. Darrigol, *Erwin Schrödinger, Philosophy and the birth of quantum mechanics,* op. cit. Also: M. Bitbol "L'alter-ego et les sciences de la nature;*Autour d'un débat entre Schrödinger et Carnap* " in: A. Soulez & J. Sebestik (eds), *Science et philosophie en France et en Autriche, 1880-1930,* To be published.

[3]E. Schrödinger, *Nature and the Greeks,* op. cit., chapter 1

[4]E. Schrödinger, *Transformation and interpretation in quantum mechanics,* (E. Schrödinger, *The interpretation of quantum mechanics (Dublin seminars 1949-1955 and other unpublished texts),* op. cit. p. 70)

Schrödinger's account of how the bits of matter of daily life, and the elements of the macroscopic world in general, must be construed quantum mechanically, makes this point very clear. According to him, the surrounding bodies are nothing other than complex observables: "Here I wish to put forth the opinion that the fundamental conceptions of wave mechanics, when regarded from the angle of the epistemological basis of the matter concept, disallows us to regard matter as constituted of particles. It has no direct relation to the particle-concept in quantum mechanics, nor indeed to the wave concept, but to the concept of observable. (...) Matter in the meaning of the philosopher really consists of (observables) - not of the particles"[1]. In 1949, Schrödinger thus conceived matter *neither* as an aggregate of particles, *nor* (in contrast with his 1926 views) as some sort of wave packet, *nor even* (clearly departing from some straightforward interpretations of the decoherence theories) as some emergent feature of a global wave-function. Matter was instead related by Schrödinger to the limiting concept of an observable. Since the construction of observables, and in particular of those observables which define the macroscopic material bodies, is admittedly conditioned by a pre-quantum knowledge (the pragmatic background of everyday life and classical physics), it can by no means be said that the quantum account of macroscopic bodies is self-sufficient. The blind spot again manifests itself at this point.

To conclude this paragraph, I think that Schrödinger's late interpretation of quantum mechanics is a remarkably well designed and convincing system of explanations, but with a missing keystone. This eagerly sought keystone is nothing more and nothing less than a proper solution of the measurement problem. Now, what Schrödinger suggested repeatedly is that the handling of the measurement problem (with its end-product discontinuities) can be postponed without any harm, that the absence of a solution to it does not make any difference, that it does not manifest itself unless one undertakes a thorough reflective investigation. In other words, the metaphor of the blind spot, which was applied by Schrödinger to the necessity of using the correspondence principle in order to find suitable expressions for the observables, appears to be even more appropriate to feature out the measurement problem in its entirety. For the measurement problem looks very much like the blind spot in our retina, which does not reveal itself by any black hole in our visual field, even though it *can* be revealed by an accurate ophtalmological investigation. The measurement problem doesn't even have to be raised when quantum mechanics is used as a *predictive* tool, for it is enough in this case to obtain the relevant overall wave function, and then calculate probabilities by applying the Born formula at some relevant stage of the

[1]E. Schrödinger, "The problem of matter in quantum mechanics", *Notes for seminar 1949,* (E. Schrödinger, *The interpretation of quantum mechanics (Dublin seminars 1949-1955 and other unpublished texts)*, op. cit. p. 98)

measurement process. It only reveals itself when a careful investigation of the possible *descriptive* meaning of the wave function is carried out.

4-4 *Neo-Schrödingerian views on the measurement problem: I- Everett's interpretation*

There exist nowadays several no-collapse interpretations of quantum mechanics which are all strikingly reminiscent of one or another feature of Schrödinger's views. Our task in the following two paragraphs is to review them carefully and to evaluate their achievements in the light of Schrödinger's own attempt at providing a satisfactory treatment of the measurement problem.

The first conception we shall examine is Everett's relative-state interpretation, for Everett himself claimed that his theory was just a development of Schrödinger's. According to Everett, the relative-state interpretation is "(...) based on pure wave mechanics"[1]. Sometimes he even calls it "the wave interpretation"[2], because "this view also corresponds most closely with that held by Schrödinger"[3]. The only difference he points to between his interpretation and Schrödinger's is that his "(...) picture only makes sense when observation processes are treated within the theory"[4]. It is only when observations processes are included within the wave-mechanical description that "(...) Heisenberg's criticism of Schrödinger's opinion - that continuous wave mechanics could not seem to explain the discontinuities which are everywhere observed - is effectively met. The 'quantum-jumps' exist in our theory as *relative* phenomena (i.e., the states of an object-system relative to chosen observer states show this effect), while the absolute states change quite continuously"[5]. Our task is then to evaluate Everett's claim of striking similarities and slight differences between his interpretation of quantum mechanics and Schrödinger's. In order to do so in an orderly way, we shall list the distinctive features of Everett's interpretation and then compare each of these features with corresponding statements arising from Schrödinger's mature position, say from 1935 to the 1950's. This list of 10 items will be given according to an approximate order of decreasing agreement between Everett and/or the Many-worlds theorists on the one side, and Schrödinger on the other.

(1) *No-collapse*. This is obviously the major point of formal agreement between Everett and Schrödinger, even though it is by no means the most specific. In Everett's interpretation, "(...) the wave function itself is held to be the fundamental entity, obeying *at all times* a deterministic wave

[1]H. Everett, "Theory of the universal wave function" in: B.S. De Witt and N. Graham, *The many-worlds interpretation of quantum mechanics*, Princeton University Press, 1973 p. 109
[2]ibid. p. 115
[3]ibid.
[4]ibid.
[5]ibid.

equation"[1]. And according to Schrödinger the law of evolution of quantum mechanics holds at all times; it cannot be infringed occasionally when a measurement is performed (see §4-3).

(2) *Conjunction.* The idea that the terms of a superposition have to be taken as coexistent, and to be expressed by a conjunction rather than a disjunction, is also common to Schrödinger and Everett. In Everett's terms, "All branches exist simultaneously in the superposition after any given sequence of observations"[2], which is tantamount to saying that "(...) all elements of (the) superposition are equally 'real'"[3]. As for Schrödinger, his clearest expression of the idea that a superposition expresses a conjunction of coexisting terms rather than a disjunctive list of probabilities is to be found in the text of the July 1952 Dublin seminar. This passage is so strikingly akin to Everett's position (and sometimes even of its many-worlds metaphoric expressions), that it is worth quoting extensively: "Nearly every result (the quantum theorist) pronounces is about the probability of this *or* that *or* that ... happening - with usually a great many alternatives. The idea that they be not alternatives but *all* really happen simultaneously seems lunatic to him, just *impossible*. He thinks that if the laws of nature took *this* form for, let me say, a quarter of an hour, we should find our surroundings rapidly turning into a quagmire, or sort of a featureless jelly or plasma, all contours becoming blurred, we ourselves probably becoming jelly fish. It is strange that he should believe this. For I understand he grants that unobserved nature does behave this way - namely according to the wave equation. The aforesaid *alternatives* come into play only when we make an observation - which need, of course, not be a scientific observation. Still it would seem that, according to the quantum theorist, nature is prevented from rapid jellification only by our perceiving or observing it. (...) The compulsion to replace the *simultaneous* happenings, as indicated directly by the theory, by *alternatives*, of which the theory is supposed to indicate the respective *probabilities*, arises from the conviction that what we really observe are particles - that actual events always concern particles, not waves. Once we have decided for this, we have no choice. But it is a strange decision"[4]. However appealing these sentences may appear to supporters of an interpretation of quantum mechanics based on the simultaneous existence of branches, or even of worlds, they must be taken with some prudence. For nowhere does Schrödinger tell us very clearly *which* wave function the statement of coexistence applies to. Is it the wave function of the object, of the composite system (apparatus+object), of the

[1]ibid.

[2]H. Everett, "Relative state formulation of quantum mechanics", in: B.S. De Witt and N. Graham, *The many-worlds interpretation of quantum mechanics*, op. cit., p. 146

[3]H. Everett, "Relative state formulation of quantum mechanics", in: B.S. De Witt and N. Graham, *The many-worlds interpretation of quantum mechanics*, op. cit., p. 116

[4]E. Schrödinger, *July 1952 colloquium*, in: E. Schrödinger, *The interpretation of quantum mechanics (Dublin seminars 1949-1955 and other unpublished texts)*, op. cit. p. 19-20

more extended composite system (observer+apparatus+object), or even of the whole universe? Only in the two latter cases would the analogy between Schrödinger's and Everett's coexistences be complete.

Things sometimes look as if Schrödinger had chosen to favour these two cases, but his way of expressing this option is quite ambiguous. On the one hand, from his standpoint, nothing prevents the application of the statement according to which terms of a superposition are to be construed as "simultaneous happenings" rather than "alternatives" to *any* wave-function. His allusion to "we ourselves" and to "nature" turning into "jelly" or prevented from "jellification", is perfectly compatible with a holistic conception of the wave function, including observers and the whole universe. On the other hand, however, in his explanation of the motives of those quantum theorists who believe that wave functions just correspond to disjunctive statements of probabilities, Schrödinger tends to refer restrictively to wave functions of *objects*. According to him the problem here is only to decide between a particle conception and a wave conception of "what we really observe", namely of objects. Besides, when he writes in subsequent paragraphs of the same text about the two modes of linkage provided by the ψ-wave, he again points implicitly towards the *system-object,* rather than towards any composite system. These two modes of linkage between events are, respectively, the transversal (wave-like) linkage and by longitudinal (particle-like) linkage.

(3) *Holism.* According to what has just been said, there appears to be a noticeable difference of emphasis between Schrödinger and Everett when holistic description is at stake. However, this difference is not very easy to appreciate in a single instance, and we thus have to evaluate it from a more general standpoint. To begin with, Everett's motivation and methods are consistently and thoroughly holistic. The motivation for his interpretation, as he states it in his paper of 1957, is cosmological: "(in the case of a closed universe), there is no place to stand outside the system and to observe it. There is nothing outside it to produce transition from one state to another". His method thus consists in including each time both the apparatus and the observer within the relevant composite system to be described by a wave function. Actually, this is just a minimal composite system. When pushed to its ultimate consequences, Everett's interpretation prompts one to look for a wave mechanical description of the whole universe[1].

As for Schrödinger, we have seen in previous chapters that he was quite familiar with holistic considerations, even in the framework of pre-quantum physics. He was also the physicist who developed the first quantum account of composite systems as early as 1927, and who emphasized the concept of *entanglement* of wave functions in a series of papers published in 1935. He did not hesitate to apply the formalism of

[1]D. Deutsch, "Quantum theory as a universal physical theory", Int. J. Theor. Phys., 24, 1-41, 1985

entangled wave functions, which was suited for describing the interaction between any two (or more) systems, to the interaction between system-object and system-apparatus. But on the other hand his use of wave mechanics was usually less daring than his own advances could have allowed. He generally limited his non-formal discussions of wave-functions to the wave-functions of *system-objects,* rather than to more holistic wave-functions including the apparatus; and he often treated the system-apparatus as a kind of background which merely determines the structure of a certain *observable*. Such was the case for instance when he wrote in 1952 "The wave-function, characterizing the state of the system, and the operator, characterizing the experimental device, together are supposed to give a certain information on the observed value"[1]. In addition, when he wished to provide a wave-mechanical description of the apparatus *also,* he tended to adopt (though admittedly as a metaphor) the simplifying and faulty device which consists in ascribing *separate* wave-functions to the object and the apparatus in order to account for the interaction between them.

This tendency to exclude (in practice) the apparatus from the wave mechanical description, or to separate its wave mechanical description from that of the object, was even stronger when the *observer* was at stake. In the 1935 cat-paper, Schrödinger found himself compelled to evoke the role of the "living subject" or of a "mental act" at the last stage of the measuring process. But he made no attempt whatsoever to encompass the said living subject within the wave-mechanical description. The problem is that even though he had just stated the concept of entanglement, even though he had insisted that the measurement problem could not be tackled properly without making extensive use of this concept, even though he recognized that it is "(...) *the* characteristic trait of quantum mechanics"[2], he was still wondering in the conclusion of the cat-paper whether it would not prove to be just a convenient trick for carrying out calculations, reflecting the non-relativistic character of quantum mechanics. Later on, when Schrödinger gave the observer a role in his account, it was always to *disentangle* a holistic wave function; not to have this observer caught up in the entanglement.

To recapitulate, holism was perfectly acceptable to Schrödinger; it figured repeatedly, in his gas theory of 1925-26, in his quantum theory of composite systems of 1935, and then in his cosmological theory of atomism in 1937-39; he even regarded it as privileged from a philosophical standpoint; but it never became a *systematic and comprehensive* tool of thought for the interpretation of quantum mechanics, as it came to be in Everett's interpretation.

[1]E. Schrödinger, *Transformation and interpretation in quantum mechanics,* in: E. Schrödinger, *The interpretation of quantum mechanics (Dublin seminars 1949-1955 and other unpublished texts),* op. cit. p. 50

[2]E. Schrödinger, "Discussion of probability relations between separated systems", loc. cit.

(4) *Relativity.* "Relativity of states" gave its name to Everett's interpretation of quantum mechanics. Its principle was stated thus: "All statements about the subsystems (...) become *relative* statements, i.e. statements about the subsystem relative to a prescribed state for the remainder (since this is generally the only way a subsystem even possesses a unique state); and all laws are correlation laws"[1]. Under closer examination, it appears that this insistence on the relativity of states and on the correlations provides the only acceptable link between the idea that many branches coexist and the fact that our lives and our social agreement about the actual surrounding macro-world are only concerned with *one* branch. As S. Saunders[2] cogently pointed out, Everett's interpretation relies on an indexical conception of actuality. This indexical character of actuality can easily be displayed through an analogy with the more familiar characteristics of time. In Everett's interpretation, the specific connection between actuality and the variously possible outcomes exhibited by the branches of a holistic superposition follows just the same pattern as the connection between *now* and the various possible tensed propositions referring to a given event. At each step of Mc Taggart's well-known regress, the contradiction between two tensed propositions such as *e is past* and *e is future* can be removed provided one says explicitly relative to *which* events f and f' each proposition is true: *e is past relative to f* and *e is future relative to f'*. Likewise, in a quantum superposition, the contradiction between the factual propositions corresponding to each term can be removed provided one says relative to which *other* factual proposition each given proposition is true: "'Observable X has value r; observable X has value s' are inconsistent. But introducing a new observable Y, we may say instead 'X has r relative to u of Y; X has s relative to v of Y' and there is no longer a contradiction"[3].

This strategy is very close to Schrödinger's in his 1935 cat-paper. In paragraph 10 of this paper, Schrödinger returns to the cat paradox that he first explained in paragraph 5. In paragraph 10, he explains the relevance of the paradox for the measurement problem, whereas in paragraph 5 he just took it as a *reductio ad absurdum* of the most straightforward realist reading of the ψ-function. He then mentions that after the entanglement between the object, the apparatus, and the cat wave functions have taken place, the "catalogue of information" of the apparatus (and of the cat) is very incomplete, for it does not even indicate which result has been recorded by the apparatus; nor does it indicate the biological state of the cat. But the global "catalogue" at least affords a list of *conditional statements* of the following form: *if* the pointer observable of the

[1]H. Everett, "Theory of the universal wave function" in: B.S. De Witt and N. Graham, *The many-worlds interpretation of quantum mechanics,* op. cit. p. 118

[2]S. Saunders, "Time and quantum mechanics", in: M. Bitbol & E. Ruhnau (eds.), *Now, time and quantum mechanics,* Editions Frontières, 1994

[3]ibid.

apparatus has value u and the cat is alive, *then* the object is characterized by the value r of observable X; *if* the pointer observable of the apparatus has value v and the cat is dead, *then* the object is characterized by the value s of observable X. The (major) difference between Everett and Schrödinger is that, according to Schrödinger, this relativization exacerbated the difficulty instead of providing any hint of a proper solution. For he insisted that after entanglement and conditionalization have occurred, it is still necessary to *decide* which one of the conditionals obtains. The reason for this discrepancy is likely to be found in the wide gap between Everett's and Schrödinger's theories of mind (see *monadology,* point (9) below).

(5) *Realism.* Realism appears to be another point of agreement between Schrödinger, the supporters of the many-worlds interpretation, and Everett. But, actually, this agreement is more verbal than genuine. Some of the most enthusiastic many-worlds theorists are metaphysical (or naïve) realists, whereas, from a metaphysical standpoint at least, Schrödinger appeared to be an *anti*-realist of the most extreme kind: namely a post-Machian "idealistic monist", according to his own wording. His methodological realist commitment in science arose from this metaphysical anti-realist ground, and it had many points in common with Blackburn's quasi-realism (see § 2-1) and also with Putnam's internalism.

As for Everett himself, he drew a close connection between 'reality' and objective description[1], which sounds quite close to what Schrödinger was inclined to think. But he also insisted that pure wave mechanics is an objective description *in so far as* the wave function is "(...) in one-one, rather than statistical, correspondence to the behaviour of the system"[2]. This idea is definitely non-Schrödingerian; at any rate it is far from Schrödinger's mature views. For in the 1950's Schrödinger held on to a statistical (or at least distributional) link between the wave-representation and the facts, through the empirical correspondence rules. And he also pointed out (see §4-2) that well-defined statistics can be considered as an objective feature of systems deriving from a certain experimental preparation, just as much as a sharp value can.

(6) *Parallelism.* One of the most interesting features of Everett's interpretation of quantum mechanics is the kind of parallelism it displays between the time-development of the wave function of a composite system (including the observer), and the sequence of experimental results as it is recorded in each observer's "memory bracket". If each memory bracket is "capable of the interpretation 'the observer has experienced the succession of events A, B, ... , C'"[3], then this parallelism can even be

[1]H. Everett, "Theory of the universal wave function" in: B.S. De Witt and N. Graham, *The many-worlds interpretation of quantum mechanics,* op. cit. p. 111
[2]ibid. p. 109
[3]ibid p. 144

identified with "psycho-physical parallelism"[1]. But this subjective interpretation of the "memory bracket" is not indispensible, and one can refer to intersubjective communication about events instead. Parallelism is not necessarily psycho-physical; it can be construed as parallelism between the evolution of the wave-function and a set of sequences of intersubjectively acknowledged facts[2]. The first interpretation of the memory bracket can be called *mentalistic,* whereas the second may be referred to as *pragmatic.*

At any rate, there is one important point which holds irrespective of whether one gives a mentalistic or pragmatic interpretation of Everett's "memory bracket". It is that even though there are as many brackets as there are terms in the superposition, each memory bracket is characterized by a content which is consistent with the following two basic epistemological requirements: reproducibility of experiments and agreement of observers about experimental results. True, reproducibility and agreement are not *causally explained* by any consideration about a transcendent object provoking *uniform* effects in the apparatus or in the brain of the observers, but they are implied by the very holistic structure of the theory supplemented with the interpretative device of "memory brackets".

This non-causal account, which was referred to as *parallelism* (see chapter 6 for more details), is a very peculiar way of dealing with reproducibility and agreement. It may seem to be totally at odds with the traditional conception of metaphysical realists according to which agreement about something is due to causal interactions between the *thing* "out there" and the receptive structures of instruments or sense organs, whose end results are either uniform experimental outcomes or sense-data. The non-causal account of reproducibility and agreement, on the other hand, sounds very much in tune with an internalist position in Putnam's sense. For, according to an internalist position, the concept of a "transcendent object causing alterations in us" is but an unwarranted extension of the immanent causal relations. Showing that the very structure of physical theories automatically ensures that facts will agree with our basic methodological requirements of reproducibility and intersubjective agreement, as can be done by using Everett's concept of "memory bracket", is thus the most a community of human scientists can hope for. We shall call this position the "structuralist" (and internalist) version of parallelism.

But actually, it was very soon considered that Everett's way of displaying reproducibility and agreement could *also* be made consistent with a metaphysically realist position, provided the metaphysical realists leave aside dualism and causal interactionism, and adopt monism and

[1]ibid. p. 117; see more comments in paragraph 6-7 of the present essay.

[2]M. Bitbol, *De l'intérieur du monde* (in preparation); M. Bitbol, "Quantum mechanics, facts, and presence", in: M. Bitbol & E. Ruhnau (eds.), *Now, time and quantum mechanics,* Editions Frontières, 1994

emergentism instead. One can for instance regard the whole universe as corresponding to a single wave function from which a certain range of macroscopical appearances emerge, each one in one world; and that the memory brackets display these emerging features as they appear in each world. Let us then call this the "emergentist" version of parallelism.

Schrödinger's attitude towards this crucial issue of parallelism is not very easy to define. Moreover, it changed during his lifetime. To begin with, the idea of causal interaction between a transcendent object and (technological or biological) receptive structures was obviously difficult to reconcile with Schrödinger's criticism of metaphysical realism and of the concept of thing-in-itself. True, he suggested from time to time a model of interaction between a wave-object and a wave-apparatus (see §4-3) which is strongly reminiscent of the causal view. But this pseudo-causal model was formulated with a lot of qualifications, as a metaphor intended as a corrective to the corpuscularian metaphor and as a tactical concession to the spontaneous metaphysics of the physicists, rather than as a description to be taken seriously.

I thus think the dominant trend in Schrödinger's thought was *parallelism,* under its two forms: emergentism and structuralism; emergentism in his earliest attempt at interpreting quantum mechanics, and structuralism in his later views. In 1926, his concept of wave-packet was just meant to display how macroscopic discontinuous and corpuscle-like features could *emerge* out of a holistic wave-like background. Later on, he developed instead a conception which is definitely closer to the structuralist version of parallelism than to the emergentist version. However, it did not prove easy for him to make this view fully explicit and developed, because the kind of parallelism it tended to promote had no special symbol (such as Everett's memory brackets) to be displayed in. Schrödinger's structural parallelism is thus dispersed and fragmentary, and we can but analyze its two main components.

The first component is the strict separation between the wave-mechanical description and the domain of (macroscopic) facts. Schrödinger advocated such a separation in what we have called his *post-modern turn.* According to his statements in *Science and Humanism,* there is on the one hand the wave-picture, and on the other hand the experimental facts about which the wave picture provides *information.* If one leaves aside the representative function Schrödinger ascribes to wave-functions, the distinction he makes is just as strong as the category distinction an instrumentalist would make between ψ construed as a probabilistic tool and the random events to which a probability is assigned. At any rate, such a radical cut enables one to characterize a structural version of parallelism as opposed to an emergentist version; for the latter would rather show how macroscopic facts *arise from* the holistic wave background.

The second component of Schrödinger's structural parallelism pertains to *any* version of parallelism. It consists in displaying a one-one correspondence, not of course directly between wave functions and facts, but between the time-development of holistic wave-functions and the *mutual correlation* of certain facts. Schrödinger's best account of this kind of correspondence is to be found in his 1935 cat-paper, where he noticed that after the measurement interaction has taken place, the overall "catalogue of information" splits into a disjunction of conditional propositions, and that *each one* of these propositions displays a strict internal correlation between facts which was not there initially (i.e. before the measurement interaction has taken place).

But did Schrödinger ever express *jointly* these two constitutive features of what we called his structural parallelism? He did so at least once, in *Science and Humanism,* expressing both separation and correspondence in a single short paragraph: "(...) what is the use of such a (wave-mechanical) description, which, as I said, is not believed to describe observable facts or what nature really is like? Well, it is believed to give us information about observed facts and their mutual dependence"[1].

In the first sentence, Schrödinger evokes *separation.* He reminds the reader of his previous statement according to which wave-mechanical description is *not* a straighforward description of facts, and *even less* of a description of what nature really *is*. In the second sentence he emphasizes that wave functions however provide:

(i) information about facts (generally of a probabilistic nature),

(ii) information about the *mutual dependence* of facts.

But saying that information about the *mutual dependence* of facts can be found in the wave function is but a loose way of expressing a one-one correspondence between the *correlation* of facts and the end-product of unitary evolution of certain wave-functions. Everett's additional contribution consisted in creating a remarkably efficient symbolism, namely memory brackets, in order to display this correspondence.

(7) *No decoherence*. Neither Everett nor Schrödinger had anything to say about the decay of interference terms in the wave-functions *of composite systems*. But they both made an assumption which ensures that measurement interactions lead to disappearance of the off-diagonal terms in the reduced density matrix *of the object*. Indeed, they both developed the wave-function of the composite system created by the measuring interaction in such a way that the relative states of the measured observable eigenstates are mutually *orthogonal;* and one can prove that, starting from an entangled wave-function for a composite system, the reduced density matrix for one sub-system is diagonal if the relative states corresponding to the other sub-system are mutually orthogonal[2].

(8) *Preferred Basis*. The designation of a preferred basis is obviously a very important problem for many-world versions of Everett's

[1]E. Schrödinger, *Science and humanism*, op. cit., p. 41

[2]B. Van Fraassen, *Quantum mechanics, an empiricist view*, op. cit. p. 204

interpretation of quantum mechanics. For if there is no *intrinsic* reason for the wave function of composite systems to develop according to a certain basis rather than according to any other one, then how can it be said that splitting of the universe into several *unambiguously defined* worlds happen spontaneously? Incidentally, this is a problem for other blends of realist interpretations of quantum mechanics as well. In spontaneous collapse interpretations, such as Ghirardi's, Rimini's and Weber's, the question arises of the "collapse basis", namely of the basis according to which spontaneous collapse occurs[1]. And in some realist versions of the modal interpretation (e.g. Dieks'[2]), one has to provide an unambiguous development of the wave function, for the wave function is supposed to describe the system *as being in a well-defined value state* corresponding to one of the terms of this development.

So, we have to evaluate briefly some recent attempts at solving the preferred basis problem, before we come to Everett's and Schrödinger's positions.

One straightforward solution relies on the condition of bi-orthogonality. According to E. Schmidt's theorem[3], each composite state-vector can be decomposed as a linear superposition of tensorial products of two state-vectors, in such a way that state-vectors belonging to the same Hilbert sub-space but to different terms of the superposition are mutually orthogonal. Dieks has therefore proposed that bi-orthogonality be *the* criterion enabling one to fix unambiguously a basis for decomposing a holistic state-vector into a linear superposition of tensorial products of two state-vectors (one for the object and the other for the apparatus). In Dieks' version of the modal interpretation, it follows that bi-orthogonality is also *the* criterion enabling one to consider that the system is some well-defined value-state corresponding to one term of the superposition[4]. Moreover, since this criterion works not only at the end of the measuring interaction process, but also at any time between the beginning and the end of the interaction, one can say, according to Dieks, that systems can always be ascribed a well-defined value-state. At each time, they can be ascribed a value-state for an instantaneous observable whose set of eigenstates is the set of orthogonal state-vectors defined by the instantaneous bi-orthogonal decomposition. The two difficulties which hinder this strategy are that the choice of bi-orthogonality has to be justified on meta-theoretical grounds, and that there are cases where the bi-orthogonal decomposition is not unique.

Another proposal was Deutsch's *kinematic independence,* according to which each term of the superposition has to evolve independently of the

[1]M. Dickson, "What is preferred about the preferred basis?", Found. Phys., 25, 423-440, 1995

[2]D. Dieks, "Modal interpretation of quantum mechanics, measurements, and macroscopic behavior", Phys. Rev. A49, 2290-2300, 1994

[3]E. Schmidt, "Zur Theorie der linearen und nicht linearen Integral Gleichungen (I)", Math. Annalen, 63, 433-476, 1907

[4]D. Dieks, "Resolution of the measurement problem through decoherence of the quantum state", Phys. Lett. A-142, 439-446, 1989

other terms[1]. Bi-orthogonality can be derived from kinematic independence, whereas kinematic independence cannot in general be derived from bi-orthogonality. In addition, kinematic independence offers a criterion of decomposition of holistic state vectors into superpositions of tensorial products of any number of state vectors, rather than only *two*. Unfortunately, this criterion shares most defects of bi-orthogonality, in so far as it (admittedly) relies on a meta-theoretical criterion, and it sometimes leads to non-unique decompositions[2]. No wonder that decoherence theories, which were initially developed in order to provide one with a self-sufficient solution to the measurement problem, have been hailed as a possible rescue for many-worlds versions of Everett's interpretation. For, as Saunders puts it "In the light of recent developments in measurement theory, in particular the theory of decoherence in open systems, (...) there are strong grounds to suppose that a 'pointer basis' (or 'preferred basis') can be derived from quantum mechanics"[3]. A demonstration of how a preferred basis can be derived from a decoherence formalism has been given by Zurek's group[4].

Two questions remain open at this stage.

(i) Why combine Decoherence with Many-worlds, rather that retaining Decoherence alone? An obvious answer is that decoherence was designed to solve the problem of the transition of the composite system (object+apparatus) from a pure state to an (approximate) mixture, whereas Everett's interpretation and/or its many-worlds versions are also able to address the problem of the transition from the (approximate) mixture to the appearance of a single well-defined outcome. Gell-Mann's and Hartle's theory of decoherent histories represents an interesting attempt at merging the two approaches into one[5]. But there are also other quite different approaches of decoherence, much less sympathetic to Everett's views. R. Omnès, for instance, insists that the only problem which has to be addressed is the one of transition from a pure state to a mixture liable to an ignorance interpretation. According to him, the problem of actuality, i.e. of appearance of a single result, which is specifically addressed by Everett, goes beyond the field of physics[6].

(ii) Is decoherence a satisfactory answer to the preferred basis problem? It is partly satisfactory because decoherence theories can be framed in such a way that they ensure uniqueness of the basis. But in so far as it does *not* dispense with metatheoretical arguments, it fails to

[1]D. Deutsch, "Quantum theory as a universal physical theory", loc. cit.

[2]S. Foster & H. Brown, "On a recent attempt to define the interpretation basis in the many-worlds interpretation of quantum mechanics", Int. J. Theor. Phys., 27, 1507-1531, 1988.

[3]S. Saunders, "Decoherence, Relative states, and Evolutionary adaptation", Found. Phys. 23, 1553-1587, 1993

[4]J.P. Paz & W.H. Zurek, "Environment-induced decoherence, classicality, and consistency of quantum histories", Phys. Rev. D48, 2728-2737, 1993

[5]M. Gell-Mann & J.B. Hartle, "Classical equations for quantum systems", Phys. Rev., D47, 3345-3382, 1993

[6]R. Omnès, *The interpretation of quantum mechanics,* Princeton University Press, 1994

represent the proper theoretical breakthrough which was the aim of every serious attempt at solving the preferred basis problem. In Saunders' terms, "unless quantum mechanics is supplemented with a new hypothesis, decoherence is approximate and interest-relative"[1]. It is approximate because interference terms do not disappear during the decoherence process, but are only made very small For All Practical Purposes. And it is interest-relative, because decoherence cannot be worked out without some supplementary assumptions which are *all* tantamount to presupposing that every physical process must eventually result in an acceptable macroscopic world for people to speak about and live in. Zurek's essential assumption, namely *coarse-graining* of the world into object, apparatus and environment, is admittedly anthropomorphic. As for Gell-Mann and Hartle, their having recourse to what they call "Information Gathering and Utilizing Systems" for setting the criteria of selection of sets of decoherent histories, is obviously meant to provide one with a physicalist equivalent of the anthropocentric account. Lockwood's reference to the "consciousness basis"[2] and Saunders' Darwinian arguments[3] have at least clearly stated the biocentric and anthropocentric features that most physicists (be they supporters of decoherence theories or of the many-worlds interpretation) have desperately attempted to avoid or to hide.

At this stage, we can come back to Everett's original conceptions about the preferred basis, and then to Schrödinger's. As we mentioned previously, Everett's strategy of decomposition of the state vector of a composite system (object+apparatus+observer), at the end of the measuring interaction, consisted in starting from the basis of eigenstates of the relevant observable (each eigenstate being a possible state of the object), and then assuming that the relative states are also mutually orthogonal. The mutual orthogonality of the relative states derives from the condition that the corresponding measurement is a "good" measurement, namely one which is capable to discriminating adjacent values of the measured variable[4].

One problem with this approach is that Everett could not justify, on a purely wave-mechanical basis, his initial preference for the observable eigenstates. The main argument he gave is clearly interest-relative, for it relies on the condition that the apparatus coordinate be *definite* in each relative state[5]. However, Everett tried to give his choice a universal significance, by generalizing the concept of a bi-orthogonal

[1] S. Saunders, "Decoherence, Relative states, and Evolutionary adaptation", loc. cit.
[2] M. Lockwood, *Mind, Brain, and the Quantum*, B. Blackwell, 1989; "'Many-minds' interpretations of quantum mechanics", Brit. J. Philos. Sci. (forthcoming)
[3] S. Saunders, "Decoherence, Relative states, and Evolutionary adaptation", loc. cit.; S. Saunders, "Time and quantum mechanics", in: M. Bitbol & E. Ruhnau (eds.), *Now, time and quantum mechanics*, op. cit.
[4] H. Everett, "The theory of the universal wave function" in: B.S. De Witt and N. Graham, *The many-worlds interpretation of quantum mechanics*, op. cit. p. 47. See also B.S. De Witt, "Many-universes interpretation of quantum mechanics" ibid. p. 157
[5] H. Everett, "'Relative state' formulation of quantum mechanics", in: B.S. De Witt and N. Graham, *The many-worlds interpretation of quantum mechanics*, op. cit. p. 143-144

decomposition (that he calls the canonical representation). He noticed that the condition of bi-orthogonality can be imposed at *any* instant during the measuring interaction, and not only at the end of this interaction[1]. This could have prompted him to consider, as Dieks did several years later, that the criterion of bi-orthogonality thus defines at each time an instantaneous observable $\tilde{A}(t)$, whose eigenvectors are those belonging to the object side of the bi-orthogonal decomposition. But Everett was quite reluctant to do so, and he even criticized explicitly any attempt at identifying $\tilde{A}(t)$ with an observable. He insisted that an observable should not depend on time, but only on initial conditions. The only real observable is therefore $\tilde{A}(\infty)$, namely the limit of $\tilde{A}(t)$ when $t \rightarrow \infty$. At any given instant t, the interaction between the apparatus and the object can but approximate the limiting conditions which makes it equivalent to a measurement of the observable $\tilde{A}(\infty)=A$. It is, therefore, clear that, according to Everett, bi-orthogonality was not sufficient *by itself* to define the sought-after preferred basis. A meta-theoretical element, enabling one to define the observable A which correspond to given initial conditions in a measurement, still had to be introduced somewhere. This meta-theoretical element was likely to be, according to Everett, a condition of *permanent* "definiteness" of the pointer states of the apparatus after a measurement.

Schrödinger expressed this absolute need for a meta-theoretical element even more strongly. True, Schrödinger was the quantum physicist who rediscovered, as early as 1935, the Schmidt theorem about bi-orthogonal decompositions[2]. He was also the first physicist who pointed out in the same paper the role of bi-orthogonal decompositions in the interpretation of entangled state vectors: "The biorthogonal development is the one which give us true insight into the entanglement. (...) one can say that the entanglement consists in that one and only one observable (or set of commuting observables) of one system is uniquely determined by a definite observable (or set of commuting observables) of the other system". But Schrödinger never tried to use bi-orthogonal decomposition of the state vector of composite systems (object+apparatus) as a criterion enabling one to define a basis of decomposition independently of our previous knowledge of observables and of their associated basis of eigenvectors. He rather insisted repeatedly, in his 1952 lectures entitled "Transformation and interpretation in quantum mechanics"[3], that the association of linear operators called "observables" with measuring devices *cannot* be obtained as an outcome of the theory itself. In lecture 4 of this series, he mentioned, in a strikingly Bohrian style, that the "assumption of correspondence" between measuring devices and

[1]H. Everett, "The theory of the universal wave function", in: B.S. De Witt and N. Graham, *The many-worlds interpretation of quantum mechanics,* op. cit. p. 54.
[2]E. Schrödinger, "Discussion of probability relations between separated systems", op. cit.
[3]E. Schrödinger, *Transformation and interpretation in quantum mechanics,* in: E. Schrödinger, *The interpretation of quantum mechanics (Dublin seminars 1949-1955 and other unpublished texts),* op. cit.

observables "(...) is a loan from classical physics. It is suggested by the fact that many of our operators are loans from classical physics, they are functions taken from classical physics, 'translated' into operators"[1]. And in lecture 7, as we already mentioned at the end of §4-3, he claimed that the association of linear operators and measuring devices has to be taken for granted, though it *"cannot"* arise from "pure mathematics"[2].

At the time when these lectures were written, Schrödinger's conviction that the observables and the associated basis of eigenstates are not derivable from the theory itself, let alone from any method using "pure mathematics", could have been considered over-pessimistic. Nowadays, however, after so many failed attempts at finding *intrinsic* criteria of decomposition of state vectors according to a preferred basis, it appears a quite reasonable attitude. Most likely, the search for intrinsic criteria for defining a preferred basis will be replaced in the coming years by a discussion of the appropriateness of meta-criteria (be they transcendental or naturalized, mentalistic or pragmatic, anthropic or Darwinian).

At any rate, in view of Schrödinger's claim that the privileged status accorded to the basis of eigenstates of an observable cannot be justified *within* the framework of the theory, it is very likely that his reluctance to embrace any version of the many-worlds interpretation of quantum mechanics in spite of coming so close to it, is partly due to his early lucid perception of the preferred basis problem.

(9) *Monadology.* As M. Lockwood[3] rightly points out, Everett's original interpretation of quantum mechanics is much closer to what one calls nowadays a "many-minds" or "many-observer states" theory[4], than to a genuine many-worlds view. "For (Everett) never speaks of dividing or differentiating worlds or universes, but only of the 'branching' and 'splitting' of 'observer states'"[5]. In his 1957 paper, Everett insisted that "throughout all of a sequence of observation processes, there is only one physical system representing the observer", thus ruling out the straightforward many-worlds interpretation of his views; yet, he said, "there is no single unique *state* of the observer" after a measurement interaction has taken place, and in *each* state, the observer has to be construed as perceiving a particular experimental outcome. As we mentioned in point (6), this link between multiplicity of observer states and multiplicity of perceptions is justified by the principle of "psycho-physical parallelism".

In Leibnizian terms, one could say that Everett chose to picture the universe as a *single world* with a plurality of observer-branch *monads* (which may each involve several observers with correlated states), rather

[1]ibid. p. 51
[2]ibid. p. 70
[3]M. Lockwood, "'Many-minds' interpretations of quantum mechanics", loc. cit.
[4]See D.Z. Albert & B. Loewer, "Interpreting the many-worlds interpretation", Synthese, 77, 195-213, 1988; D. Z. Albert, *Quantum mechanics and experience*, Harvard University Press, 1993
[5]M. Lockwood, "'Many-minds' interpretations of quantum mechanics", loc. cit.

than as a plurality of *worlds*. As it was the case for monads, observer-branches can be said to have access to the universe from their particular viewpoint, and to lack "windows" to communicate with each other. And, as it was the case also with monads, one may speak of a mutual agreement between the observer-branches in spite of their lack of communication, provided an external viewpoint (God's viewpoint or the theoretician's viewpoint) is adopted. The only important difference from Leibnizian monads is that observer-branches do not agree with each other about *single events,* but only about *statistics.* Agreement about statistics is the only point about which some kind of "pre-established harmony" between most observer-branches would have to be invoked (though this is obviously not the kind of vocabulary most present-day philosophers would be eager to promote).

However, both this monadological conception of the world and the physicalist conception of mind which is more or less explicitly associated with it, are definitely foreign to Schrödinger's thought.

In his *Mind and Matter,* Schrödinger stated a problem which has obvious relations with what is at stake in Everett's interpretation of quantum mechanics. This problem is: "(...) the arithmetical paradox; the *many* conscious egos from whose mental experiences the *one* world is concocted"[1]. The very formulation of the problem seems to require a monadological solution. But the terms in which Schrödinger presents this kind of solution leaves no doubt about his absolute lack of sympathy for it: "One way out is the multiplication of the world in Leibniz's fearful doctrine of monads: every monad to be a world by itself, no communication between them; the monad 'has no windows', it is 'incommunicado'. (...) I think there are few to whom this suggestion appeals, nay who would consider it as a mitigation at all of the numerical antinomy"[2].

In addition, Schrödinger was extremely opposed to any version of a physicalist conception of mind. For, as he wrote in a paper of 1946[3], *the Mind is basically Subject, and as a subject it cannot be taken as an object at all.* In this text, his conception of the Mind appears to stem from Schopenhauer's abstract concept of the pure knowing subject, and to have affinities with Wittgenstein's metaphor of the eye in the visual field in the *Tractatus* (5.633). But a quite different influence on Schrödinger's conception of the Mind is also manifest in certain texts: it is the influence of the original Vedantic doctrine of identity according to which the Mind (my mind) is just "(...)identical with the whole and therefore cannot be contained in it as a part of it"[4].

So, it can be said that Schrödinger wavered between two possible explanations of the necessary *absence* of the Mind in our world picture:

[1]E. Schrödinger, *Mind and Matter,* op. cit. p. 128
[2]ibid. p. 129
[3]E. Schrödinger, "Der Geist der Naturwissenschaft", Eranos Jahrbuch, 14, 491-520, 1946
[4]E. Schrödinger, *Mind and Matter,* op. cit. p. 128

either the Mind plays the role of the (self-invisible) eye in this world picture *(Wittgenstein),* or it is just *identical* with this world picture *(Advaita Vedanta).* But whatever option is adopted, it has the twofold consequence that Mind cannot be an object of (scientific) study, and that the Mind is functionally *unique.*

Thus, Schrödinger's solution of the "arithmetic paradox" could only be spiritualist and monist. When he stated what this solution should be according to him, Schrödinger referred explicitly to the doctrine of the Upanishads according to which there is only one Mind, not one for each body, let alone one for each possible brain state. In order to illustrate the kind of mind-body relation he was contemplating, he quoted the following sentence by a Persian mystic of the middle ages, Aziz Nasafi: "The spiritual world is one single spirit who stands like unto a light behind the bodily world and who, when any single creature comes into being, shines through it as through a window"[1]. Such a conception has obviously nothing to do with Everett's combination of monadology and latently physicalist conception of the Mind.

One must note, nevertheless, that although Schrödinger's monistic philosophy of mind is clearly different from Everett's monadological treatment of minds, it is not incompatible with Everett's theoretical framework as such. As I showed a few years ago[2], it is perfectly possible to reconcile Schrödinger's cluster of monistic conceptions of mind with something like the many branches theory, provided some slight interpretative moves are made.

One has only to suppose that Everett's holistic wave function describes every *thing* , but not *everything.* In this case, one may consider that there is *something* which remains definitely outside the quantum description, just as it remains outside any description of the objective world. This *something* is the Mind in a very abstract sense; that is Schrödinger's Mind identified with the pure placeless, timeless and disembodied knowing subject (as in his paper of Eranos Jahrbuch of 1946[3]), or Wittgenstein's "metaphysical subject", which is just as absent from the objective world as the eye in the visual field. This being granted, the next two steps consist in formulating a perspectivist analysis of physical systems (the physical systems display themselves through a multiplicity of aspects seen from the different standpoints an observer may adopt), and in making a far-reaching substitution: here, it is the abstract Mind which is going to play the role usually ascribed to an observer, whereas the observer's body becomes part of the described physical system. Carrying out this substitution analogically, one would say that the composite system (object+apparatus +observer's body) gives rise to "aspects" that the Mind perceives by changing its point of view relative to the system, just as an

[1]E. Schrödinger, *Mind and Matter,* op. cit. p. 129

[2]M. Bitbol, "Perspectival realism and quantum mechanics", in: P. Lahti and P. Mittelstaedt, (eds.), *Proceedings of the symposium on the foundations of modern physics 1990,* World Scientific, 1991

[3]E. Schrödinger, "Der Geist der Naturwissenschaft", loc. cit.

ordinary observer can see several aspects of an object by moving around it.

But this seems absurd. If the Mind is placeless, how could it change its *point* of view? If it has no sense organs, how could it perceive at all? Fortunately, Everett's formalism offers a very straightforward generalization of the concept of point of view. Once the wave function of the overall 'real system' has been defined, it displays, at any given time, the set of memory states which result from the interaction between the object in the restricted sense, the apparatus, and the observer's brain. Rather than saying that this interaction results in a branching off of the world or of the observer's mind, we could rather consider that it confronts the One Mind with all its own *possible states* in a given situation. Take then the concept of "point of view" in its most figurative sense, the one you use when you notice, for instance, that your point of view on life is determined by your past experiences. This kind of "point of view" represents in some way no less than your personal identity. Now, in this figurative sense, the set of memory contents displayed in Everett's expression of the state vector of the composite system can be identified with the set of all the "points of view" Mind has the possibility to adopt on the system.

At this point, we must stress one important difference between this position and Everett's. Everett considers, in good agreement with his physicalist conception of the mind, that in each branch there is a (relative) state which *describes the observer as definitely perceiving a particular object state*. In each branch, there is something like a physicalistically construed mind which perceives the world from his own point of view. As previously emphasized, we may say that Everett's branches represent a quite exact equivalent of Leibnizian *monads*. But in the monistic-spiritualistic conception, even though there are still many points of view, there are not as many perceptions, let alone as many minds, as points of view. There is only One Mind which may adopt any one of the available points of view. Here, one cannot speak of a plurality of monads construed as *substantialized* points of view with a physicalistically based awareness of their own, but rather of a plurality of structural positions the Mind may happen to occupy. To use the beautiful metaphor Schrödinger borrowed to the Persian mystic Aziz Nasafi, each particular Everett's branch, with its memory content, can also be considered in such a conception as a window through which the One Mind has the possibility to shine. Let us call this combination of Everett's formalism and Schrödinger-like monistic-spiritualistic outlook "the One Mind - many points of view interpretation".

However, even with the previous qualifications, the idea that Everett's formalism can be interpreted as a technique for displaying the whole set of figurative points of view the One Mind may adopt when faced with a composite system including an observer's body, is not devoid of difficulties.

It is at least clear that these difficulties are definitely distinct from those faced by the many-worlds conception of the same formalism. Whereas the many-worlds interpretation sounds like an imaginative fantasy whose challenge is to justify the necessity of generating many unobservable situations, "the One Mind - many points of view interpretation" threatens other elements of the set of our most intimately rooted beliefs. These are the problems we shall now address.

To begin with, there seems to be something odd in the contention that Everett's writing of the state of a compound system displays a set of "possibilities", namely the set of the *possible states* of the One Mind. After all, a debate took place between physicists who contended that a measurement yields transition from several possibilities to a single actuality, and Everett, according to whom *all* the elements of the superposition are *actual* relative to the corresponding observer's state. But to understand Everett's position, one must recognize that his primary concern was not to eliminate the concept of "possibility" from the interpretation of quantum mechanics. It was rather to avoid considering any physical *transition* from the possible to the actual. And, since actuality is considered as the most obvious characteristic of an experimental result, Everett had but one solution to eliminate the necessity of considering a transition: it was to spread out actuality somehow onto every term of the superposition.

Now in "the One Mind - many points of view interpretation", the last stage of an experiment, namely its epistemic level, amounts to the following event: Mind identifies itself to one of those points of view on the compound system which were made available by the measuring interaction. This identification is definitely not a physical process. It does not involve anything which *changes* the state vector describing the real system. It is not tantamount to a *physical* transition from the possible to the actual. Thus, although the modal concept of "possibility" has not been eliminated from "the One Mind - many points of view interpretation", and although actualisation retains a special status in this interpretation (for it means identification of the One Mind to one of the available points of view), no physical transition from the possible to the actual is involved.

However, at this stage, the reasoning remains incomplete. For, if Mind identifies itself *irreversibly* to one of its possible points of view, what difference other than purely verbal is there between the usual physical actualisation and this peculiar type of actualisation we have called "identification" of the One Mind to a certain point of view? Don't they both ascribe a particular status to one term in the superposition? To clear up the non-trivial difference between the two concepts, we need a precise spatio-temporal analysis of "identification". This analysis is likely to eliminate radically the possibility of speaking of "identification" as if it were a "process" able to *occur* reversibly or irreversibly. But before performing this analysis, we must study another related difficulty.

A problem in "the One Mind - many points of view interpretation" is that the perspectival description of the world it involves lacks a meta-level from which the points of view can be described. Even though the theoretician ascribes himself a (purely formal) meta-position, his picture does not include any "point of view of points of view" which would be able, as the Leibnizian God, to encompass the whole series of monads. Indeed, as soon as the Mind identifies itself with a point of view, it can but identify itself with a *particular* one among those made available by the time development of the composite system which includes the observer's body. However, this being granted, the mere insistance on the "particularity" of a point of view sounds artificial. Since no point of view is available from which all the other points of view would be seen as equivalent, the point of view Mind adopts, when adopted, is not one among others; it is *the* point of view, self-referred to as *my* point of view.

Let us retain that the Mind, having no point of view of its own, can but adopt *particular* points of view and identify itself completely with each of them. The Mind is by itself point-of-view-less, just as it is placeless and timeless. The aporia is then the following: the Mind is not within the world since, even if it can identify itself to any available point of view which partake of the world, it does not *reduce* to this point of view. Nor does the Mind stand outside the world, since it has no point of view of its own, independent from the points of view the world can offer. The Mind can only be considered as an empty space in the triadic relation: "point of view of (   ) on a 'real universe'". One can thus see in what sense the Mind may be said to retain its necessity, even though the 'real universe' admittedly gathers all that falls under the categories of knowledge: the Mind holds a key *role* in the very *constitutive relations* of this knowledge. It has a *functional* status.

Let us now come back to the problem of the spatio-temporal analysis of "identification". Let us suppose that Everett's holistic wave functions represent a faithful description of the objective world. In a perspectivist framework, this means that they correctly list the available points of view at a given time and that they enable one to calculate the set of possible points of view associated to a particular compound system at time t, given this set at time t'. The last step of an experiment, namely awareness, or "identification" of the Mind to a particular point of view, cannot however be included in this description. "Identification" *to* a given point of view does not pertain to the description *of* points of view, and moreover, since it involves a timeless entity (the One Mind), it has nothing to do with any temporal process. Using once more, but in a slightly different version, the metaphor Schrödinger borrowed to Aziz Nasafi, it is appealing to fancy the Mind as a beam of light (of heavenly, non-disturbing light, of course!) illuminating a particular point of view, whereas the others remain in the darkness. But this is again misleading. No objective element can give any distinct character to a point of view the Mind has adopted, since every relevant objective element is already included in the spatio-temporal description afforded by the wave function. In fact, the Mind is the only entity concerned by "identification".

The strangest point is that, in this conception, nothing prevents the Mind from identifying itself to this and that point of view, here and there, before and after. The Mind has no spatio-temporal location, and it cannot be aware of any "change" occurring while it performs the odd trip we have just sketched. Indeed, if it identifies itself to a given point of view, this implies that it adopts the whole associated memory content. It can never remember a "previous" point of view, for all its available "memories" are confined to the particular point of view it occupies "presently". Of course, the words "previous" and "presently" have no other significance than that of elements in an analogical picture of what we have called *identification*. Mind has no history; and present, past and future are meaningless for it. Having a history would mean holding *traces* of a past. But all the possible traces are included in the description of the objective world, where they appear as (figurative) points of view. The only history Mind can have pertains to the point of view it identifies itself to.

The difference between physical actualisation and mental actualisation or "identification" is thus considerable. Physical actualisation is an irreversible process, by which one of the possible points of view is selected and given a distinctive status, so that any further description of possible points of view is dramatically affected by it. "Identification" has no import on physics at all, it does not modify the becoming of the set of possible (figurative) points of view. Again, it is pure awareness. "Identification" provides a connection between the objective world unfolding in space-time, and the placeless and timeless Mind.

But this connection is of a very special nature. For instance, it would be absurd to state that "identification" of Mind with a given point of view has occurred at a given time; for temporal location pertains to the objective world. But it would be no less absurd to infer from this impossibility that "identification" *never* occurs, since awareness would then prove impossible. How can we reconcile these two constraints bearing on the time of identification, if we are to remain in the framework of Schrödinger's monistic-spiritualistic conception? In his philosophical writings Schrödinger has given a very precise statement of his doctrine about the relation between the Mind and time. This statement can be used in order to clarify the temporal aspects of the "One Mind - many points of view interpretation of quantum mechanics". In chapter 4 of *Mind and Matter*, one finds a reflection on what we have called the *timelessness* of Mind: "Mind is by its very nature *a singulare tantum*. I should say: the overall number of minds is just one. I venture to call it indestructible since it has a peculiar time-table, namely mind is always *now*" [1]. Then, in the framework of this doctrine, one can give a definite

[1]E. Schrödinger, *Mind and Matter*, op. cit. p. 145

(though very "peculiar") answer to the question of the time of *Identification* of the Mind with a particular point of view: Identification occurs *now*. A "now" which, provided it is not restricted to the token-reflexive component of its meaning, cannot be ascribed any proper location in time, even though it bears *relations* (past, present and future) with locations in time[1].

To conclude this digression on the "One Mind - many points of view interpretation" we must compare it with an interpretation which has been proposed by E. Squires and which *looks* quite similar[2]. There are however major differences between these two interpretations.

In order to deal with the fact that, in pure unitary quantum mechanics, the outcome of a measuring interaction is a superposition of eigenstates, whereas only one of them is actually observed, E. Squires assumes that universal "consciousness" or "mind" selects freely *one* of the terms of the superposition. But, by enabling the popular fuzzy components of the concept of "mind" to intrude into the analysis, the author is led to predict paranormal effects. To begin with, according to him, particular observers can consult the universal mind about other observers' states of "mind": this is "quantum telepathy". Secondly Squires' "universal mind" can perform specific choices instead of random selections of eigenstates: this yields "quantum psychokinesis". But if, following Schrödinger, one reaches enough rigour in the definition of Mind, any spurious justification of paranormal effects becomes pointless. In one of the restrictive acceptions of the word "Mind" Schrödinger uses, namely that of a pure *knowing* subject, or a pure pre-requisite for knowledge, there is nothing to be *known* in Mind. Mind has no memory content of its own. Thus no observer's counterpart (or "point of view", in the One Mind-many points of views interpretation) can consult Mind directly, if he wants to know another counterpart's memory content; for as soon as Mind has identified itself to some point of view, it has completely forgotten the memory content associated to any other point of view and completely adopted the memory content associated to its present point of view. No ground is therefore left for Squires' "quantum telepathy". In addition, the concept of choice is meaningless for Mind if Mind is construed as a pure knowing subject with no biography and no criteria of choice of its own. Its "choices", if any, are bound to be blind. Hence, the absence of any ground for Squires' "quantum psychokinesis".

From a philosophical standpoint, Squires' conclusions follow from a very straightforward version of dualism[3]. By contrast, the "One Mind - many points of view interpretation" is related to a softened version of dualism, which we could call "functional dualism". In such a doctrine, the

[1]M. Bitbol, "Now and Time", in: M. Bitbol & E. Ruhnau (eds.), *Now, Time and Quantum mechanics*, op. cit. See further developments in chapter 6 of the present essay.

[2]E. J. Squires, in: M. Cini and J.M. Lévy-Leblond (eds.), *Quantum theory without reduction*, Adam Hilger, 1990

[3]E.J. Squires, "How to test for Cartesian dualism by quantum experiments", in: P. Lahti and P. Mittelstaedt, (eds.), *Proceedings of the symposium on the foundations of modern physics 1990*, op. cit.

"knowing subject" has no content, but it still retains a *function:* it is the *to whom* a result appear, or the *by whom* a physical structure of some brain becomes *experience*.

Let me add a last warning. My aim here was not to endorse the daring view of quantum mechanics called the "One Mind - many points of view interpretation", but rather to show the possibility of framing an interpretation which be both isomorphic to Everett's interpretation and compatible with Schrödinger's monistic-spiritualistic doctrine of mind. The "One Mind - many points of view" interpretation is precisely of this kind. The deliberately restricted conclusion I wish to draw from it is then the following: disagreement with Everett on the theory of mind is *not* sufficient in isolation to explain why Schrödinger did not formulate anything like Everett's interpretation of quantum mechanics in spite of his coming so close to it. A proper explanation of Schrödinger's abstention must include his early perception of the preferred basis problem (point 8), and his own formulation of the probability problem which is to be discussed hereafter (point 10).

(10) *Probability.* The most important difference between Everett and Schrödinger is certainly to be found in their attitude towards the probability problem. Here, Schrödinger appears to be more aware of the difficulties than Everett.

According to Everett, quantum mechanics does not have to postulate its interpretation (in the restricted sense of *probabilistic correspondence rules),* for it arises quite naturally from its own formalism. The sequences of results in the memory bracket of each observer-branch, he maintained, are generally boun to obey the statistics which are to be expected from applying the Born's rule. In the conclusion of his 1957 paper, he emphasized that this ability to derive a probabilistic scheme from the theory itself was the very point on which his approach was superior to the orthodox interpretation: "Objections have been raised in the past to the conventional interpretation or 'external observation' formulation of quantum theory on the ground that its probabilistic features are postulated in advance instead of being derived from the theory itself. We believe that the present 'relative-state' formulation meets this objection, while retaining all of the content of the standard formulation". But as Albert, Loewer, and Lockwood[1] rightly point out, Everett's account of probabilities is inherently circular, for the sequences of results in each observer branch are constructed in such a way that their statistics agree automatically with the square modulus of the corresponding coefficient in the superposition.

There has therefore been much work done since Everett's first paper towards solving the apparent conflict between the priority the relative state interpretation gives to the pure unitary deterministic evolution of a

[1]See D.Z. Albert & B. Loewer, "Interpreting the many-worlds interpretation", loc. cit.; M. Lockwood, "'Many-minds' interpretations of quantum mechanics", loc. cit.

holistic wave-function, and the associated probabilistic statements. In the framework of the many-worlds interpretation, Deutsch[1] for instance proposed that each (Everett-De Witt) world is in fact an infinite bundle of (Deutsch) worlds whose measure is proportional to Born's probability.. This move is sufficient to reconcile quantitatively the idea of equiprobability of worlds with Born's rule. But it does nothing to clarify the connection between the unitary evolution and the *appearance* of a certain discrete distribution characterized by statistical regularity, from the point of view of somebody living in *a given* (Deutsch) world. Albert's "many minds" conception, according to which there are infinitely many minds adopting stochastically one observer branch or another is a possible dualist answer to that challenge. However, as Lockwood pointed out, this conception also raises many difficulties (about personal identity, for instance), besides its relying on a metaphysical tale. Further work is needed to adapt the kind of probabilistic solution it provides to an acceptable metaphysical (or altogether non-metaphysical) outlook. An interesting and sound mentalistic variety of this kind of solution was proposed by Lockwood[2], and a pragmatic variety was suggested by myself[3].

Here again, in the probabilistic correspondence rule problem as in the preferred basis problem, we *cannot* avoid introducing some extrinsic meta-theoretical elements in our reasoning. But this was exactly what Schrödinger claimed. As we have seen, the concept of *fact* was often treated by him as a pre-scientific entity, which must be linked to the theory by means of an interpretative probabilistic scheme (or correspondence rules), whose major features must be asymptotically *compatible* with the theory, but which has no immediate counterpart in the formalism. Everett himself recognized that his relative state formulation of quantum mechanics "(...) can be said to form a *metatheory* for the standard theory"[4], but, unlike Schrödinger, he tended to anticipate a situation where the metatheory somehow partakes of the theory. Schrödinger's extreme *reluctance* to accept Rosenfeld's contention "(...) that a mathematically fully developed, good and self-consistent physical theory, carries its interpretation in itself"[5] obviously does not fit at all with Everett's attitude, according to which "Any interpretative rules can probably only be deduced in and through the theory itself"[6].

[1]D. Deutsch, "Quantum theory as a universal physical theory", loc. cit.

[2]M. Lockwood, "'Many-minds' interpretations of quantum mechanics", loc. cit.

[3]M. Bitbol, *De l'intérieur du monde* op. cit.

[4]H. Everett, "Relative state formulation of quantum mechanics", loc. cit. in: B.S. De Witt and N. Graham, *The many-worlds interpretation of quantum mechanics,* op. cit. p. 149

[5]E. Schrödinger, "Might perhaps energy be a merely statistical concept?" in: E. Schrödinger, *Gesammelte abhandlungen,* op. cit. vol. 1, p. 510

[6]H. Everett, "Relative state formulation of quantum mechanics", loc. cit. in: B.S. De Witt and N. Graham, *The many-worlds interpretation of quantum mechanics,* op. cit. p. 149

4-5 *Neo-Schrödingerian views on the measurement problem: II-'Modal' and 'Critical' interpretations*

There is another cluster of no-collapse interpretations which are quite close to Schrödinger's views. These interpretations are usually referred to as "modal interpretations" or as "critical interpretations". The "modal interpretations" were promoted by Kochen, Dieks, Healey, Van Fraassen[1], and the "critical interpretation" was formulated by H. Wimmel[2]. As we shall see, these interpretations of quantum mechanics are in *some* respect closer to Schrödinger's interpretation than Everett's because they combine rejection of the collapse of the wave function with full recognition that the meta-theoretical status of the empirical correspondence scheme is irreducible. In some other respects, however, they depart drastically from Schrödinger's view, because they are associated either with a flatly anti-realist conception of scientific theories or with a variety of realism which is definitely non-Schrödingerian.

Let us begin with Van Fraassen's version of the modal interpretation, before we develop Wimmel's "critical interpretation". The very heart of Van Fraassen's interpretation is to be found in his penetrating criticism of Von Neumann's position. According to Van Fraassen, one of the most questionable aspects of Von Neumann's interpretation of quantum mechanics is his identification of two things: possession of a value of some observable by a system, and probability one, as calculated by application of Born's rule on the state vector, of obtaining this value *if* a measurement is performed on the system. In other terms, Van Fraassen criticizes Von Neumann's identification of "value-attributing and state-attributing propositions"[3].

The problem is that Von Neumann did not only considered that a system can be said to possess a value of a certain variable when it is in an eigenstate of the corresponding observable. He *also* accepted, conversely, that if the outcome of a measurement is uncertain, if the state vector is *not* an eigenstate of some observable, then the system does *not* possess any value of this observable. But in this case, there appears to be a serious conceptual discontinuity between the case where the probability of an outcome is equal to one and the case where the probability is smaller than one (even by an infinitesimal amount). When the probability is equal to one, the system is supposed to be characterized by a well-defined value of the observable, whereas when the probability is not equal to one, the same observable is supposed to have no value whatsoever. In order to remove this discontinuity, there are two possible strategies one might adopt.

[1]S. Kochen, "A new interpretation of quantum mechanics", in: P. Lahti & P. Mittelstaedt (eds.), *Symposium on the foundations of modern physics,* World Scientific, 1985; D. Dieks, "Modal interpretation of quantum mechanics, measurements, and macroscopic behavior", Phys. Rev. A49, 2290-2300, 1994; R. Healey, *The philosophy of quantum mechanics,* Cambridge University Press, 1989; B. Van Fraassen, *Quantum mechanics, an empiricist view,* Oxford University Press, 1991

[2]H. Wimmel, *Quantum physics and observed reality,* World Scientific, 1992

[3]B. Van Fraassen, *Quantum mechanics, an empiricist view,* op. cit. p. 276

Either one retains the possessed value assumption for the case of certainty and tries to extend its characteristics to the other cases, or one tries instead to make sense of a situation where the probability ascriptions which are derived from a state vector, are *not* equivalent to value ascriptions, *even* in the case of probability one.

The first strategy was promoted by Einstein. In the Einstein-Podolsky-Rosen paper, it was emphasized that when a certain outcome of an experiment performed on some system can be predicted with probability one, then there exists an "element of reality" associated with that system. It was also implied that *this* is merely a paradigmatic case, and that in many other cases an element of reality could still be attached to the systems, without being ascribed the probability one by the purely statistical formalism of quantum mechanics. The hidden-variable theorists then took up very seriously this implication, and tried to generalize the pattern of sharp-value ascription.

Van Fraassen's interpretation comes much closer to the second strategy. Van Fraassen insists on the fundamental difference between probability ascription (even if it be equal to one) and value ascription, and on the corresponding distinction between quantum dynamical states and experimental events: "(...) a *state,* which is in the scope of quantum mechanics, gives us only probability for actual occurrence of *events* which are outside this scope"[1]. Dynamical states, and dynamical states *only,* are described by quantum mechanics; events are an extra-theoretical element *concerning which* probability calculations using the state-vectors are performed. Accordingly, no (experimental) event whatsoever can be ascribed the capacity of modifying or interrupting the continuous development of the state vector according to the Schrödinger equation: "The transition from the possible to the actual is not a transition *of* state (...)". Van Fraassen's account of a measurement involves a *parallel* description of dynamical state evolution and value ascriptions rather than an interruption of dynamical state evolution by the experimental acquisition of values. It associates a deterministic continuous change of the dynamical state, and an indeterministic discontinuous link between the preparative input and the experimental output. Let us call IN the preparative input, OUT the output, W the dynamical state at t and W' the dynamical state at t'. Then, the typical measurement process has the form:

$$\text{“IN,} \quad W \quad \rightarrow \quad \text{OUT,} \quad W'\text{”}$$

"(...) The evolution of W into W' is deterministic, in accordance with Schrödinger's equation (...), and without acausal jumps or collapses. But W' only tells us what is the possible and probable character of OUT (including *some* necessities of course), and does not fully determine it"[2]. Another distinctive feature of Van Fraassen's interpretation is that, even

[1]ibid. p. 279
[2]ibid. p. 277

though he thinks the correct treatment of the measuring interactions involves holistic state vectors for the composite systems (object+apparatus), he does not so much insist on it as on the state of the object itself (namely the improper mixture obtained by reduction of the overall pure state).

This account of Van Fraassen's modal interpretation thus displays at least four important points of convergence with Schrödinger's mature treatment of the measurement problem:

(i) Dissociation and parallelism between the wave-mechanical description (dynamical states) and the domain of experimental facts (value states);

(ii) The meta-theoretical status of the observable basis and of the probabilistic correspondence rules;

(iii) No collapse (universality of the evolution of dynamic states by the Schrodinger equation);

(iv) The use of holistic wave-functions, but focusing on the state of objects.

There is however one obvious difference between Van Fraassen and Schrödinger. It bears on the issue of realism. Van Fraassen's constructive empiricist position does not enable him to see the state vectors as much more than mathematical tools for probability calculations, or than "bookkeeping devices". The very name of his interpretation (the "modal" interpretation) implies that state vectors are nothing else than formal elements able to display the *possibilities* that our understanding associates with *actual* empirical facts for the sake of prediction. By contrast, Schrödinger's tendency in the 1950's was to endow the wave-functions with some sort of ontological status. At first sight, Van Fraassen's position and Schrödinger's thus seem to be irreconcilable, because they respectively embody an anti-realist and a realist outlook. But things are not so simple. As we have shown in previous chapters (especially §1-5 and §2-1), Schrödinger's entity realism in physics was *grounded* in a definitely anti-realist outlook. It was much closer to Blackburn's quasi-realism, which has also some strong affinities with anti-realism, than to any kind of naïve realism. The difference between Van Fraassen's position and Schrödinger's towards the wave-functions, has thus nothing to do with a metaphysical dispute. It merely illustrates the difference between somebody who contents himself with a reflective analysis of the procedures of experimental sciences, and somebody who tries to use this reflective analysis as a *preliminary move,* allowing him eventually to shift the focus of the physicist's attention from little bodies to distributional entities like wave-functions. In other terms, it illustrates the difference between genuine anti-realist reductionism and an attitude which, in spite of its anti-realist departure point, considers intentional directedness towards proper entities as an essential component of scientific work.

In view of this analysis of the meaning of Schrödinger's methological realism, Schrödinger's views must be considered, in some respects, as being much closer to Van Fraassen's empiricist version of the modal interpretation than to Dieks' realist version. For Dieks considers that there is a single value ascription associated with each linear superposition; his realism is predominantly realism about value ascriptions rather than realism about the linear superpositions themselves. This kind of realism is definitely more akin to the hidden variable kind of realism about properties, than to Schrödinger's combination of (quasi-)realism about the wave-function and criticism of the realist "belief" according to which systems *possess* properties (see § 4-1).

Hermann Wimmel has recently developed another conception of quantum mechanics which has many features in common with the modal interpretation, but which also has additional points of affinity with Schrödinger's views. To begin with, his conception emphasizes more than any other the dissociation between the domain of the ψ-states and the domain of facts. According to Wimmel, "(...) ψ-functions and operators do not by themselves define factual values, possessed by the system, of corresponding dynamic variables"[1], even if the system is in an eigenstate of some observable. He also insists on the necessity of avoiding any expression which would suggest that properties are already out there in systems, waiting to be measured: "Reference to the quantum object proper, in this case the atom, is only made by stating that a particular record can be associated with one of the atom's spin eigenfunctions and/or eigenvalues. The measured atom is not said to possess (or have possessed) the associated spin eigenvalues, contrary to the Copenhagen interpretation. Neither is the atom said to have acquired the spin eigenfunction associated with the event that has occurred"[2]. This clarification is welcome, for it manages to dispense with the very formulations which made most blends of popular Copenhagen-like interpretations latently self-contradictory. In these interpretations, manipulation of the concept of "disturbance" of objects by measurement suggested that there is a sense in which objects can be said to have properties *before* any measurement; and association of state vector projection with the vague idea that attribution of state is equivalent to value attribution when the probability of this value calculated from this state is equal to one, suggested that there is a sense in which objects can be said to have properties *after* the measurement. But on the other hand any consistent talk about properties was prohibited, in so far as it could not generally be "verified". In Wimmel's formulation, this conflict is avoided from the outset by ruling out its first term.

In good agreement with the idea that factual values are not directly defined by a (possibly eigen-) ψ-state, the two domains of ψ-state

[1]H. Wimmel, *Quantum physics and observed reality*, op. cit. p. 15
[2]ibid. p. 21

evolution and value-ascriptions are carefully distinguished throughout Wimmel's treatment. Their relations are construed in terms of parallelism rather than in terms of periodic intervention of one of them into the other: "(...) the interpretation has to use two incongruent models in parallel, namely the quantum formalism, or ψ-model, which is not a description of facts, and a fact-describing (realistic) model of facts or F-model, that contains, and works with, directly observable facts and events"[1]. This crucial point of parallelism between the ψ-model and the F-model is then contrasted with the Copenhagen conception of the measuring process, which, in spite of Bohr's holistic moves, contained so many remnants of the dualist and causalist picture of the interaction between the object and the apparatus. It cannot be said, for instance, that the ψ-model is specifically designed to make predictions on the (microscopic) object, whereas the F-model is specifically attached to the (macroscopic) apparatus. There is no "cut", not even an arbitrarily located cut, between the portion of the measurement chain where the quantum formalism applies and the portion where classical concepts, characterized by their affinities with the structures of ordinary language, are to be used instead. "The ψ-model is, from the outset, taken to apply universally (...). The ψ-model covers both the 'quantum system' and the 'experimental apparatus', which thus constitute a formal unity. (...) There is no question of 'interaction' between the two models; this is neither necessary nor conceptually possible. The two models are so to speak *'connected in parallel',* while the aggregate of quantum system and (classical) apparatus of the Copenhagen interpretation creates the impression of being, so to speak, *'connected in series', and that in an undefined way"* [2]. On the one hand, dissociation of the ψ-model and of the F-model is so absolute that any influence of one of them on the other is precluded. The idea that events could in some way emerge from or be produced by the ψ-model is no more plausible in this perspective than the idea that the occurrence of facts produces sudden temporal changes (namely collapses) of the ψ-function. And on the other hand there must exist rules which ensure that the global wave-function is able to "mirror" factual reality, i.e. to reflect somehow the reproducibility and permanence of (intersubjectively recognized) records[3]. These rules, or "translation modes" between the ψ-model and the F-model, are of two types. There is a F→ψ translation mode which corresponds to the definition of the ψ-function to be associated to each given *preparation.* And there is a ψ→F translation mode which is probabilistic, and which reduces to the Born rules for results of *measurements.* The overall structure of this interpretation scheme is displayed in the following diagram.

[1]ibid. p. 25
[2]ibid. p. 31
[3]ibid. p. 41

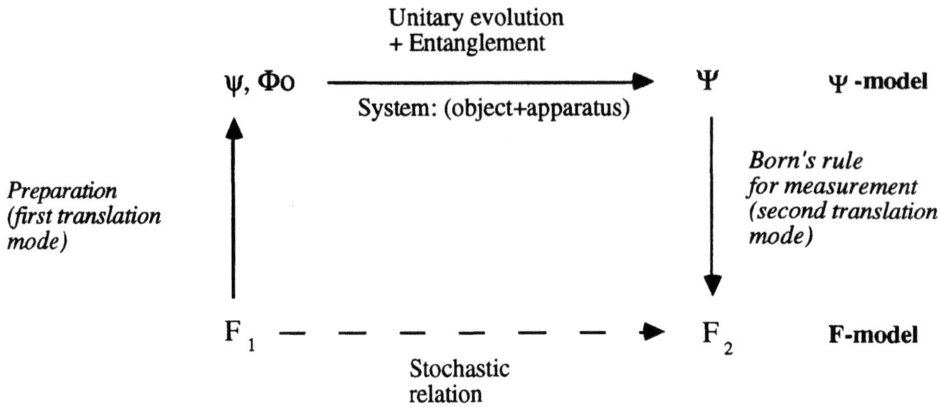

In this account of the "critical interpretation" of quantum mechanics, we already recognize several steps of Schrödinger's thought in the 1950's: (i) rejection of his early emergentist (wave-packet) view of the relation between wave-functions and corpuscle-like facts; (ii) holistic description of the measurement process, involving entanglement between the ψ-function of the object and the ψ-function of the apparatus; (iii) rejection of wave-packet collapse; and (iv) post-modern dissociation between ψ-functional description and experimental facts.

Another major point of convergence is the idea that the observable basis of eigenvectors and the empirical correspondence rules are definitely meta-theoretical. They pertain to the relation between the ψ-model and the F-model, *not* to the ψ-model itself.

The observable basis, to begin with, cannot be derived from the ψ-model according to Wimmel: "Wave functions and (density) operators do not predict what *type of factual states or factual events* will occur or be observed in a physical situation. (...) Wave functions and density operators must be used according to qualitatively different modes of application, depending on the physical system and the experimental situation. In particular, mutually independent, irreducible modes of translation between the ψ-model and the F-model must be used in different situations in order to cover all possible cases". Each observable basis of eigenvectors represents a mode of translation, appropriate to a given experimental situation, between the ψ-model and the F-model, in the direction ψ→F. As such, it cannot arise either from the ψ-model or from the F-model. Its relational status makes it clear why it has to be considered, in Schrödinger's terms, as the "blind spot" of the quantum theory.

Once the observable basis, and the "type" of factual events to be expected, is determined by an analysis of the experimental situation, the next step is to define the (probabilistic) correspondence rules between the ψ-model and the F-model *as generally as possible,* namely to make the ψ→F translation mode apply not only to *one* particular fact, but to any *sequence* of facts. However, a difficulty arises at this point. Since we have previously dismissed the possibility that the unitary evolution of ψ-functions be interrupted by an element of the F-domain, we cannot use the computational device of the reduction of the wave-function; we cannot use this convenient tool in order to evaluate the probability of the outcomes which may arise from the *next* experiment, once the outcome of the *previous* experiment is known. Reduction of the wave-function is especially useful to account for experimental reproducibility and probability evaluations for sequences of experimental outcomes, and some kind of substitute has to be found for it.

Wimmel thus proposes to use the same strategy as Mott in 1929[1] (see §2.6 of the present essay). Mott treated the problem of the appearance of α-rays tracks in a Wilson chamber. But in so doing, he did not consider the wave-function of an α-particle in isolation; he did *not* employ the usual trick of *reducing the wave function* each time the α-particle ionizes a molecule of water, and then calculating the evolution of the corresponding wave-packet. Instead, he described the α-particle and two molecules of water by a single holistic wave-function progressively undergoing entanglement, and then calculated the *joint probability* of a two-ionizations event. This joint probability is such that the second ionization (and the subsequent drop formation) is likely to occur on a line joining the parent nucleus and the first ionization, thus giving rise to the appearance (or to the "illusion", as Schrodinger's puts it) of a particle track.

In order to predict the probability of *sequences of events,* possibly arising from the interaction of the object and several pieces of instrumentation, Wimmel similarly considers the global wave function of the object *and* these pieces of instrumentation, then develops this wave function according to a composite observable basis, and finally calculates the joint probability of the sequence considered as a whole by applying the generalized Born rule to the coefficient of the resulting entangled linear superposition. Just as in Schrödinger's mature conception of quantum mechanics, the evolution of the wave function is increasingly holistic but continuous, and the only probabilistic element comes in when the *final link* between the entangled wave function and the experimental facts is at stake. At no point between the preparation of the experiment and the final outcome (which here amounts to a track), does one have to suppose that quantum jumps or localized particle collisions occur. All the consequences of quantum mechanics, *including* those which were initially interpreted in terms of discontinuous processes (e.g. the photoelectric effect, and Planck's radiation law), can be derived in this way.

Wimmel's interpretation of quantum mechanics also gives some weight to Schrödinger's attempt at pushing the measurement problem to the edges of the theory. For here, the measurement problem is not only

[1]N.F. Mott, "The wave mechanics of α-ray tracks", Proc. Roy. Soc. Lond., A126, 79-84, 1929; A.B. Pippard, "The interpretation of quantum mechanics", Eur. J. Phys., 43-48, 1986

pushed to the edges; it is deliberately bypassed, through the formalized dissociation between the ψ-model and the F-model. No transition from the holistic and continuous ψ-model to a fact ever has to occur, *even* at the very end of the measurement process. Only a translation between the two models has to be performed, for the sake of those who happen to live in an environment of events. Of course, this kind of agnosticism sounds quite frustrating when it is stated so bluntly. But I think that the Schrödinger/Modal way of ignoring the measurement problem, and of developing the idea of a translation scheme between two parallel processes, namely ψ-process and F-process, will be more readily accepted as soon as a proper philosophical discussion of it is provided. This discussion is clearly lacking in Wimmel's book; and it was only outlined by Schrödinger. I will thus try to undertake such a discussion in chapter 6.

The major difference between Wimmel's interpretation on one side and both Everett's and Schrödinger's on the other side, obviously bears on realism. In the same way as Van Fraassen, who considers that the ψ-functions are a mere modal clothing on a backbone of experimental outcomes, Wimmel insists that the only reality is the "observed reality". His account of Everett's interpretation is thus a balanced mixture of approval of its structural features and criticism of its general outlook. To begin with, Wimmel approves some aspects of Everett's relative state interpretation, for it leads one to use the same kind of formula for treating multiple events as the Mott-Wimmel formula. It also implies discarding completely the collapse of the wave-function. Thus, according to Wimmel, "It may have been the unnecessary 'many-worlds' costume and other shortcomings of their treatment that prevented the correct parts concerning the refutation of ψ-reduction from being recognized in this case as well"[1]. But, under the heading "other shortcomings", Wimmel later criticizes Everett for having missed "(...) the necessary distinctions between, on the one hand, facts and events and, on the other, ψ-information (...)"[2]. This criticism is mostly irrelevant, for the distinction between ψ and F series is perfectly displayed in Everett's notations and comments, by means of a separation between the time-development of the wave-function and the content of the memory brackets. However, it is also true that Everett's realist conception of the wave function made him try to *append* the F-series to the ψ-series. This did not prevent one from seeing the difference between the two series, but it could have led some physicists to think that the content of the memory bracket is a mentalistic epiphenomenon with respect to what happens in the real world, whose state is described by a universal wave function including the bodies of the observers. By contrast, Wimmel does not try to include the observer's state in the holistic wave function he uses, and he emphasizes an anti-realist conception of the ψ-function: "(...) it is best to abstain from

[1]H. Wimmel, *Quantum physics and observed reality*, op. cit. p. 46
[2]ibid. p. 68

wishing to see in wave functions and operators, as it were, a second 'reality'. Rather it appears that one should understand wave functions solely as a means of formulating certain auxiliary conditions that have to be obeyed by the world of (directly or classically observable) facts and events"[1]. Though Wimmel discusses extensively the case of adiabatic measurements (see §4-2) which give direct and non-perturbative access to the distributional characteristics of the wave functions, he does not think that even this can constitute a sufficient reason for endowing wave functions with some sort of reality. According to him, then, the adiabatic measurements have to be regarded only as "quasi-measurements".

But what applies to the difference between Everett's and Wimmel's interpretations does not necessarily apply to the difference between Schrödinger's and Wimmel's interpretations. There are more structural similarities between Schrödinger's and Wimmel's position than between Everett's and Wimmel's. This close kinship with an overtly anti-realist interpretation of quantum mechanics can be seen as one more symptom of the philosophical specificity of Schrödinger's realism of ψ-waves.

Let us recall that Schrödinger was just as eager as Wimmel to emphasize the distance between the ψ-description on the one side, and facts as well as "nature as it *is*" on the other. He could also accept, as an anti-realist like Wimmel, that ψ-waves are but "constructs". The major difference however is that this did not prevent him, unlike Wimmel, from thinking that the physicist's intentional gaze had to be directed towards something. The organized tension towards future information, the frame of expectation of subsequent experimental confirmations which is associated with this intentional gaze, must be maintained if possible - displaced if necessary, but not suppressed according to Schrödinger. This is what we have repeatedly referred to as Schrödinger's quasi-realist, or phenomenological, stance. The only difficulty was to choose a proper object of intentional directedness. Now we know that according to him, ψ-waves, not the traditional entities of physics (namely particles) are the best candidates. His reasons for such a choice were the following (see chapter 5 for further analysis of points (i) and (ii), and §4-2 for further analysis of point (iii)):

(i) the things and facts of everyday life, which are the paradigm entities of intentional directedness, are no *less* artificial than ψ-functions; according to Schrödinger they and ψ-functions are equally to be considered as "constructs";

(ii) but constructs which enable one to perform daily or experimental actions so efficiently must be taken as seriously as the hypothetical transcendent entities of metaphysical realism. One has no reason to discard this *seriousness* which is the right ethical attitude for a working scientist to adopt, and to promote some kind of anti-realist reductionism instead;

[1]ibid. p. 62

(iii) in the situation governed by quantum mechanics, the statistical or distributional characteristics of the ψ-function fulfill the conditions of stability, reproducibility, reidentifiability, which make objective knowledge possible, whereas the isolated experimental facts, which are supposed to characterize hypothetical bearers of properties such as particles, do not.

CHAPTER 5

THE "THING' OF EVERYDAY LIFE[1]

As we emphasized in previous chapters, Schrödinger's rejection of corpuscularian entities, and his adoption of wave entities instead, can only be understood in the light of his general views on ontology. Our aim in this chapter is to provide one with an exposition and a critical assessment of Schrödinger's conception of the objects (or "things") of everyday life, and then to connect it back with the ontological issues raised by quantum mechanics. This connection is by no means artificial, for Schrödinger developed his ideas about the "things" of everyday life in permanent relation with his reflection on quantum mechanics. In most texts, quantum mechanics even appears as the true motivation for a (usually short) discussion about general ontology.

Schrödinger gave his best definition and his most precise description of what he calls a "real object", in four texts whose dates of publication range between 1928 and 1954[2]. The crucial question he tried to answer was the following: can one still consider that quantum mechanics describes the behaviour of "real objects"; and if the answer to this question is positive, which kind of objects do the quantum laws rule? In Schrödinger's own wordings, as they were used during his 1928 conference, the new situation created by quantum mechanics prompts one to ask: "Do these atoms and electrons, etc., really exist and, if so, are they in precisely the configurations we attribute to them? Is their existence, as many declare, as definitely guaranteed as the objects of my environment which can be touched and handled?"[3].

The very form of Schrödinger's question about atoms and electrons implies that a preliminary enquiry about the objects "which can be touched and handled" must be conducted. The "things" of everyday life have the special status of a set of objects whose existence is merely presupposed by our daily actions and speech. It is only by comparison to them, and to the unproblematic attitude we adopt towards them, that any question about the "existence" of the objects of modern physics can make sense. Schrödinger's digression about the familiar, visible and tangible objects had the advantage of offering a frame of reference, which, by comparison, made it easy to pick out the *lacunae* which affects the putative objects of quantum mechanics. It was only then, during a second phase of his reflection, after he had clarified the status of the familiar "things" of

[1]A former (and shorter) version of this reflection on Schrödinger's conception of the "thing" of everyday life can be found in: M. Bitbol, "Esquisses, forme, et totalité", in: M. Bitbol and O. Darrigol (eds.), *Erwin Schrödinger, Philosophy and the birth of quantum mechanics*, op. cit.

[2]E. Schrödinger, "Conceptual models in physics and their philosophical value"; Francfort, December 8, 1928, in: *Science and the human temperament*, op. cit., p. 119-132; *Nature and the Greeks*, (Conferences delivered in Dublin in 1948), op. cit.; "What is an elementary particle?", Endeavour, **9**, 109-116, 1950. E. Schrödinger, *Science, philosophy and the sensates*, (chapter 3 of a series of lectures which were intended to be given in Harvard, as William James Lectures, in 1954; in: E. Schrödinger, *The interpretation of quantum mechanics (Dublin seminars 1949-1955 and other unpublished texts)* op. cit.

[3]E. Schrödinger, "Conceptual models in physics and their philosophical value", loc. cit. p. 119

the mesoscopic scale, that Schrödinger felt he was ready to address the problem of the status of the micro-objects.

5-1 *The three features of objects*

In several texts, Schrödinger begins his exposition stating concrete examples of "things-of-our-environment". These familiar (and at times literary) sentences are obviously meant to provide the reader with a network of familiar and reassuring reference-marks, before the difficult ontological investigation starts. The range of his familiar illustrations is quite broad. He chooses either to comment on a fruit basket[1], or an "iron letter-weight in the shape of a Great Dane"[2], or his own arm[3], or even "Marble Arch"[4].

Then he organizes his thought along three principal lines:

(1) The object of everyday life only appears to us from one of its sides, or through one of its sensorial aspects. Marble arch only shows me one of its two fronts, and if I am not looking presently at the fruit basket which is in the room, the only element I perceive of it is its fruity fragrance. The objects only shows us one of its profiles or aspects; its presentation to us is *perspectival* and *incomplete*.

(2) *Form,* in the broader sense, is what confers individuality and permanent identity on entities given to the senses. Form is what gives these entities their distinguishing features and a certain amount of stability.

When form happens to be inaccessible at every instant, or when it is not sufficient to single out a certain object, such an object can be identified by relying on the continuity of its trajectory, which is a spatio-temporal configuration, namely a historical variety of form.

(3) Let us suppose that the elementary components of matter cannot be individualized, either by instantaneous form or by historical form. Their various combinations however give rise to complex observable forms which enable one to distinguish them from one another. These organized wholes, or their characteristic observable form, must be considered as the only real, individuatable, and permanent, objects.

Perspectivism, emphasis on *form,* and *holism,* are the three main features of Schrödinger's conception of real objects. We are then going to analyze these three features in turn, putting them back in the context of some relevant philosophical traditions.

[1]ibid.
[2]E. Schrödinger, *Science and Humanism,* op. cit. p. 18
[3]E. Schrödinger, *Nature and the Greeks,* op. cit. Conclusion
[4]E. Schrödinger, "Indeterminism in physics", in: *Science and the human temperament,* op. cit. p. 62

5-2 *The aspects and the "thing"*

Let us begin with perspectivism. The real object only presents to us one of its sides or one of its aspects. It is only given partially to our sight or to any other sense. This remark is by itself commonplace. But it still has to be accomodated into a comprehensive view of the world. According to what Schrödinger calls the "popular view" of the world, it is usually considered as the effect or our adopting a particular standpoint in space and time. In the framework of thought of the "popular view", indeed, our seeing only one aspect of the thing is a mere consequence of the relative situation of a human body (our body) and the object which is in front of it. Each spatial position of the perceiving body relative to the perceived object defines the perspective according to which a profile of the object is projected. And any motion, either of the human body or of the object, any obstacle between the body and the object, modify the perceived profile according to a set of geometric and kinematic rules.

The "popular" account of the perspectival features of our perception of objects as it has just been summarized, however suggests the very method one could use in order to reverse it. As the French mathematician of the seventeenth century Gérard Desargues demonstrated in his treatise on perspective, if it is possible to derive the profiles of a three-dimensional body by application of the rules of projective geometry, then it must also be possible to reconstruct the object from certain ordered sets of profiles. Space itself, which, in the "popular view", provides a passive frame for the positions and relative displacements of the human body and the objects, could be construed in a Leibnizian spirit as the system of all possible viewpoints.

Such a complete reversal, which is tantamount to starting from the profiles in order to get the objects, rather than considering that the profiles are simply subjective views of a world of objects and human bodies already 'out there', will be referred to as the "Ptolemaic revolution". This (somewhat ironical) expression must not be taken as expressing any conflict between Schrödinger's perspectival conception of the bodily objects and Kant's "Copernican revolution". It rather aims at underlining the opposition of Schrödinger's perspectivist views towards a physicalist and/or naturalistic outlook which is much more common nowadays and which presents itself as "Copernican"[1]. It also evokes a parallel with a set of iconoclastic ideas Husserl developed during the last phase of his career. In a text whose short title is "Earth does not move" ("Die Ur-Arche Erde bewegt sich nicht")[2], Husserl argued that, from a phenomenological perspective, the earth had to retain its pre-Copernican

[1]A. Shimony, "Scientific inference", in: *Search for a naturalistic world view I,* Cambridge University Press, 1993, p. 183

[2]Part of: E. Husserl, *Grundlegende Untersuchungen zum phänomenologischen Ursprung der Räumlichkeit der Natur,* in: M. Farber, *Philosophical essays in memory of E. Husserl,* Harvard University Press, 1940

status of an absolute framework of motion. Of course, he did not wish to dispute the *scientific* value of Copernicus' and Newton's achievements. He rather tried to underline, in a provocative way, that what he calls *the sediments of the life-world,* are primary with respect to any further scientific representation. Now, this *life-world* has many features in common with what J. Searle refers to as the "Background" and that he defines as "(...) the set of non intentional or pre-intentional capacities that enable intentional states to function"[1]. Therefore, what had to retain a privilege, according to Husserl, was not the earth as an object of intentional directedness (a certain planet), but the earth as a *background* of ordinary human life.

By insisting on the unique status of the earth, Husserl was only drawing in a provocative way the ultimate consequences of his own conception of philosophy. For a philosophy based on the method of the "phenomenological reduction" was to be conceived as a systematic attempt at bracketing the intentional gaze of the "natural attitude" (or "popular view"), and redirecting attention towards the elements of the background: the Earth as "archè" rather than the Earth as a planet; the (transcendental) self rather than the (empirical) subjects; the meanings (or noema) rather than the referred to objects. And also (but in the spirit of Heidegger's philosophy, not Husserl's), the things as "ready-at-hand" (Know-how), rather than the things as objects of knowledge (know-that).

It is thus only because people adopt spontaneously the "natural attitude" which considers that aspects are subjective views of objects, that the opposite attitude which considers aspects as primary, seems to be "egocentric" or "anthropocentric". From the position that Husserl tried to promote, this judgment would have to be modified, and the criticisms would rather bear on the tendency to draw dogmatic conclusions from the "natural attitude". Indeed, if the attention was permanently and consistently redirected towards the background meanings, the "natural attitude", with its object-centration which most likely generates a substantialist metaphysics, would appear as a most risky attempt at hypostasizing the ideal coexistence of an unbounded set of aspects.

This reversal of criticism is not meant to convey the idea that the perspectivist conception of the "thing" of everyday life is the best possible approach, let alone that it is unproblematic. Commenting on Schrödinger's own variety of perspectivism, we shall rather list an impressive number of difficulties which undermine seriously this conception. But one must not dismiss perspectivism from the outset with arguments which presuppose that the "natural attitude" cannot even be challenged; one has to evaluate it according to fair standards.

Schrödinger performed what we have called the "Ptolemaic revolution" for his own sake. He expressed his position very clearly in several texts,

[1]J. Searle, *The Construction of Social Reality,* Allen Lane, The Penguin Press, 1995

which we shall have to study in detail in order to display his acute perception of the problems. His views must also be compared and sometimes contrasted with a series of other perspectivist and/or phenomenalist conceptions which include Berkeley's, Mill's, Mach's and Russell's. It will finally be necessary to evaluate them in the light of Husserl's phenomenology as well.

Let us begin the presentation of Schrödinger's version of phenomenalism with a long quotation from *Nature and the Greeks:*

"(...)envisage any one of the material objects around us, for example my arm and hand. As a material object, it is composed, not only of my own direct sensations of it, but also of the imagined sensations I would have in turning it round, moving it, looking at it from all different angles; in addition it is composed of the perception I imagine you to have of it, and also, if you think of it purely scientifically, of all you could verify and would actually find, if you took it and dissected it, to convince yourself of its intrinsic nature and composition"[1].

The characteristic features of a perspectivist and/or phenomenalist construal of the "thing" of everyday life are almost exhaustively defined in the previous sentences. But some imprecisions in Schrödinger's vocabulary point towards the difficulties one is likely to meet when this conception is worked out.

Consider for instance Schrödinger's use of the words "sensation" and "perception" in very similar contexts. True, Schrödinger does not make the elementary mistake which would consist of taking sensation and perception as synonyms. He even stresses the distinction in a later sentence of the same text, by writing about the "percepts *and* sensations" (not of the percepts *or* sensations) which are "(...) included in my speaking of this arm as of an objective feature of the 'real world around us'". However, the intervention of sensation and perception in succession or in conjunction raises the question of what is initially "given", before the things can be constituted out of it. Is this supposed primeval "given" made of pure elementary sensations, or has it the complex structure of perception? Can perception concern *only* the shape and the *qualia* which compose a profile? Or is perception already pervaded by intentional directedness towards an object, thus making circular the idea that objects are made of, or constituted by, percepts?

The second problem is about the status of the constituted object, as it is raised by Schrödinger's repeated use of the verb "to compose". Has the object to be taken litterally as a *compound* of sensations and percepts? Is the object at least an *organized* compound of sensations, as it is suggested by the metaphor of a building composed of bricks arranged in a certain order that Schrödinger used in earlier paragraphs of *Nature and the Greeks?* Or does the object represent something beyond the sum total of its appearances: an invariant, an organizing principle (of the appearances),

[1]E. Schrödinger, *Nature and the Greeks*, op. cit., p. 92

or, may be, as in Husserl's phenomenology, the "objective unity" of various noetical phases[1]?

The third difficulty arises from the systematic use of the conditional mood, which intervenes in hypothetical or counterfactual propositions. Something like: *If I turned my arm round (but, actually, I am not doing this), I would have such and such sensations.* Why does one have to use *possible* elements in the constitution of an object which is supposed to be *actual?* Is it really appropriate to ground categorical statements on hypothetical material? What are the relations between (i) the present tense of the conditional mood, in the counterfactual propositions used by Schrödinger, (ii) the past tense of the indicative, corresponding to memorized perceptions, and (ii) the future tense of the indicative, corresponding to perceptual expectations? Do Schrödinger's counterfactual propositions make sense without some presupposition of identity (in this case the identity of *the arm)* irrespective to the actual or possible position of my eyes relative to it? And aren't we then caught once again in a circular process of definition?

These are the questions which will be addressed in forthcoming paragraphs.

5-3 *The "elements" of the construction (Mach, Russell, Schrödinger, Husserl)*

According to Schrödinger himself, his conception of the real objects of everyday life stems from the empiricist tradition, and more especially from the positivist doctrines of the nineteenth century. In paragraph 7 of his paper "What is an elementary particle?", he even avoids using the first-person mode of discourse, and rather traces his ideas about material object back to several other philosophers: "I wish to set forth a view on matter and the material universe, to which Ernst Mach, Bertrand Russell, and others were led by a careful analysis of concepts". These explicit references to Mach and Russell must however not be taken too seriously, for Schrödinger did not content himself with repeating the doctrines of his predecessors; he elaborated and refined them.

The major characteristic feature that Mach's positivism[2] shares with classical empiricism is the radical distinction it draws between the experience of sensible qualities, and the perceptual or intellectual "interpretations" of this experience. Only sensorial experience can be ascribed, according to Mach's positivism, the status of factual material; only atomic sensations are truly *given*. The perceptual or intellectual components of knowledge are only considered as an artificial structure superimposed onto the brute data, or as useful fictions which can be modified according to the needs. In particular, Mach separated carefully the sensations, or elements, from the act of organizing them into

[1]E. Husserl, *Ideas*, op. cit. §98
[2]E. Mach, *The analysis of sensations*, Dover, 1959

reasonably stable complexes called "material objects". Sensations are allegedly known by direct acquaintance, whereas their familiar gatherings and orderings can only be justified as a realization of the ideal of "economy of thought". Something quite similar can be said of Russell. For even though Russell entertained the possibility of taking perception, "(...) that integral experience of things in their environment"[1], as the basis of his analysis of knowledge, he ended up giving a reductive account of perception, in terms of sensations. According to him, perception is only a set of sensations which one associates by habit with expectations of future sensations. "What is called perception differs from sensation by the fact that the sensational ingredients bring up habitual associates - images and expectations of their usual correlates - all of which are subjectively indistinguishable from the sensations"[2]. The privilege he confered upon sensations and images had the consequence of more or less reducing to sensation every other element traditional psychology locates in the "mental universe". According to Russell, all the constituents of mental life were constructs made out of ultimate atomic elements, namely sensations and their imaginative reproductions: "I propose to argue that thoughts, beliefs, desires, pleasures, pains and emotions, are all built up out of sensations and images alone, and that there is reason to think that images do not differ from sensations in their intrinsic character"[3].

But Schrödinger was not so comfortable with sensualistic reductionism. The building materials of the world, according to him, include "(...) sense perceptions, memory images, imagination, thought"[4], on an equal footing, without any attempt at minimizing the role of thought or at reducing it somehow to sensations and images. This original emphasis of Schrödinger's phenomenalism on *thought* is sometimes concealed when he says he is just reporting on Mach's and Russell's ideas in his own terminology, but it reappears as soon as the latter terminology is carefully analysed. According to Mach and Russell, says Schrödinger, "(...) a piece of matter is the name we give to a continuous string of events that succeed each other in time, immediately successive ones being as a rule closely similar"[5]. But then come some indications about the meaning Schrödinger ascribes to the word "event". From these indications, we see that, in Schrödinger's conception, the basic constitutive elements do not reduce to bare sensations as in Mach, or to sensations plus images of sensations as in Russell. Schrödinger's "events" already mix up sensitive and imaginative components, into a "complex" whose structure is likely to be of an intellectual origin: "The single event is an inextricable complex of sensates, of associated memory images, and of expectations associated with the former two"[6]. Thus, in his version of phenomenalism, the constituents

[1]B. Russell, *The analysis of Mind*, G. Allen & Unwin, 1971, p. 157
[2]ibid.
[3]ibid. p. 121
[4]E. Schrödinger, *Nature and the Greeks*, op. cit., p. 92
[5]E. Schrödinger, "What is an elementary particle?" loc. cit.
[6]ibid.

of the *real objects* are not just actual and imagined sense-data, but rather perceptual *complexes* which are so elaborate, so "inextricable", that they make any further analysis into elementary data quite difficult and probably pointless.

To recapitulate, according to Schrödinger, real objects are *two-level* complexes. They are complexes of events, which are themselves inextricable complexes of sensorial, imaginative, and intellectual components. Increasing priority is given to the hierarchies of these complexes over what the complexes unite. And the complexes eventually acquire complete *autonomy* with respect to their constituents: "After a certain wealth of association has come to outshine the core of sensates, the latter is no longer needed to keep the complex together. It persists even when the contact of our senses with the object temporarily ceases. And more than that: the complex is latently conserved even when the whole string is interrupted by our turning away from the object to others and forgetting all about it"[1]. The "complex" here obviously holds the role of the *substance,* i.e. the role of something permanent over and above the presence or absence of experiential content at any given time; but it is construed as an invariant structure organizing the sequence of actual and possible experiential contents, rather than their underlying substrate.

Let us remark at this point that the distinctivee features of Schrödinger's conception of the real object which have just been discussed make it much closer to Husserl's phenomenology than to any empiricist or positivist doctrine. In paragraphs 19 and 20 of his *Ideas,* Husserl compares his approach to classical empiricism. He underlines some striking similarities, and also some important differences. Both empiricism and phenomenology are motivated by a severe criticism of tradition, habits and prejudices. They both aim at "(...) being guided by the facts themselves, getting away from talk and opinion back to the facts"[2]. But to the empiricist, the facts themselves reduce to the facts of pure experience; they reduce to sensations. Whereas, in phenomenology, the facts (Sachen) are not only the facts of nature. The kind of "immediate seeing" of facts a phenomenologist has to consider does *not* reduce to the sensory seeing of experience; it extends to "seeing in general as primordial dator consciousness of any kind whatsoever"[3]; it includes both experience and intuition of essences. Let us now come back to Schrödinger. Schrödinger's insistance on encompassing thoughts into the domain of the constitutive elements out of which a real object is made, and also his making "complexes" autonomous with respect to any experiential content, imply the same kind of half-acceptance and criticism of the empiricist tradition as Husserl's. On the one hand, empiricism is praised for its deconstructive undertaking, which leads to a thorough analysis of the common-sense conception of the world and of material

[1]ibid.
[2]E. Husserl, *Ideas,* op. cit. §19
[3]ibid.

objects. And on the other hand, the empiricist's conception of the product of the analysis is criticized either explicitly (by Husserl) or implicitly (by Schrödinger). According to both Husserl and Schrödinger, this product must be much more elaborate than just sensations. Husserl calls it "Noema" and Schrödinger calls it "event". Husserl's noema is a complex "essence" which, according to H.L. Dreyfus' sound synthesis[1], gathers three functions: it has to pick out an object (it has an *aboutness),* it describes the object under a given *aspect,* and it includes the additional *expected* aspects the object picked out *could* exhibit. As for Schrödinger's "event", it is an inextricable perceptual complex of sensations and expectations which has obviously a lot more in common with Husserl's "noema" than with the bare sensations of the empiricist and positivist traditions.

To summarize, Schrödinger's allegiance to the positivist tradition is deep, as he himself acknowledges, but it is by no means uncritical. It can best be understood in the light of Husserl's following remark: "If by 'positivism' we are to mean the absolute unbiased grounding of all science on what is 'positive', i.e. on what can be primordially apprehended, then it is we who are the genuine positivists"[2].

5-4 *Are the "basic data" really basic?*

Mach construed the real objects or "bodies" as *complexes* of elements[3]. It is then likely that Schrödinger borrowed the concept of "complex" from Mach. But, as we have already mentioned, Schrödinger used Mach's concept of "complex" extensively, applying it not only to the constituted objects but also to the constituting events. And he also took this concept so seriously as to endow the complexes with complete autonomy with respect to their elements. This insistance on complexes rather than on constituents may have originated in Schrödinger's perception of a loophole in the doctrine of the *Analysis of sensations*. The problem is that Mach's account of the relation between these complexes and the elements which compose them makes quite difficult to see why the elements should be given any priority over the complexes. On the one hand Mach expresses himself as if the elements came first, before the complexes at any rate: "colors, sounds, temperatures, pressures, spaces, times, and so forth are connected with one another in manifold ways. (...) Relatively greater permanency is exhibited, first, by certain complexes of colors, sounds, pressures and so forth, functionally connected in time and space, which therefore receive special names and are called bodies"[4]. And on the other hand, he mentions that, in some way, the complexes are apprehended as wholes, apparently

[1]H.L. Dreyfus (ed.), *Husserl, intentionality and cognitive science,* MIT Press, 1982, (introduction by H.L. Dreyfus)
[2]E. Husserl, *Ideas,* op. cit. §20
[3]E. Mach, *The analysis of sensations,* Open court, 1914, p. 5
[4]ibid. p. 2

endowed with "properties", before they are analysed into elements: "The complexes are disintegrated into elements, that is to say, into their ultimate component parts"[1]. Are the elements then pre-given and later connected in order to make complexes, or are the complexes pre-given and later disintegrated in orders to reveal elements?

The dilemma is that the sensation-element, which is *in principle* to be construed as the primeval and unconditioned datum out of which everything else is made, appears *de facto* as the elusive end-product of a very elaborate process of abstraction. The modern version of phenomenalism, which combines the basic outlook of classical empiricism with the characteristic features of the logic-linguistic turn, has often been criticized on this ground. The difficulty, here, is about the status (elementary or elaborate) of the so-called "basic" sentences, or protocol sentences, rather than on the status of plain sensations. Carnap's version of phenomenalism, as it has been developed in *The logical structure of the world* [2], has been an easy target for this kind of criticism[3].

Let us now give more details about how "simple sensations", or "basic sentences" stating sensations, may arise as the end-products of a process of thought which takes its departure point from some sort of "complex". There are at least three of these processes of thought:

The first one is the so-called *phenomenological reduction,* which starts from the perceived "thing" of the *natural attitude* and progressively goes back to the pure sensible matter (or "hyletic data") through two stages: the structured directedness that Husserl calls the *noema,* and the stream of experience that he calls the *noesis*. Characteristically enough, it is not in the first pages of his *Ideas* that Husserl states what he means by *hylè,* but rather in the second half of the book, after he has performed his series of reductions (or "bracketings"). There, he underlines that the hyletic contents are merely "present" in ordinary perceptual experience, whereas their being *objectively apprehended* is characteristic of the "analysing experience" of the philosopher[4].

The second process uses linguistic analysis. As W. Sellars (after Wittgenstein) pointed out, the discourse on private sensations cannot be primitive. It is one of the most elaborate kinds of discourse, because it is based on a background acceptance of ordinary language and of reference to public objects. "(...) The training of people to respond conceptually to states of themselves which are not publicly observable requires that trainer and trainee alike (...) share both the intersubjective framework of public objects and the intersubjective theory of private episodes (...)"[5].

The third process is typical of the scientific outlook, and it is this one which was insisted upon by Schrödinger. The sensation of yellow, for

[1]ibid. p. 5

[2]R. Carnap, *The logical structure of the world,* Routledge & Kegan Paul, 1967

[3]N. Goodman, "The significance of Der logische Aufbau der Welt", in: P.A. Schilpp (ed.), *The philosophy of Rudolf Carnap,* Open court, 1963

[4]E. Husserl, *Ideas,* op. cit. §98

[5]W. Sellars, "Phenomenalism", in: *Science perception and reality,* Routledge & Kegan Paul, 1963

instance, can be construed as the (somewhat mysterious) end-result of the interaction between an electromagnetic wave of appropriate wavelength, and a vast neuro-physiological system which goes from the retina to the cortical neurons of the brain. "If you ask a physicist what is his idea of yellow light, he will tell you that it is transversal electromagnetic waves of wavelength in the neighbourhood of 590 nm. If you ask him: but where does the yellow come in? he will say: in my picture not at all, but these kinds of vibrations, when they hit the retina of a healthy eye, give the person whose eye it is the sensation of yellow"[1].

The paradox of these "sensations" which are both immediately present and very remote (not to say inaccessible) as soon as one tries to transform them into an object of discourse or of scientific inquiry, was central in Schrödinger's thought. We shall discuss two varieties of the paradox. One of them arises from a direct discussion of *immediacy,* and the other one has to do with philosophical *foundationalism.*

First variety. According to Schrödinger the sensual qualities are *manifest,* but they have been removed from our picture of the world in order to make it completely objective. Their being absent in the objective picture of the world is by no means contingent; for their withdrawal is just an important part of the *definition* Schrödinger gives of the process of objectivation. However, conversely, "Once we have removed (sensual qualities) from our 'objective reality' we are at a desperate loss to restore them. We cannot remove them entirely, because they are *there*, we cannot argue them away"[2]. The temptation, when one faces this dilemma, is to invent (or reinvent) dualism: "(...) we have to give them a living space, and we invent a new realm for them, the mind, saying that this is where they are, and forgetting the earlier part of the story (...)"[3]. But of course, this "forgetful" attitude is not satisfactory either, and Schrödinger criticizes it eagerly in chapters 3 and 4 of his *Mind and Matter* [4].

Second variety. The sensual qualities partake of the *foundations* of any objective description, even though they cannot be a component of the resulting picture of the world. One must keep in mind "(...) the two general facts (a) that all scientific knowledge is based on sense perception and (b) that none the less the scientific view of natural processes formed in this way lack all sensual qualities and therefore cannot account for the latter"[5]. Schrödinger traces back this version of the paradox to Democritus B125, whose first part can be taken (and has often be taken) as an early statement of physicalist reductionism, but whose second part is a skeptical undermining of the first part: "'Ostensibly there is colour,

[1]E. Schrödinger, *Mind and Matter,* op. cit. p. 166

[2]E. Schrödinger, *William James lectures (c. 1954),* 3rd lecture, (in: E. Schrödinger, The interpretation of quantum mechanics (Dublin seminars 1949-1955 and other unpublished texts), op. cit. p. 145)

[3]ibid.

[4]E. Schrödinger, *Mind and Matter,* op. cit. p. 127-128; See paragraph 6-6 for further comments on Schrödinger's attack against dualism.

[5]E. Schrödinger, *Mind and Matter,* op. cit. p. 177

ostensibly sweetness, ostensibly bitterness, actually only atoms and the void'. To which the senses retort: 'poor intellect, do you hope to defeat us while from us you borrow your evidence? Your victory is your defeat'"[1].

Actually, this impossibility of fitting immediacy with objective knowledge, or to reconcile the ground of *qualia* with the allegedly grounded bodily objects, is a small part of a much more general difficulty which has to do with time. As Hegel noticed, the bare fact of immediate certainty is the poorest kind of truth. In order to say something, in order to communicate, one has to go beyond this Now which "(...) has already ceased to be when it is pointed out"[2]; one has to look for invariants, independent from any spatial, temporal or personal point of view. The dilemma thus takes the following form: on the one hand the resulting objective picture is essentially timeless, it is separated from the instability of the Now, and on the other hand one still has good reasons, in the framework of what Derrida calls the "metaphysic of presence", to think of our experience *here and now* as of the ground of the objective picture. In M. Merleau-Ponty's terms, "It is true that we should never talk about anything if we were limited to talking about those experiences with which we coincide, since speech is already a separation. (...) The fact remains, however, that the primary meaning of discourse is to be found in that text of experience it is trying to communicate"[3].

One cannot avoid taking this major paradox very seriously into account. Does it undermine phenomenalism so deeply that the whole undertaking has to be given up? Is it still reasonable to think that the bodily objects of everyday life are made out of these "private sense-data" whose concept emerges as an elaborate by-product of *public* language? And can one accept the idea that even though discourse *about* the sense-data requires so many mediations, still the sense-data themselves remain the *immediate* ground from which everything else stems?

A classical defense of phenomenalism consists in distinguishing between the "context of discovery" and the "context of justification". True, one could say, the context of discovery of what we call sense contents is a global background knowledge of familiar objects, and the sense contents are in this sense secondary. But in the context of justification the only acceptable criteria are the sense contents themselves, which thus have to be taken as primary[4]. This argument is quite weak, however, because it is tantamount to endorsing Mach's dilemma, and trying to establish a hierarchy between its two horns. The argument accepts both that sense-data arise from a "disintegration" of our experience of things, and that the things are made of sense-data by way of "connection", but it endows the second process with an epistemological priority over the first one.

[1]ibid.
[2]G.W.F. Hegel, *The phenomenology of Mind*, Trad. J.B. Baillie, Mac Millan, 1910, p.98
[3]M. Merleau-Ponty, *Phenomenology of perception*, Routledge & Kegan Paul, 1962, p. 337
[4]W. Sellars, "Phenomenalism", in: *Science perception and reality*, op. cit.

Another defence of phenomenalism, promoted by Goodman, is much more subtle. To begin with, according to Goodman[1], *even* if it were true *in principle* that one cannot grasp the elementary constituents which are required by phenomenalism, it would not be a good reason to give up the phenomenalist program of constitution of the world. Provided one focuses attention on the *structures* which arise from the constitution rather than on the constituent parts themselves, this program can be pursued without harm. And provided one considers the outcome of this program only as a rational reconstruction rather than as a revelation of the actual process of *genesis* of familiar objects, no confusion is to be feared. Actually, what makes Goodman's argument quite strong, much stronger than any previous defense of phenomenalism at any rate, is that he does not side with classical phenomenalists when metaphysical questions are at stake. He rather dissolves the metaphysical background of the dispute between phenomenalists and physicalists. Phenomenalists, everyday life physicalists, and modern physicalists, hold explicitly or implicitly that their entities (namely sense-data, mesoscopic bodies, or elementary particles respectively) are the basic stuff out of which the world is made. So, they try to display a one-way mode of translation of every proposition into basic propositions bearing on their favourite entities. But if one does *not* consider that there is a description which is more fundamental than another, then it follows that the task is rather to promote mutual intertranslatability than one-way translations. "The perceptual is no more a rather distorted version of the physical facts than the physical is a highly artificial version of the perceptual facts"[2]. For according to Goodman, there are no ultimate, version-independent, facts. In addition, if one is to remain consistent, it cannot even be said that Goodman's versions are versions *of the world,* because there is no version-independent world. Goodman's versions *are* worlds.

It is very likely that the phenomenalist program can only be saved at this cost, namely by renouncing any claim to metaphysical priority for sensa-data or protocol sentences over other entities or sentences. For if no privileged metaphysical status is ascribed to them, one has no longer to rely on the slippery argument of their *immediacy* in order to promote the idea that they are to be taken (or at least that they *can* be taken) as the basic entities. It is enough to point out that they hold the role of nodal points in a certain network of entities and relations which constitute one acceptable Goodman's "version". In addition, if no privileged metaphysical status is ascribed to sense-data or protocol sentences, the previous argument about their derivatedness does not weaken any longer their being ascribed a logical priority in a certain version. The circumstance that sense-data arise *de facto* as remote and distorted by-products of the (large scale) physicalist description of the world, then

[1]N. Goodman, "The significance of Der logische Aufbau der Welt", in: Schilpp (ed.), *The philosophy of Rudolf Carnap,* op. cit.
[2]N. Goodman, *Ways of worldmaking,* Hackett Publishing Company, 1978, p. 93

appears as a mere indication of their particular position in a system of intertranslation rules within which *we,* speakers of the everyday language or specialists of macro-physics, happen to occupy a biased position, and not as a compelling argument against their being ascribed a fundamental status in some Goodman's "version" or language game.

Furthermore, arguing away the "elements" of the phenomenalist language game by pointing out that they depart considerably from the picture of the world promoted by modern *micro*-physics would be quite unfair. It would sound as if the phenomenalist language game held a very special position, out of the main road represented by modern physics. But this is not true. If one had to point towards a language game, or a set of language games, which really represent the main road and which play a central role in our lives, it would not be the language game of micro-physics. It would be *ordinary speech;* or ordinary speech associated with those elements of (classical) physics which happen not to disrupt its structure. For, as Bohr pointed out, the language game of modern physicists happens to be dependent (and *must* be dependent, for the sake of unambiguous communication) on a description of apparatuses and pointers which makes use of ordinary language and of classical physics. In this perspective, the language game of modern specialists of micro-physics appears to be somehow grounded on an underlying layer of ordinary speech, just as much as the phenomenalist language game. In addition, the (non-mathematical part of the) language game of modern physicists appears as far apart from ordinary speech, as complicated and as clumsy when compared with it, as the phenomenalist language game. Thus, both language games, that of modern physics and that of phenomenalism, have many features in common. Both language games start from ordinary speech. Both language games manipulate entities which are far apart from the "things" of everyday life and speech: in modern physics, the entities are mathematical symbols such as operators and vectors in a Hilbert space, or strange half-corpuscular half-undulatory non-individual "particles"; and in the framework of phenomenalism, the entities are sensations, percepts, or 'events'. Finally, both language games claim to be able to account retroactively for the appearance of a macro-world of "things" by means of their entities.

But there is also a major difference. Modern physicists are sometimes trying desperately to recover some sort of isomorphism between their secondary language game and the forms which are presupposed by the primary language game of ordinary speech (even at the cost of introducing paradoxes). They are often tempted by minimizing the distance between their entities and the kind of objects (namely material bodies) which are manipulated in everyday life, and which were idealized by classical physics. They tend not to content themselves with *starting* from ordinary speech for the sake of stating experimental conditions; instead, they look for some sort of residual isomorphism between the entities which are the *final* outcome of their theoretical investigations and the typical presuppositions of ordinary speech. Let us give two examples. Heisenberg's "uncertainty" relations were greeted by many physicists in 1927, because they allow one to determine the range within which it is *still* possible to speak *approximately as if* the basic elements of the world were corpuscles with a trajectory. As for modern hidden variable theories, they represent an attempt at modifying drastically the characteristics of classical corpuscles, rather than renouncing the idea that the world is made of corpuscle-like entities.

By contrast, phenomenalists deliberately accept (and even amplify) the distinction between their basic "elements" (sensations or percepts) and the "things" of everyday life.

Thus, even if the phenomenalist program cannot be fully worked out, even if it stumbles over many difficulties, it retains a non-negligible capacity to awaken both philosophers and some modern physicists from their excessive fascination for the familiar world as it is shaped by the presuppositions of ordinary speech and actions. It does so by means of two of its features:

(i) The accepted distance between the basic entities of phenomenalism and everyday modes of speech suggests that it is at least conceivable to recover the familiar world of "things" by using "elements" which bear no resemblance *at all* with these things. Transposed to the problem of microphysics, this is tantamount to say that the appearance of a macroscopic world with material bodies in it can perfectly be accounted for by entities which are very far apart from corpuscules (say by an appropriate manipulation of some global wave-function). This simple remark may prevent physicists from being trapped into traditional linguistic forms and representations which are mainly adapted to our mesoscopic environment, and which must therefore be taken *only* as a *departure point,* not as a *final* result of the microphysical investigations. The characteristics of the microscopic world call either for clearly stated limitations of the field of applicability of the traditional linguistic forms and practices (Bohr), or for new categorial forms (quantum logic), or even for a completely new set of entities (Schrödinger).

(ii) The strategy of phenomenalism consists in deflecting attention from its familiar focus. Whereas the "natural attitude" directs attention towards the surrounding material objects, phenomenalism is associated with a kind of impressionist bracketing of this usual outlook. Attention is now redirected towards the immanence of pure phenomena, and the familiar things are supposed to be nothing more than "logical constructions"[1] out of the phenomena. This procedure tends to promote a very critical attitude concerning the intentionally aimed at objects. It precludes the usual hypostasis of "things" into "substances", and then into underlying "substrata". Similarly, redirecting our attention from the objects of experimental investigation to the instruments and procedures of this

[1]B. Russell, Mysticism and logic, op. cit. p. 149

investigation, can represent a good preliminary move in order to reconsider critically the nature of the hypothetical objects of microphysics.

As I. Berlin pointed out, one of the best reasons to defend phenomenalism is that it is "(...) recommended for its therapeutic properties as an antidote to metaphysical hanckering after non-sensible substrata"[1]. These therapeutic virtues would be especially useful in modern physics, for physicists are sometimes too much in a hurry when they try to figure out the kind of objects whose properties are supposed to be made manifest by the available experimental phenomena. True, as we have noticed, phenomenalism is undermined by major difficulties. But the efficiency of the therapy doesn't even require that phenomenalism be a completely credible alternative to the material object language. A therapy is not a comprehensive substitute for life; it is just a partial substitute or a temporary assistance. Once the therapy has been successful enough, one can usually drop it.

As we shall see in subsequent paragraphs, Schrödinger's combination of phenomenalist premises and realist attitudes can ultimately be justified this way. Even though Schrödinger did not renounce so easily the metaphysical content of phenomenalism in his philosophical writings, he behaved in his scientific work as if he was *only* considering phenomenalism as a tool for freeing the physicist's intentional gaze from the spell of traditional systems of entities, and redirecting it towards new systems of entities. His choice of a therapeutic metaphor in similar circumstances clearly support this interpretation: "(Philosophical positivism) is a salutary antidote against the rashness with which scientists are prone to believe that they have understood a phenomenon, when they have really only grasped the facts by describing them"[2].

5-5 *The construction of objects and the unconscious*

Let us now follow Goodman's suggestion and focus our attention on both the strategies of constitution and the resulting structures, rather than on the constitutive "elements".

Schrödinger uses a very powerful metaphor in order to work out the phenomenalist strategy of constitution of the objects of daily life. He says that the same "building material", or the same "bricks" (namely the "elements") constitute the world and the mind. According to their *ordering* they constitute various material objects, or various mental states or events. But, even though there are bricks on the one side and buildings (including the material objects) on the other side, there are no *conscious* builders or architects. Schrödinger considers that the processes of association and expectations which lead to the construction of a "thing"

[1] I. Berlin, *Concepts and Categories*, Oxford University Press, 1980, p. 34

[2] E. Schrödinger, *Nature and the Greeks*, op. cit. p. 89

out of the elementary building material are mostly *unconscious:* "We expect certain tactile sensations if we touch (the fruit basket), sensations of taste if we bite through a fruit, a crackling of the basket if we press it together. We are usually not aware of all these expectations; we focus them unconsciously into what we call a fruit basket which really exist"[1]. By holding this position, he was not arguing for any original view. He was just developing systematically a conception which originated in classical phenomenalism. J.S. Mill describes "irresistible" associations of one sensation to another prompted by "habit"[2], and E. Mach evokes "(...) the partly instinctive, partly voluntary and conscious, economy of mental presentation and designation (...)"[3] which makes us speak as if there were permanent entities called bodies, rather than mere sensations.

But Schrödinger does not content himself with calling in the *unconscious* when the constitution of bodies of everyday life is at stake. He provides us with a kind of genetic account of how an adult has come to associate present elements with expectations in order to form the idea of a body. According to Schrödinger, the ordinary notion of real objects is a result of our striving for invariance, which has become so natural that we are not even surprised by the fact that "(...) we continually 'see' features that we do not see"; that for instance "Looking at my table lamp I see that its socle is a black, shining square, though it actually appears as an irregular quadrangle, not altogether black, because one part reflects the bright sky"[4]. The systematic research of invariants is "natural" to us adults because "(...) in its most important early stages we are not aware of it"; because it is the "(...) continuation of a behaviour that we have adopted from earliest babyhood, have developed to high perfection, and use every awake minute of our life for orientating ourselves in our daily surrounding"[5]. The invariant notions of existing things were formed "(...) by experience, by the experimental science of the baby and the small child"[6].

It is especially important to notice that Schrödinger does not try to find any conscious equivalent of the kind of procedure which is the adult end-product of the baby's striving towards invariance. Striving towards invariance is *not* an *intellectual procedure* which we just happen to be forgetful of. As we noticed above, Schrödinger says it is a *behaviour;* It must be called "(...) a behaviour, not a method (full awareness sets in only in the scientific stage)"[7]. Schrödinger even insists that we must not be impressed by the circumstance that he himself describes the genesis of our adult conception of the world in adult terms. The very fact we describe

[1]E. Schrödinger, "Conceptual models in physics and their philosophical value" loc. cit.

[2]J.S. Mill, *An examination of Sir W. Hamilton's philosophy,* Longman & Green, 1865, chapter XI

[3]E. Mach, *The analysis of sensations,* op. cit.

[4]E. Schrödinger, *Science, philosophy and the sensates,* in: E. Schrödinger, *The interpretation of quantum mechanics (Dublin seminars 1949-1955 and other unpublished texts)* op. cit. p. 147

[5]ibid. p. 146

[6]ibid. p. 149

[7]ibid. p. 146

something "(...) creates the wrong impression as if we were speaking of a conscious process that involves a lot of reasoning, while it is a behaviour that we develop spontaneously and inadvertently"[1].

To summarize, the point Schrödinger is so strongly emphasizing is that what we first acquire during our early childhood is not a *knowing that,* but a *knowing how.* In the controversy about the status of the background of practices of our adult speech and actions[2], which took place between Husserl and the cognitivists on the one hand, and Wittgenstein, Heidegger and Searle on the other hand, Schrödinger clearly sides with the latter. For he considers that this background is non-representational (it is just a *behaviour,* not a method), and that moreover it was not obtained by representational methods at an early stage of the development of human beings (Schrödinger says it was *not* a conscious process, it did *not* involve reasoning, it developed "inadvertently").

In his *Mind and Matter,* Schrödinger develops his concept of *knowing how* quite extensively, tracing it back to both our phylogenetic and ontogenetic past. About the phylogenetic component of our *knowing how,* he expresses his agreement with some speculations of R. Semon, a biologist and evolutionist of the beginning of the century whose central concept is referred to as *the Mneme.* About the ontogenetic part, Schrödinger gives some indications of his own. He especially insists in *Mind and Matter* that the acquisition of *knowing how* is not *only* a characteristic of early childhood, but that it can also happen during our adult life. During adult life, consciousness may intervene during a short period, and then withdraw when the resulting *knowing how* becomes successful enough: "consciousness is associated with the learning of the living substance; its *knowing how (Können)* is unconscious"[3]. Or, in other terms: "(...) consciousness is the tutor who supervises the education of the living substance, but leaves his pupil alone to deal with all those tasks for which he is sufficiently trained"[4].

Such a genetic account of the notion of "thing" can of course be criticized according to two lines of argument at least. The first line of argument goes against the old-fashioned picture of a baby who receives passively a host of flickering sensations, before he progressively shapes out invariants and behaves accordingly. There have been considerable controversies in recent years about the original findings of Piaget, according to whom the notion of a permanent spatio-temporal object is a late acquisition of the baby rather than something which is innate or acquired very early. But at any rate, nobody denies that newborns show something one can but call an *intentional behaviour*[5] almost immediately after their birth. The most obvious and the most primitive of this

[1]ibid. p. 146

[2]See introduction by H.L. Dreyfus in: H.L. Dreyfus (ed.), *Husserl, intentionality and cognitive science,* MIT Press, 1982

[3]E. Schrödinger, *Mind and Matter,* op. cit. p.105

[4]ibid. p. 103

[5]See D. Dennett, *Consciousness explained,* Penguin, 1993

intentional behaviours is the newborn's creeping and making his head oscillate towards her mother's breast. In view of this very early onset of intentional behaviours, it becomes quite unlikely that the intentional unity of objects, as it is implied by our adult language and actions, results from former organization of passively received sensations into approximately invariant complexes. Nothing supports the idea that the intentional unity of everyday life objects is a late synthesis making us oblivious of an allegedly passive origin of knowledge. Its true forerunners are likely to be layers of less and less differentiated intentional unities, the most primitive one being the maternal source of satisfaction of the needs. Instead of a transition from passive reception to directed activities oriented by recognition, the ontogenesis of objects would then have to be construed as iterative changes of the aim(s) of intentional directedness. This series of changes would reach social stability as soon as linguistic abilities are acquired.

The idea that the evolution of early epistemic abilities involve successive changes of intentional directedness with a social component, rather than individual transition from non-intentional receptivity to intentional attitudes, fits quite well with Schrödinger's own evaluation of the significance of the quantum revolution. We have already mentioned that, as a physicist, he usually behaved as if he considered that the phenomenalist outlook was nothing more than a useful intermediate therapy able to promote a socially acceptable transition from one system of intentional entities to another. In his philosophical work, and especially in his account of early childhood, however, Schrödinger stayed close to the classical empiricist tradition; he stuck to the idea that babies first collect elementary data and that they begin their striving for invariance *alone*. But he then also recognized the importance of a final social step in our forming the invariants we call "things": "(Forming invariants) begins within the sensory complex of the individual, but very soon extends to forming mutual invariants, in common to the individuals that are in social contact"[1]. He further insisted on the social components of our concept of a real world surrounding us, when he attacked the idea of adopting positivism, with its *methodological solipsist* components, as a *permanent* attitude: "The conception of a world that really exists is based on there being a far-reaching common experience of many individuals, in fact of all individuals who come into the same or a similar situation with respect to the object concerned. (...) This proper basis of reality is set aside as trivial by the positivists when they always want to speak only in the form: if 'I' make a measurement then 'I' 'find' this or that (and that is to be the only reality)"[2].

[1]E. Schrödinger, *Science, philosophy and the sensates*, in: E. Schrödinger, *The interpretation of quantum mechanics (Dublin seminars 1949-1955 and other unpublished texts)* op. cit. p. 146

[2]E. Schrödinger to A. Einstein, November 18 1950, in: K. Przibram, *Letters on wave mechanics*, (Tr. M.J. Klein), op. cit. p. 37

Schrödinger held two slightly different attitudes towards phenomenalism, according to whether he was dealing with the objects of physics or with the things of everyday life. But in the light of the previous distinction between individual invariants and social invariants, it appears that these two attitudes were perfectly compatible. In his genetic account of the "things", which involve both individual and social processes, he endowed phenomenalism with a foundational status. And in his reflections about modern physics, which only bear on socially accepted invariants, he confined phenomenalism in the role of a temporary "antidote" helping one to cure epistemological illusions or ingrained ontological habits.

The second line of argument one may develop against Schrödinger's empiricist ontogenetic account of the notion of "thing" is that it seems to *presuppose* the "popular view" of the world. For instance, when Schrödinger writes about the baby's learning how to to disregard perspective distortions and variations of illumination in order to unite the aspects into the idea of a permanent "thing", he adds: "it is astonishing how quickly the baby does learn this with such emotionally stressed objects as the milk-bottle, the comforter, the rattle, and his own limbs, particularly the interesting toes"[1]. Thus everything looks as if we had the baby on the one side, the objects on the other side, and a causal relation between the two. The objects cause sensations in the baby's mind, and the baby learn how to organize them into permanent entities. But the ambition of a truly phenomenalist account is to display the *genesis* of this popular view of the world which involves a strong dualist commitment. How can one take the phenomenalist seriously if he takes the popular view as a basis of his genetic account? Here again, the whole undertaking is threatened by circularity. But Schrödinger was perfectly aware of this risk. He was aware that his own exposition would prompt irresistibly the reader to think that the baby is just *discovering* his environment of things; that the baby is not *building* it out of the bare material of sensory complexes. He thus added the following corrective remark at the end of his genetic account: it is "(...) *not* (relevant) to say: oh well, that is just only the way by which the child learns 'what the world really is like'. The latter is true, but trivial. For by 'what the world really is like' we mean the notion that we, the ordinary man or woman, have formed when we were small. That any small child going through similar experiences reaches the same aspect, is trivial and does not clinch the inevitability of this aspect"[2]. According to these sentences, the difficulty is methological rather than metaphysical. We are trying to catch the genesis of our adult view of the world from the standpoint of adult life and speech. No wonder that we use the categories which pertain to this mature view when we try to account for what happened before it.

[1]ibid.
[2]ibid.

The general lesson to be learnt from Schrödinger's exposition is therefore that the task of revealing the initial steps of our acquiring the current adult view of the world is at the same time hopeless and indispensible. On the one hand it is hopeless (in Wittgenstein's *Tractatus* sense) because we are trying to account for a background *knowing how* in terms of a mature *knowing that;* because we are trying to speak about what comes before speech; and because we are attempting to explain the *constitution* of things whereas we are caught up into a universe of discourse and actions which already presupposes the said things. On the other hand it is indispensible (as a therapy), because if we do not undertake it, we are likely to remain so completely trapped into a system of prejudices which is only acceptable at the familiar mesoscopic scale, that we are unable to cope with the new situation created by modern physics.

Now Schrödinger's strategy, as he sketched it in the last quoted sentences, consisted in alleviating the hopelessness of the genetic account of the "thing" by transferring the burden of proof and the charge of circularity onto the upholders of the "popular view". For, he says, when it is claimed that the baby learns 'what the world really is like', it is only meant that the baby learns the notion that other people have formed when they were babies. At no point does one go beyond the domain of accessible knowledge, and reach some transcendent realm. Even the fact that any reasonably fit baby finally reaches the same invariants, 'does not clinch the inevitability' of these invariants. It does not determine them as securely, at any rate, as if they were *things-in-themselves* that are completely independent from any process of individual and social construction. If anything, our final agreement about a universe of "things" proves the inevitability of a certain kind of *relationship* between ourselves and the world we partake of, *given our phylogenetic, ontogenetic, and cultural background;* it does not prove the existence of an *absolute* world-structure.

In other terms, the widespread agreement about the things of everyday life should not be taken as a sign of their independent existence, but rather as an indication that no question about their existence can be raised meaningfully within a form of life which presupposes their availability. True, it is not impossible to *state* such a question. But when it is stated, it must not be taken at face value. One should not consider that it is directed towards the things, but rather towards the very background of human life and communication. According to Carnap[1], this kind of question is by no means a theoretical question about the things, but rather a practical question about the opportunity of altering the form of life which takes them for granted. In Carnap's classification, a question about the existence of things is an "external" question bearing on the appositeness of the

[1] R. Carnap, "Empiricism, Semantics, and Ontology", in: R. Carnap, *Meaning and Necessity,* The University of Chicago Press, 1956

background, not an "internal" question which only makes sense against this background when it is accepted.

5-6 *The "thing" and the future*

Using the word "construction" when the early constitution of the familiar world of "things" is at stake, suggests that the outcome of this activity is somehow static; that it is a well rounded *construct.* After all, in the very first sentence of *Mind and Matter,* 'the world which surrounds us' was identified with a "construct". But, as Schrödinger soon recognized, the way real objects of everyday life present themselves to us does not fit at all with the representation that they are static end-products of a process of construction. Perceiving a real object, according to Schrödinger, is tantamount to associating unconsciously with it an indefinite number of *expectations* concerning *future* explorations or experiments. Besides the "momentary sensational core"[1], there is, associated to it, the *anticipation* of "occasional occurrence"[2] of further sense-perceptions. The present sensional core is extrapolated by "images and expectations"[3] which refer to the future. The visual picture of a fruit basket, for instance, projects itself into a host of future possibilities of tactile, olfactive, auditive and gustative, sensations.

True, this remarkable unfolding of protentions based on an unconscious work of imagination is characterized by *order.* According to Schrödinger, we anticipate the occasional occurrence of virtual sensations *"(...) in definite relationship to one another"* [4]. For, as M. Merleau-Ponty puts it: "The unity of the object is based on the foreshadowing of an imminent order which is about to spring upon us a reply to questions merely latent in the landscape"[5]. In the Kantian tradition, order (especially causal order) is indeed a crucial ingredient for the constitution of objects, for it is a condition of possibility of objective knowledge in general as opposed to dreams. But besides this importance of order, another distinctive feature of the anticipations which are involved in the perceptions of familiar objects is their *predictive openness.* The expectations do not pretend to be exhaustive, let alone certain. The order they suppose is therefore only programmatic. No wonder that the key-expression of Schrödinger's perspectivist account of the real objects of everyday life is *"And so on":* "And so on. There is no end to enumerating all the potential percepts and sensations on my and on your side that are included in my speaking of this arm as of an objective feature of the 'world around us'"[6]. Even the primeval "event", i.e. the inextricable perceptual "complex" whose aggregation with other events into a "string" is constitutive of the material

[1]E. Schrödinger, "What is an elementary particle?" loc. cit.
[2]E. Schrödinger, "Conceptual models in physics and their philosophical value" loc. cit.
[3]E. Schrödinger, "What is an elementary particle?" loc. cit.
[4]E. Schrödinger, "Conceptual models in physics and their philosophical value" loc. cit.
[5]M. Merleau-Ponty, *Phenomenology of perception,* op. cit. p. 17
[6]E. Schrödinger, *Nature and the Greeks,* op. cit. p. 93

bodies, encompasses endless "expectations". The constituent (the "event") bears the same mark of incompleteness as the constituted object itself.

By insisting so much on expectations, on anticipations, and more generally on incompleteness, when the constitution of the real objects of everyday life is at stake, Schödinger manifested his critical appraisal of the empiricist and positivist tradition he was admittedly relying on. This aspect of his conception of the "thing" of everyday life indeed makes it definitely closer to Husserl's and Heidegger's phenomenology than to Mach's and Russell's phenomenalist constructivism, even though Schrödinger never referred explicitly to phenomenology. As a matter of fact, the emphasis on the future, on hypothetical statements about the future rather than on counterfactual statements about the present, is one of the main characteristics which differentiate phenomenology from most versions of phenomenalist constructivism. The expression "and so on" or "and so forth", associated with the thesis that the idea of a "thing" implies endless possibilities, is typical of Husserl's statement of his conception of the material bodies of everyday life in the last paragraphs of *Ideas:* "We grasp (the idea of a Thing) in the free process of running through the possibilities, in the consciousness of the limitlessness of the development of intuitions of the same order. We thus grasp at first the idea of the Thing empty of all intuitional content, and of this individual thing as something which is given 'just so far' as the agreeing intuition 'reaches', but remains at the same time determinable *'in infinitum'*. The 'and so forth' is an absolutely indispensible phase in the thing-noema, and we have a clear insight of its necessity". A few lines below, Husserl insists that: "All components of the Thing-idea are themselves ideas, each implying the 'and so forth' of 'endless' possibilities"[1].

When Schrödinger's uses a metaphor, namely that of the building material or of the "bricks" out of which the "world which surrounds us" is made, he comes quite close to the phenomenalist constructivist doctrine. But when he leaves aside metaphors and tries to describe what happens when we perceive a familiar object, he speaks of indefinite progression of anticipations, of presumptive synthesis of aspects into material objects, and of aspects which already convey expectations. In such descriptions, he reaches almost the same conclusions as the phenomenologists. In particular, as we suggested in former paragraphs, his events (which encompass anticipations) are definitely much closer to what phenomenologists call the *profiles* surrounded by a *"horizon"*[2] *of expectations,* than to the bare "elements" of the empiricist and positivist tradition.

[1]E. Husserl, *Ideas,* op. cit. §149

[2]E. Husserl, *Ideas,* op. cit. §27: "What is perceived (...) is partly pervaded, partly girt about with a dimly apprehended depth or fringe of indeterminate reality". It is this fringe that is called "horizon" (e.g. at the end of §47); See also M.Merleau-Ponty, *Phenomenology of perception,* op. cit. p. 69.
H. Kuhn ("The phenomenological concept of 'horizon'", in: D. Farber, "Philosophical essays in memory of E. Husserl", op. cit.) gave the following short definition of Husserl's concept of horizon: "'horizon' is but another name for the totality of serial potentialities involved in the object as noema".

It is only when Schrödinger attempts to account for the objects of scientific knowledge, not only the "things" of everyday life, that he feels compelled to come again much closer to phenomenalist constructivism than to phenomenology; much closer to a conception of material objects which identify them with permanent orderings of presently given elements and present counterfactual possibilities than to their phenomenological description as profiles surrounded by ever-developing horizons of expected possibilities. It is when the objects of scientific knowledge, and especially the objects of physics, are at stake, that he has to take seriously the phenomenalist identification of real objects with static constructs made out of an infinite set of possible percepts and/or possible experimental outcomes.

Through its extensive use of mathematics, physics has indeed undergone an unprecedented reversal of epistemological hierarchies. Perceiving a "thing" in our direct environment usually means feeling a contrast between the vividness of the presently apprehended "core", aspect, or profile, and the vagueness of the horizon of expectations which surrounds it. By contrast, most objects of physics cannot be perceived directly; furthermore, practical circumstances sometimes prevent one from apprehending them indirectly through the use of instruments; and finally there are even some objects whose allegedly intrinsic features are *in principle* inaccessible to any direct or indirect investigation (see e.g. Bohm's particles whose trajectories cannot be investigated experimentally without being modified). Accordingly, in the physicist's notion of his objects, the role of possibilities exceeds by far the role of any kind of perceptual or experimental actuality. Whereas the *permanence* of the "thing" of everyday life was somehow a substitute for its perceptual *presence*, be it in memory or in imagination, the permanence of the objects of physics tends to rely less and less on presence (or on actuality) and more and more on formal interpolations between possibilities.

Schrödinger fully recognized this tendency towards abstraction, namely towards pushing the present *actuality* to the margins of scientific knowledge. According to him, the basic reason for this process is *objectivation*. For what objectivation tends to exclude from the picture of the world, or to confine in the neighbouring realm of the "Mind", is precisely the "sensational core" which is the central component of the *actual* perception of familiar things. Accordingly, objective knowledge involves a progressive departure from the perceptual order of priorities, which puts actuality above possibility, and apprehended aspects above expectations.

As long as the "things" of everyday life were concerned, it was still acceptable to make a kind of compromise between actuality and (expected) possibilities. But in the domain of fully constituted objective knowledge (in physics, for instance), possibilities became overwhelmingly more important than actuality, and anticipations turned out to be predominant over immediate apprehension. "As our familiarity with a piece of matter grows, and in particular as we approach its scientific aspect, the range of expectations in regard to it widens, eventually to include all the information science has ascertained, e.g. melting point, solubility, electric conductivity, density, chemical and crystalline structure, and so on. At the same time, the momentary sensational core recedes in relevance (...)"[1]. The progressive regression of the momentary sensational core is all the more significant since it prepares its (almost) complete disappearance. The last (ideal) stage of the process of objectivation has been reached when one can construe present perceptions as a contingent, and possibly very indirect, mean of ascertaining the permanence of objects, instead of seeing permanence as a daring extrapolation of presence. As Schrödinger puts it: "After a certain wealth of association has come to outshine the core of sensates, the latter is no longer needed to keep the complex together. It persists even when the contact of our senses with the object temporarily ceases. And more than that: the complex is latently conserved even when the whole string is interrupted by our turning away from the object to others and forgetting all about it"[2]. This process of increasing autonomy of the structural "complexes", which is characteristic of objective knowledge in general, has been amplified by science and formalized through a set of conservation laws: "Science has substantiated it; though the appearance in bulk may change, the ultimate constituents of matter do not"[3].

5-7 *Possibilities and infinities*

As we suggested in the former paragraph, one has to fulfill a crucial condition in order to weaken (or to abolish) the dependence of the "complex" towards the "momentary sensational core". This condition is the following. The mere *endlessness* which characterizes the expectations or the "horizons" associated to perception, has to be replaced by an ordered *infinity* of co-existent aspects ideally encompassed by the "complex". For whereas in the phenomenological horizon-structure the sensational core of the percept still retains a privilege (namely that of being the departure point of expectations), it is put exactly on the same footing as the possible aspects when it is integrated with them in an ordered set of co-existent aspects. Even its complete absence (the set of coexistent aspects being composed only of *possible* aspects as long as no actual perception occurs) would make little difference for the complex.

There is a Kantian approximate equivalent of this change: it is the transition from the temporal synthesis of transcendental imagination to the (essentially timeless) unity of apperception which is provided by pure understanding. There is also a Husserlian equivalent of the transition from endless expectations to infinite numbers of aspects: it was referred to as

[1] E. Schrödinger, "What is an elementary particle?" loc. cit.
[2] ibid.
[3] ibid.

the *substruction,* or *the infinitization,* and was considered by Husserl, in his *Krisis der europaïschen Wissenschaften,* as the most typical and at the same time most dangerous move of modern science. Husserl denounced its growing influence in our cultural outlook. For then, he said, we are seriously at risk of becoming forgetful of the fact that, in spite of its remarkable developments, science ultimately originates in the world-life of everyday *finite anticipations.*

Infinitization, namely replacement of expectations by actual infinities, was fully worked out by Russell between 1913 and 1915, in a series of texts[1] which were quite familiar to Schrödinger. He prepared this replacement by introducing the concept of *sensibile* (plural: *sensibilia).* In his definition, *sensibilia* are entities "(...) which have the same metaphysical and physical status as sense-data, without necessarily being data to any mind"[2]. Just as sense-data give rise to sensations whenever a subject is aware of them, sensibilia would give rise to sensations *if* a subject were aware of them. They are *possibilities* of promoting sensations. With this kind of entity, it becomes conceivable to overcome the essential incompleteness of any perceptual apprehension of material bodies, and to define these bodies in terms of a complete set of co-existent aspects: "The 'thing' of common sense may in fact be identified with the whole class of its appearances - where, however, we must include among appearances not only those which are actual sense-data, but also those 'sensibilia', if any, which on grounds of continuity and resemblance are to be regarded as belonging to the same system of appearances, although there happen to be no observers to whom they are data"[3].

But of course, this is only one possible version of the phenomenalist conception of the "thing" of everyday life. In his version, Russell sides unambiguously with Mill, against Berkeley and Mach. Except for the interesting distinction he makes between (objective) sense-data and (subjective) sensations, Russell's concept of *sensibile* is remarkably close to Mill's idea of "(...) possible sensations; sensations which we are not feeling at the present moment, but which we might feel, and should feel if certain conditions were present (...)"[4]. And his identification of the 'thing' to a class of possible appearances is quite similar to Mill's definition of matter as "permanent possibility of sensation"[5]. By contrast, E. Mach strongly criticised "Mill's possibilities of sensations" and replaced the notion of an object made of such possibilities by the notion of an object made of "functional relations of the (sensational) elements"[6].

Russell's choice to define objects in terms of classes of *sensibilia,* rather than functional relations of sensations, as Mach proposes, is obviously

[1]B. Russell, *Our knowledge of the external world,* Open Court, 1914; "The relation of sense-data to physics", in: *Mysticism and logic,* Unwin Paperbacks, 1986
[2]B. Russell, "The relation of sense-data to physics", in: *Mysticism and logic,* op. cit.
[3]ibid.
[4]J.S. Mill, *An examination of Sir W. Hamilton's philosophy,* op. cit. chapter XI
[5]ibid.
[6]E. Mach, *Analysis of sensations,* op. cit. p. 363

related with his reflection on Cantor's theory of classes. The major analogy he uses in order to illustrate his constructivist conception of the "thing" of everyday life is accordingly borrowed from mathematics. Russell notices that, instead of inferring irrationals as the supposed limit of series of rationals which have no *rational* limit, we nowadays define irrationals as a certain *class* of ratios[1]. But this involves replacement of an endless development, leaving "(...) the existence of irrational merely optative", by a class which involve an *actual infinity* of ratios. A corrective is that such classes need not be ascribed any metaphysical status, but "(...) can be regarded as symbolically constructed fiction"[2]. Similarly, in the case of our conception of the "thing", endless developments of anticipations should be replaced by symbolically constructed classes encompassing actual infinities of sensibilia. The consequence of such constructivist definitions is that one drops any motivation for associating an element of transcendence to the defined entities. Whereas the irrationals seemed to be something beyond the rationals when they were defined as inaccessible limits, they now appear as unusually large classes of rationals. The difference between irrationals and rationals reduces to the difference between a class and its members. Similarly, the difference between the things and its appearances reduces to the difference between a class and its members (the sensibilia). No equivalent of the transcendent substance underlying its appearances is left; not even something like Husserl's purported "X"[3].

Russell's conception however raises at least two difficulties.

The first difficulty is about the concept of *construction*. The word "construction" lies just at the intersection of two domains of meanings. One of them is more appropriate to deal with the objects of scientific investigation; the other one points rather towards more familiar operations, which could be applied to our conception of the "thing" of everyday life. To begin with, the connotations of overt (intellectual) *activity,* which are associated with the word "construction", clearly makes it more apt to refer to what scientists do in their laboratories, than to what a man-in-the-street does when he perceives and recognizes a material object of his environment. The idea that perception involves some kind of explicit intellectual process has been the target of many cogent criticisms, from the *Gestalt* psychologists to Wittgenstein. True, there is still a possibility to rescue the applicability of the word "construction", even in the case of the "thing" of everyday life. It is to adopt an ontogenetic approach. For then, one might say, the notion of "thing" can be considered as a product of a partly unconscious process of construction taking place during early childhood. And, one might add, this initial unconscious process of construction is unconsciously recapitulated by the adult each time perception of a familiar object occurs. But here another

[1]B. Russell, "The relation of sense-data to physics", in: *Mysticism and logic,* op. cit.
[2]ibid.
[3]E. Husserl, *Ideas,* op. cit. §135

problem arises. The kind of "constructions" Russell was pointing towards were "logical" or "symbolic" constructions, not concrete constructions made brick after brick, element after element, like those which the passive pre-logical baby of the empiricists is likely to perform. If one insists too much on ascribing a concrete content to the concept of construction, in such a way that it becomes able to encompass the alleged empirical and unconscious genesis of the notion of "thing", then this concept is likely to lose part of its ability to accomodate the case of scientific objects. Indeed, any construction performed concretely by a finite being, either consciously or unconsciously, is bound to be finite and to fall short of the requirements of a scientific construct. As Goodman noticed in his discussion of Carnap's constructivism, many philosophers have accused this doctrine of being doomed to finiteness. And, if it is doomed to finiteness, it cannot really be completed in such a way that it accounts for the infinite world of mathematics and mathematical physics[1]. Carnap's constructs, which are both incomplete and uncompletable, are very unlike Husserl's noema which are incomplete but completable and even revisable.

To summarize, either the concept of "construction" is ascribed a purely intellectual, symbolic, and abstract, meaning, and its use must be restricted to the case of scientific objects; or one tries to endow this concept with a more concrete, genetic, and finite, meaning, in order to accomodate in it the empiricist conception of the "thing" of everyday life, and it is then unlikely to be adapted to scientific objects any longer.

Phenomenalist constructivism has been an attempt to extend the process of a symbolic construction of the objects of scientific enquiry to the entities (namely the "things") any human activity, including experimental science, happens to presuppose. But by undergoing such an extension, it has weakened its ability to account for scientific objects without reaching a fully convincing position about the "things" of everyday life. We shall see in paragraph 5-8 how Schrödinger exploited this ambivalence, in order to promote the "therapeutic" efficiency of phenomenalist constructivism in the interpretative problems of quantum mechanics.

We now come to the second difficulty which hinders Russell's conception of the "thing". How are we to define the class of appearances which has been identified with the "thing of common sense"? Are we to choose an extensional or an intensional definition? The most straighforward, though obviously circular, method, would be to state that the thing is the class of *its* appearances, and that appearances belong to the class if they are appearances *of* the object. Schrödinger uses essentially this method when he writes in *Nature and the Greeks* that a material object "(...) is composed not only of my own sensations of *it,* but also of the imagined sensations I would have in turning *it* round, moving *it,* looking at *it* from all different angles"[2]. But of course, this procedure is utterly unsatisfactory. In order to avoid self-reference, the general

[1]N. Goodman, "The significance of Der logische Aufbau der Welt", loc. cit.
[2]E. Schrödinger, *Nature and the Greeks,* op. cit. p. 92; my italics

strategy one could follow amounts to providing the relevant class with an extensional definition. But a simple enumeration of the appearances which constitute the class will not do. For the class is generally infinite. We thus arrive at what is perhaps the last possibility of rescuing Russell's constructivism. It consists in indicating a rule which enables one to *generate* all the possible aspects belonging to the class, given one aspect (or a sub-set of aspects). The object can then be conveniently identified with this generative law rather than with the generated class. This is certainly what Russell had in mind when he underlined that his definition of an object involved "logical" or "symbolic" constructions, not concrete constructions made element after element. This is also the kind of conception of the "thing" M. Merleau-Ponty overtly advocated: "Our perception ends in objects, and the object once constituted appears as the reason for all the experiences of it which we have had or could have. For example I see the next door house from a certain angle (...). The house *itself* is none of these appearances; it is, as Leibniz said, the 'geometral' of these perspectives and of all possible perspectives, that is the perspectiveless position from which all can be derived, the house seen from nowhere"[1]. Then, according to this view, the "thing" is not an appearance, it is not something beyond the appearances, it is the generative law of the class of "its" appearances, namely the "geometral" of all possible perspectives.

Now, if one analyses carefully Schrödinger's genetic account of the "thing" of everyday life, it is almost exactly in these terms that he expressed his position. In his description of the way children form their notion of the surrounding objects, he did not insist on the child's aptitude for forming vast classes of appearances, as this would have been implied by a straightforward extensional conception of the constructivist thesis. He rather pointed towards the opposite trend, namely the trend towards elimination or oblivion of more and more appearances; for the process of forming invariants he insists upon involves an increasing ability to neglect a host of irrelevant variations. One of the possible invariants, probably the most primitive, is the "typical" profile of the object; say, for instance, the façade of the house if the object is the whole house. Hence this "automatic correction"[2] we perform, even when the object is far away or when it is seen under an unfamiliar angle, in order to impose on it a typical size and shape. But the maximum generality is reached when the invariant is able to generate all the aspects encompassed by the "thing". At the end of his study of our "striving for invariance", Schrödinger is thus tempted by merely *identifying* the "thing" with the invariant: "The most

[1]M. Merleau-Ponty, *Phenomenology of perception,* op. cit. p. 67; see also P. Heelan, *Space perception and the philosophy of science,* University of California Press, 1983, p. 134

[2]E. Schrödinger, *Science, philosophy and the sensates,* in: E. Schrödinger, *The interpretation of quantum mechanics (Dublin seminars 1949-1955 and other unpublished texts)* op. cit. p. 147

fundamental invariants that we form at an early stage *are* the *things* in our surroundings, including our own body"[1].

In so far as Schrödinger considers the invariant as the generator of aspects, one could say he just identified the "thing" with the generative law of every possible appearance belonging to a certain class, thus adopting a purely formal and deductivist conception of the "thing". The matter is not so simple, however. There are still many inductivist connotations in the way Schrödinger accounts for the genesis of the notion of invariant. For instance when he says that "We *gradually learn* to disregard these changes or rather to unite the various aspects into the idea of a thing which does not change"[2], the emphasis is put not so much on the derivatedness of the aspects, as on their progressive synthesis into the idea of an invariant "thing". But Schrödinger does not come so far from the intellectualist and deductivist implications of the concept of a "geometral" allowing one to generate perspectives, when he notices that the invariant becomes very soon predominant over the aspects, and that it somehow becomes autonomous. According to him, the invariant holds the role of an ideal norm, by comparison to which every change has to be evaluated. Once a child has acquired the notion of invariant, this child may oppose the *true* changes of shape of an object to the merely perspectival changes, provided he is "(...) helped by the principles of rigid geometry"[3]. This tendency towards priority of the invariant over any range of possible variations, is confirmed by a note of approval *("approbatio")* which Schrödinger wrote in the margin of page 209 of his copy of M. Born's *Natural philosophy of cause and chance*. There, at the end of a genetic account of the "thing" of everyday life which has much in common with Schrödinger's, M. Born explains that a set of observed profiles "(...) is defined by assuming a definite invariant or *gestalt,* not the other way round".

As we shall see in the next paragraph, both Schrödinger's tendency to establish a close relationship between the "thing" of everyday life and the object of scientific inquiry, and his epistemology, which often comes very close to Popper's, predisposed him to provide his concept of invariant with a very strong deductivist component.

5-8 *The "thing" as theory, and the theory as expectation*

Schrödinger's investigation about the "thing" of everyday life was construed by him as little more than an introduction to a serious analysis of the object of modern physics. This is shown by the fact that his best and most developed versions of this investigation are to be found in texts where they served as introductions to a critical examination of the concept of elementary particle. As a consequence, Schrödinger wavered between a

[1]ibid. p. 148
[2]ibid. p. 146-147
[3]ibid. p. 148

quasi-phenomenological attitude, which is more adapted to the case of familiar "things", and a constructivist attitude, which fits better with the case of objects of science. He especially wavered between endlessness and infinity, between succession of expected appearances and co-existence of profiles, between simple aggregates of aspects and "geometrals" able to generate these aspects. His vocabulary fluctuates accordingly. In his 1928 conference[1], for instance, the "thing" is indifferently identified with a "complex", with a "frame", and with a "scaffolding". He writes about "(...) independent complexes which we call existing objects"; or about an object which is "(...) nothing more than a frame which serves to unite certain sense-perceptions"[2]. And he raises the question as to whether the pictures by which we try to visualize the micro-structure of matter do "(...) resemble (the palpable objects around us) in being the scaffolding for a series of perceptions, which can be conceived, if not actually experienced"[3].

The status of Schrödinger's concept of "complex", to begin with, is quite ambiguous. On the one hand, it is suggested that a complex is nothing more than a string of connected perceptions. In addition, even though in a complex, the actual perceptions are supplemented by possible perceptions, they are not supplemented by an *infinity* of such perceptions; only by "*a number of* virtual perceptions"[4]. If we retain these indications, we must say that a "complex" is a finite aggregate of actual and possible perceptions. But on the other hand, as we have already pointed out, Schrödinger's complexes progressively acquire a status of their own, quite independently of the perceptions they serve to connect. In "What is an elementary particle?", the complex finally stands on its own feet, even when the original string of perceptions and anticipations is interrupted. The complex, which was first defined as an ordered set *of* perceptions, then becomes identical to a *frame* or a *scaffolding,* namely to a structure which is able to *accomodate* perceptions though it is not *made of* perceptions. In its initial version, the complex was dependent on the perceptions it incorporated: if one perception had been added to it or substracted from it, it would have changed. But the end-product of the process of abstraction, be it the purely structural component of the complex, or the frame, or the scaffolding, is completely free of any link with particular perceptions. It has a lot in common with a scientific theory or a scientific construct.

This convergence of Schrödinger's phenomenalist notion of "thing" with his conception of scientific constructs, is already manifest in his 1928 paper. There, Schrödinger used the same word ("scaffolding") when he referred to scientific models as when he dealt with "things" of everyday life. While the "thing" is a "(...) scaffolding for a series of perceptions

[1]E. Schrödinger, "Conceptual models in physics and their philosophical value" loc. cit.
[2]ibid. p. 119-120
[3]ibid. p. 130
[4]ibid p. 120

(...)", it is also "(...) the ultimate purpose of all schemes and models to serve as scaffolding for any observations that are at all conceivable"[1].

In his text of 1954 intended for William James lectures[2], Schrödinger developped this convergence of the notion of "thing" and of the concept of scientific construct. He stated explicitly that our common conception of "what the world really is like", and of the "thing" of everyday life which partakes of it, is a kind of scientific theory: "I range it with scientific constructs". Of course, at this stage, one could criticize Schrödinger for having just blurred the difference between the "thing" and the scientific constructs, without giving enough justifications for his move. But Schrödinger did not content himself with dropping distinctions that so many other epistemologists have carefully laid down. He rather used his comparison between "things" on the one side and scientific constructs or theories on the other side, to promote mutual clarification of the two concepts. Firstly, clarification of the notion of "thing" by current epistemological reflections on scientific theories; and secondly, clarification of the status of scientific theories by comparing them with a frame of thought which is deemed to be their ancestor, namely our "popular view" of the world.

Let us begin with the first kind of clarification. Schrödinger developped quite systematically his identification of our "popular view" of the world to a scientific theory, even though the said theory admittedly dates from an archaïc phase of the development of the child. He used expressions like "(...) the experimental science of the baby and small child (...)", and he established a very narrow connection between thing-like invariants and the principles of conservation which, according to him, "(...) refer to the eventual scientific formulation of virtually the same aspect"[3]. It then appears that Schrödinger essentially adopted the hypothetico-deductive conception of the "things" of everyday life that W. Sellars defines as the *Neo-Lockean approach* to phenomenalism. Namely one in which "(...) the framework of physical objects is analogous to a theory" enabling us to impose some order onto the flux of our sense-impressions. According to such a version of phenomenalism, "(I)t is *reasonable* to suppose that physical objects exist, although we do not directly perceive them, because the hypothesis that there are such things enable us to understand why our sense contents occur in the order in which they do"[4].

One effect of this conception is to weaken the position of the "things" in the network of "certainties" (in Wittgenstein's sense) we presuppose both in ordinary life and in science. For, if the notion of "thing" is *only* a scientific construct, if our popular conception of "what the world really is

[1]ibid. p. 131

[2]E. Schrödinger, *Science, philosophy and the sensates,* in: E. Schrödinger, *The interpretation of quantum mechanics (Dublin seminars 1949-1955 and other unpublished texts)* op. cit.

[3]ibid. p. 149

[4]W. Sellars, "Phenomenalism", in: *Science perception and reality,* op. cit. p. 85

like" is to be considered as a primitive scientific theory, "This makes it liable to, and capable of, being subjected to revision and changed and improved, as all scientific theories are"[1]. The main consequence of this position has already been developed in paragraphs 4-1 and 4-2 of this essay. Since the concept of material body, which arises from the popular notion of "thing", is liable to revision, then there is absolutely no reason to hold on to it at any cost. The concept of material body is to be put on exactly the same footing, according to Schrödinger, as the axioms of a theory; and one should therefore drop it unhesitantly as soon as it imposes tensions and paradoxes in our account of experimental phenomena. Even Schrödinger's urge for pictures could not persuade him to be more conservative about the concept of material body, for this concept is only one of the possible modalities of picturing, not an integral part of it.

We come now to the second kind of clarification. This time, it means clarification of the status of scientific theories and objects of science by their being compared to the familiar "things" which surround us. The objects of scientific inquiry are often described by Schrödinger as simple extrapolations, or as sophisticated equivalents, of the "things" of everyday life. Just after having defined a familiar "existing *object*" as a complex of actual and virtual observations, he goes on writing: "I believe that, with respect to objects of science, we cannot really attribute another meaning than the one just indicated to the concept of 'really existing'"[2]. This continuity from the primitive "thing" to the object of science manifests the common *elementary* or *sensational* ground on which, according to Schrödinger, both of them are built: "(...) all our knowledge about the world around us, both that gained in everyday life and that revealed by the most carefully planned and painstaking laboratory experiment, rests entirely on immediate sense perceptions (...)"[3]. This remains true, even though the dependence of the scientific objects on perceptions is much more remote than the corresponding dependence of the "things", and even though scientific objects depend on abstract *relations* between perceptions rather than directly on perceptual *contents* [4]. In "What is an elementary particle?", Schrödinger expressed even more directly his view of the unshakeable dependence of science on the immemorial modes of exploration and ordering which led us to adopt our *popular view* of the world. This time, he did not only point out that science ultimately relies upon the same sensational material as everyday experience, but also, more generally, that its very methods of investigation protract the mode of orientation in daily life: "Physics takes its start from everyday experience, which it continues by more subtle means. It remains akin to it, does not transcend it generically, it cannot enter into another realm".

[1]E. Schrödinger, *Science, philosophy and the sensates,* in: E. Schrödinger, *The interpretation of quantum mechanics (Dublin seminars 1949-1955 and other unpublished texts)* op. cit. p. 149
[2]E. Schrödinger, "Conceptual models in physics and their philosophical value" loc. cit.
[3]E. Schrödinger, *Mind and Matter,* op. cit. p. 166
[4]E. Schrödinger, *Science, philosophy and the sensates,* in: E. Schrödinger, *The interpretation of quantum mechanics (Dublin seminars 1949-1955 and other unpublished texts)* op. cit. p. 133-140

This being granted, the major distinctive features of the "thing" of everyday life are likely to be transmitted to the objects of science, not to say to the scientific theories. In this paragraph, we shall insist on one of these features, namely *openness*. As we noticed formerly, the "thing" presents itself to us under one of its profiles. Since this is *only* a profile, not the object in its (ideal) entirety, it bears the mark of incompleteness. The virtual percepts which are synthetized, together with the actual ones, into the concept of a "thing", are only *predictable* (and they are usually *unconsciously* predicted). They are only *anticipating* a future confirmation. The possibililities are not yet fixed; they are not freed from any link with time; they are not included into a block-framework which would imply that they are all simultaneously co-existing. The thing is thus an open entity whose profiles give promise of something new to perceive, and which is averse to the very idea of completion. It is a *project* of order, not a well rounded ordering scheme. None of the possible re-orderings of the "thing" is to be understood as the last one.

Now, it is exactly this feature of "openness" that Schrödinger considered as the most characteristic of science in general[1]. An openness which extends to the whole set of principles and presuppositions which underly the scientific undertaking, not only to each scientific construct or each scientific theory separately.

As early as 1935, Schrödinger stated a conception of science wherein the major characteristic feature of scientific theories is *falsifiability*. His position was quite close to Popper's, whom he had met in Oxford in 1935[2], and whose *Logik der Forschung* had been published in 1934. He formulated a fully hypothetico-deductive view of scientific theories and indicated, in a very Popperian spirit, that refutation of a theory by experimental results is to be taken as an occasion for further advances rather than as a failure: "At least one then knows where the arbitrariness lies and where the improvement must be made in case of disagreement with experience: in the initial hypothesis or model. For this one must always be prepared. If in many various experiments the natural object behaves like the model, one is happy and thinks that the image fits the reality in essential features. If it fails to agree, under novel experiments or with refined measuring techniques, it is not said that one should not be happy. For basically this is the means of gradually bringing our picture, i.e. our thinking, closer to the realities"[3].

However, the last sentence is quite ambiguous. The expression "closer to the realities" could be interpreted in two ways at least. Either one thinks that it implies the possibility of reaching a *true* theory, and a perfect *correspondence* with reality, though admittedly in the (very) long term. Or one rather takes it litterally as meaning that final truth and

[1]E. Schrödinger, *Nature and the Greeks*, op. cit. p. 4
[2]K. Popper, *Unended quest*, Routledge, 1974, p. 108
[3]E. Schrödinger, "The present situation in quantum mechanics", in: J.A. Wheeler & W.H. Zurek (eds.), *Quantum theory and measurement*, op. cit. p. 152

correspondance are excluded; that theories could only come *closer and closer* to the realities without ever reaching them. The difference is essential, and Schrödinger usually supported the second interpretation. In the same paper of 1935, he precluded any possibility of a final completion of scientific knowledge: "(...) it is quite probable that the adaptation of thought to experience is an infinite process and that 'complete model' is a contradiction in terms, somewhat like 'largest integer'"[1]. In *Science and Humanism,* he became even more critical, for he *also* insisted that the old belief in asymptotic *convergence* of scientific theories was likely to be unacceptable nowadays: "While asserting that any model we may conceive is sure to be deficient and would surely be modified sooner or later, one still had at the back of one's mind the thought that a true model exists - exists so to speak in the Platonic realm of ideas - that we approach to it gradually, without perhaps ever reaching it, owing to human imperfections. This attitude has now been abandoned"[2]. And the reason why it has been abandoned, according to Schrödinger, is the big conceptual leap one had to make in order to accomodate the phenomena of microphysics. This leap between macro- and micro-physics is so big, actually, that any confidence in smooth convergence of theories towards reality has been destroyed. Hence a quite radical consequence: if one asks whether a model of these phenomena is at all capable of being either true or false, Schrödinger's answer is the following: "Probably it is not. Probably we cannot ask for more than just adequate pictures (...)"[3].

In all these criticisms of the very idea of a final *truth* or ultimate *correspondence with reality* of scientific theories, one feels that Schrödinger remains somehow reluctant. "This attitude has now been abandoned"; but *must* it be abandoned? No model of the phenomena of microphysics is *probably* capable of being either true or false; but are we *sure* of this? In these prudent formulations one may read a remote hope of a completion of science, and one must not be surprised that this hope manifested itself at a certain period of Schrödinger's life. However aware of the philosophical problems Schrödinger may have been, he did not hesitate to claim, in a daily newspaper interview of 1947, that he had discovered the "Final" laws of physics. And the corresponding technical paper was entitled "The final affine laws"[4]. The temptation was so great, due to the mathematical elegance of Einstein's unified field theories, that even Schrödinger could not resist.

In 1947, Schrödinger was (provisionally) caught in a process of fascination which he himself described quite accurately. He described it in his 1935 cat-paper, some years before 1947, and then, even more strongly and critically, in the first chapter of *Nature and the Greeks* which he wrote in 1948. Let us state the reason for this fascination in general

[1]ibid.
[2]E. Schrödinger, *Science and Humanism,* op. cit. p. 25
[3]ibid. p. 24
[4]E. Schrödinger, "The final affine laws", loc. cit.

terms, before we come to Schrödinger's own formulations. Scientific theories, and especially theories of classical physics *look complete*. They bring expectations together into a systematically ordered set, and they transform them into coexistent possibilities in a timeless scheme. Their *openness,* if any, is not intrinsic. It is not manifested in their very structure; it has the extrinsic epistemological status of a possibility of *falsification* (according to Popper), or of a possibility of *being replaced by another theory* if they belong to a regressive "research program" (according to Lakatos). In other terms, even though the research programs themselves are open, this is not the case of the nucleus of theories which these programs incorporate. Hence the temptation of projecting the characteristics of the hard core theories onto the research program, indeed onto the whole scientific undertaking. And hence also the hope for a final theory which will be able to freeze any past, present, and future, expectation, into a definitively timeless structure.

This mechanism of projection is remarkably similar to the one Schrödinger criticized when he inquired into the origin of the ingrained determinist beliefs of the classical physicists: "Perhaps the method (of the mechanical model) is based on the belief that somehow the initial state really determines uniquely the subsequent events, or that a *complete* model, agreeing with reality in *complete exactness* would permit predictive calculation of outcomes of all experiments with complete exactness. Perhaps on the other hand this belief is based on the method"[1]. So, the belief in determinism *seems* primitive, and it *seems* to be the reason why so many physicists looked for a *complete* model which would allow one to predict any conceivable experimental event. But actually, it is quite possible that everything has to be taken upside-down, and that what made philosophical determinism appear so "natural" is the very mathematical method used by classical physicists, together with their increasing confidence in its success. Similarly, one might say, the belief that the undertaking of science can be brought to an end through the possession of some final *true* theory *seems* to be primitive and it *seems* to be the most basic motivation of scientists. But it is also possible that everything has to be taken upside down, and that what makes the belief in a final true theory so "natural" is the closed structure and timelessness of the nucleus of most theories. There is a perfect analogy between the reason which promoted fascination for determinism and the reason which promoted fascination for ultimate truth of scientific theories. Actually, this is more than an analogy. In Schrödinger's quoted sentences, the two points, namely determinism and final truth, are closely connected. For, according to Schrödinger, perfect predictability was traditionally considered as a characteristic feature of a model "agreeing with reality in *complete exactness*", i.e. a model which can claim final truth.

[1]E. Schrödinger, "The present situation in quantum mechanics", loc. cit. p. 153

To be accurate, one must say that the urge for (ideal) determinism and for ultimate truth is much older than classical science. As Schrödinger pointed out, the claim that it is possible to "(...) *close* the disconcerting 'openness' of the outlook gained from experience alone (...)"[1] is an immemorial characteristic of theological thinking. After the birth of modern science, this claim was either taken as a motivation to bring science to completion by using extrinsic dogmatic material, or somehow imposed directly onto science itself. The first trend is typical of religion, in its attempts at gaining respectability of the very same grounds as science; and the second trend is typical of *scientism*. Our interpretation of the origin of the beliefs bearing on determinism and final achievement of scientific knowledge must thus be reformulated. Fascination for determinism and ultimate truth is certainly more than a by-product of the methods of science. But by the very (Platonic) perfection of the structures they use, sciences tend to give credence to (or to revive) claims which are typical of pre-scientific thought. It is then the role of the philosopher, or of a philosopher-scientist like Schrödinger, to remind us from time to time that the "openness" we try to reduce, and whose sought reduction is such a strong motivation for research, is inseparable from the very presuppositions of empirical knowledge. And that, therefore, the idea of completeness of empiric sciences involves a contradiction in terms.

To conclude this paragraph, we now have to raise a question which was latent in the previous comparisons of the "thing" of everyday life with a scientific construct. According to Schrödinger, the notion of "thing" is falsifiable *because* it can be identified to a (very primitive) scientific theory. But conversely, the entire scientific undertaking is considered by him as starting from everyday experience, "(...) which it continues by more subtle means"; science prolongs the (partly unconscious) investigation which led us to adopt our "popular view" of the world, and this means that it incorporates one of the most distinctive feature of the entities of this "popular view", i.e. openness (the "and so forth" of anticipations in perception of "things"). At first sight, this procedure of mutual justification looks quite circular.

We have already addressed a little part of the charge of circularity by noticing that tracing back the origins of science to our most ingrained (pragmatic) methods of investigation does not account for the falsifiability of each particular scientific theory. It rather points, broadly speaking, towards the incompletability of the scientific undertaking considered as a whole.

Another way of addressing the charge of circularity is to notice that the falsifiability of scientific theories does not only take over the openness of the manipulated and perceived "thing"; it rather amplifies it. True, in everyday life it is always possible to recognize the failure of a particular anticipation, for instance the illusory nature of a certain perception. Yet

[1] E. Schrödinger, *Nature and the Greeks*, op. cit., p. 4

the general scheme (or "popular view") according to which the surrounding world is made of "things" or material bodies interacting in ordinary space and time is never questioned. Whereas, in science, a whole theory, together with its most fundamental axioms, can always, in principle, be replaced. So, what Schrödinger tried to project onto our "popular view" of the world was more than the simple *openness* which the scientific undertaking has inherited from its ontogenetic roots in daily investigations and perceptions. It was the kind of radical revision of beliefs which is typical of scientific revolutions and which has no manifest equivalent in our way of dealing with the view of the world presupposed by daily actions and speech. What Schrödinger insisted upon when he compared the notion of "thing" to a scientific construct is not only that any expectation associated with the perception of a particular thing can be deceived (which is the usual *openness* of our familiar knowledge of the world), but that this very notion of "thing" should not be sheltered; not any more, at any rate, than the most fundamental axioms of a (falsifiable) scientific theory.

But is this possible? Are we free to dispense with the very notion of "thing"? In principle, we are not. For the notion of "thing" is not just a notion among many others. And propositions which state the existence of material bodies or "things" are not just ordinary propositions. As Wittgenstein noticed in *On certainty,* "(...) the *questions* that we raise and our *doubts* depend on the fact that some propositions are exempted from doubt, are as it were hinges on which those turn"[1]. A proposition which states the existence of material bodies is precisely this kind of hinge on which our discourse turns, because anything we can say when we use ordinary language takes it directly or indirectly for granted. We should even say that propositions which state the existence of material bodies are non-sensical because, in the language game we play, there is no possibility of stating their negation. This is the reason why we have never learned them: "Children do not learn that books exist, that armchairs exist, etc. etc. - they learn to fetch books, sit in armchairs, etc., etc."[2]. Propositions which state the existence of material bodies do not belong to the normal use of ordinary language, but rather to the atmosphere of philosophical perplexity which surround it. Doubt about a fraction of our picture of the world is always possible but it cannot go as far as threatening the very framework of presuppositions which underlies any formulation of doubt. And the notion of "thing" just happens to be part of this framework. "Doubting and non-doubting behaviour. There is the first only if there is the second"[3].

Now, the whole language-game of experimental science, including experimental micro-physics, is merely an extension of the language-game of everyday life (Schrödinger insisted on something quite similar when he

[1]L. Wittgenstein, *On certainty,* Basil Blackwell, 1974, §341
[2]ibid. §476
[3]ibid. §354

said that physics cannot transcend our everyday experience). It is thus part of the language game of experimental science not to allow doubt about the existence of the material objects which constitute the experimental devices. Then, even if the structure of a theory which accounts satisfactorily for the available experimental data gives one strong motives for questioning the notion of individual, reidentifiable, localized substrate of properties, which is constitutive of the notion of "thing", one cannot just drop it. For if one tries to dispense with this notion, one undermines the very experimental grounds of the doubts which bear on it. Universal skepticism, in experimental science as in everyday life, is self-defeating.

So, what is the right attitude to adopt in these difficult circumstances where experimental results (of micro-physics) seem to threaten the presuppositions to which they owe their significance? There are at least three possibilities. We shall call them, respectively, conservative monism, loose dualism, and ontological parallelism.

(i) To begin with, *conservative monism* aims at preserving at any cost the traditional notion of material body, even in the microscopic domain. M. Born advocated such a position (see paragraph 1-2 of this essay), but he had to introduce many qualifications to his corpuscularian views in order to fit the constraints imposed by Bohr's complementarity and Heisenberg's uncertainty relations. D. Bohm then fully worked out the consequences of preserving the corpuscularian categories in micro-physics. He showed that this kind of ontological conservatism is by no means impossible, but that it imposes an impressive list of conditions[1] which some could find difficult to accept.

(ii) Let us then come to *loose dualism*. In the present context, *dualism* is tantamount to accepting a separation between a domain (the macroscopic world) wherein the traditional notion of material body operates, and a domain (the microscopic world) wherein it collapses. *Loose* dualism means recognizing that the exact location of this separation can fluctuate, and that it is in principle arbitrary. Bohr's position essentially identified itself with loose dualism. This position met a series of problems which are all linked to the fact that it projects on the scale of magnitudes (micro-macro) a necessary compromise between the methodological requirement of maintaining the notion of "thing" as a framework for describing experimental arrangements, and the full recognition that this framework has come under very strong pressure in quantum mechanics.

(iii) Finally, let us examine *ontological parallelism*. An ontological parallelist fully acknowledges the role of the notion of "thing" in everyday life. And he also recognizes that the most natural attitude to adopt in *experimental* physics is to extend the corpuscularian categories

[1]*Non-locality* and *contextualism* of properties are conditions imposed on hidden variable theories of any kind. They are imposed on theories which take *particle* properties as their hidden variables (or "beables"), but they also apply to the theories which take *field* properties as their "beables". See D. Bohm & B.J. Hiley, *The undivided universe*, Routledge, 1993. Other conditions are more specific of the corpuscularian varieties of hidden variable theories. See H.R. Brown, C. Dewdney, & G. Horton, "Bohm particles and their detection in the light of neutron interferometry", *Foundations of physics*, **25**, 329-347, 1995

throughout the scale of spatial magnitudes (even if it means using them with a lot of qualifications which are typical of the language of modeern physicists); for this attitude allows one to establish full categorial continuity between the presuppositions which underlie the use of instruments and those which underlie discourse about the purported objects of experiments. Yet, the parallelist also warns one against any ontological hypostasis of this "natural attitude" of the experimenter. He even thinks that experimental evidence is enough to claim that corpuscularian notions are ontologically unacceptable *throughout the scale of spatial magnitudes*. Even though they are *in practice* indispensible or useful almost everywhere, they are *in principle* inadequate everywhere. The only thing one has to check in order to reconcile these two conflicting statements is that a correct (universal) theory involving *no corpuscularian entities at all* displays reasonable *compatibility* with the presuppositions of the experimenter's actions and speech in the laboratory.

This latter approach is quite close to Hume's and Wittgenstein's *skeptical solutions to the skeptical doubts* [1]. Here as in Hume's case, the notion of "thing" is *nowhere valid*, but everything looks as if this notion were *in practice* acceptable almost *everywhere*, either by direct (careless and inattentive[2]) acquaintance in our immediate surroundings, or by approximative continuity in the microscopic domain (provided one takes into account Heisenberg's uncertainty relations). Schrödinger was certainly, among the founders of quantum physics, the thinker who came closest to this position. On the one hand he extensively documented the all-pervasive importance of the notion of "thing" in everyday life, and on the other hand he believed this notion could and should be subject to experimental falsification. He nevertheless failed to make completely clear the idea of *double universality* (practical universality for the notion of "thing" and *quasi-realist* ontological universality for the entities of the most advanced theory of modern physics) which is characteristic of what we have called "ontological parallelism".

5-9 *Realism and morals*

The double claim that (i) the notion of "thing" is to be classified with scientific constructs, and that (ii) scientific research is an extrapolation of the mode of investigation of ordinary life, provides us with interesting new elements for our inquiry into the meaning of Schrödinger's scientific realism.

The first part of the claim, i.e. that "things" are analogous to scientific theories, has usually been associated with an attack on realist interpretations of scientific theories. As Sellars pointed out, the "new

[1]See S. Krikpe, "Wittgenstein on rules and private language", in: I. Block (ed.), *Perspectives on the philosophy of Wittgenstein*, Basil Blackwell, 1981

[2]D. Hume, *A treatise of human nature*, (ed. L.A. Selby-Bigge), Oxford University Press, 1960, Book I, Part IV, Chapter II.

phenomenalists" who support a hypothetico-deductive conception of both "things" and scientific theories, usually consider at the same time that "(...) theoretical entities are 'calculational devices' and do not exist in the full-blooded sense in which observables exist"[1]. According to them, theories are only formal schemes, and theoretical entities are just convenient devices in order to perform as economically as possible the task of deriving observational consequences from observational premises. But then, the framework of "things" of our "popular view" of the world, which has been identified with a primitive scientific theory, must also be reduced to a mere conceptual tool which enables one to behave efficiently and to predict correctly future observations from present ones. If identification of the framework of "things" with a scientific theory occurs in the context of an anti-realist conception of scientific theories, then even our "natural attitude" in everyday life comes under suspicion. Thus, most defences of realism had to deal jointly with "things" and scientific theories.

One of these defences is M. Born's. In his paper "Physical reality", M. Born discusses the ideas of H. Margenau, who "(...) advocates the standpoint that reality consists of two layers: the immediate data of the senses, and 'constructs'; the latter include things of everyday life as well as scientific constructs (...)"[2]. Against this conception, Born emphasizes both that sense impressions are not the primary data (for the true primary data of perceptions are 'Gestalts' assimilated to invariants), and that scientific constructs can be ascribed the character of "real things" provided they are invariant in regard to the relevant transformations.

This being granted, Schrödinger's scientific realism seems to become more difficult to understand than ever. After all, as Popper complained at length in his "Unended quest", Schrödinger accepted that sense impressions are (in principle, even if not in fact) primary data. True, he did not reject the idea that we perceive 'Gestalts' (or 'events' which are very elaborate 'complexes'), nor did he underestimate the role of invariants; but he still separated sensations construed as truly "given" from thought and imagination which are supposed to order them. Furthermore, he also accepted that both the "things" of everyday life and the scientific entities are only *constructs,* and that the former are primitive instances of the latter. Schrödinger's position is thus remarkably close to that of Sellar's new phenomenalists, and to that of Margenau as criticized by Born. Yet, the general conclusion he drew from this position was almost *exactly opposite* to theirs. He did not insist, as they did, that since things are merely scientific constructs, and since scientific constructs are merely formal devices for prediction of observable events, therefore things are only primitive devices for prediction of observable events. He rather emphasized a reasoning which consists in taking everything the other way round. In contradistinction to new phenomenalists, Schrödinger

[1]W. Sellars, "Phenomenalism", in: *Science perception and reality,* op. cit. p. 86
[2]M. Born, "Physical reality", loc. cit.

pointed out that since scientific constructs are extensions of the procedures of orientation of our everyday life, and since these procedures of orientation are so important to us that we call their invariants "real objects", there is no reason *not* to endow a certain class of scientific entities with the status of real objects.

We must remember that, in Schrödinger's eyes, a construct such as the "thing" of everyday life, "without (which) we cannot achieve a single step in practical life"[1] must not be "questioned in the least". Accordingly, even though "(...) speaking and thinking in terms of what is really going on in the physical world" is just a "convenient habit"[2], we have no reason to give up this habit, even in science. For, he says, if we give it up without any counterpart, we are in great danger of letting the individual observations "(...) fall apart without any connections"[3].

The symmetry of Schrödinger's double claim that "things" are to be classified with scientific constructs, and that science extrapolates the mode of inquiry of ordinary life, is thus only apparent. It is the *second* part of the claim, not the first, which finally determined his epistemological position towards the entities of modern physics.

But why is it so? Why did Schrödinger not follow the slope of *scientific* anti-realism which is the most natural counterpart of his phenomenalist outlook? Or, once again, why did he associate a metholological realist attitude in physics with a definitely anti-realist view of our familiar environment? We can gain insight into the main motivations of Schrödinger's choice from an analysis of I. Berlin's cogent criticism of several consequences of phenomenalism[4]. According to I. Berlin, a characteristic feature of phenomenalism is that it aims at replacing categorical statements (about "things" and their properties) by bundles of hypothetical statements about *sensibilia* or possibilities of observations. But by doing so, phenomenalism mixes up two things: the categorical statement itself and the description (by means of hypothetical statements) of the conditions without which we would not be justified in asserting it. Stating the list of its conditions is by no means equivalent to stating it. For categorical statements, which, when used in isolation, are usually considered as typical locutionary acts, have also an illocutionary connotation and a perlocutionary effect. Their illocutionary connotation is due to the fact that stating them amounts to a promise: the promise that the speaker would show the reasons he has to believe them if one asked for further justifications. As for their perlocutionary effect, it comes from the circumstance that they "(...) tend to direct attention to - invite us to look for - things and events in a way in which other kinds of expressions do not. This is felt more clearly about expressions containing

[1]E. Schrödinger, *My view of the world,* op. cit. p. 64 see §1-5

[2]E. Schrödinger, "What is an elementary particle?"

[3]E. Schrödinger to A. Einstein, November 18 1950, in: K. Przibram, *Letters on wave mechanics,* (Tr. M.J. Klein), op. cit. p. 38

[4]I. Berlin, "Empirical propositions and hypothetical statements", in: *Concepts and Categories,* op. cit.

demonstratives like 'this', or 'that', or 'here', but applies no less to existential propositions without demonstratives which identify something in the time series"[1]. In short, categorical statements about "things" and their properties incite people to inquire; or at least they incite people to dispose themselves to inquiry. They prompt one to adopt the right attitude for research. On the contrary, as long as we content ourselves with stating hypothetical propositions such as *if somebody did D and adopted position P, then he would have sensation S,* we do not prompt anyone to direct attention towards something; we do not incorporate in our statement the (implicit) perlocutionary force which would push our interlocutor to start looking by himself for the described conditions. "(I)n themselves, hypotheticals do not 'point'; otherwise they would cease to be hypothetical, they would lose their conditional, non-actual-fact-asserting force"[2].

Thus, categorical statements "point", and hypothetical statements do not "point". Categorical statements tend to shape regulative targets for investigations, whereas hypothetical statements can only initiate infinite regresses of justifications. Categorical statements presented as true (or merely asserted) are part of the normative background that regulates everyday action and science as material practices[3]; but hypothetical statements, if isolated and completely devoid of any implicit categorical elements, are likely to be associated with practices lacking the direction provided by appropriate regulative principles. This does not mean that a theory involving *dispositions* can never be connected with a properly regulated scientific activity. But in order to fulfill such a methodological condition, a theory must be expressed in terms of categorical statements *about* entities which *have* dispositions, or at least about the dispositions themselves, rather than directly in terms of an open hierarchy of hypothetical statements. True, dispositions are by definition dependent on a ground-level of hypothetical statements; yet this does not prevent one from introducing an elaborate kind of directedness, by a discourse pointing towards dispositions taken as second-level entities.

Actually, the case of dispositions is exemplar, for it shows that the pointing ability of categorical statements does not depend on an ontologically categorical background (which may well be lacking in the case of dispositions), but rather on the associated illocutionary act force.

We could say in other words that any attempt at reducing categorical statements to bundles of pure hypotheticals is *eo ipso* an attempt at dropping the *moral content* which is associated with the assertion of a categorical statement. Let us explore this idea a little further. Categorical statements connotate the seriousness of the speaker when he asserts them; and they prompt the hearers to start serious investigations directed

[1]ibid. p. 45
[2]ibid. p. 47
[3]See R. Harré, *Varieties of realism*, Basil Blakwell, 1986, p.91

towards a precisely defined purpose. Now *seriousness* and *systematic directedness* of attention or action are just the basic features of a moral attitude, according to a modern thinker such as I. Murdoch. In a platonician context, this idea would be (too) straighforwardly expressed thus: "An object of serious thought must be something real, serious thinking is moral truthful thinking, goodness is connected with reality, the supremely good is the supremely real"[1]. This way of putting things is difficult to accept, because it is tantamount to accepting a naïve version of idealism. One can nevertheless retain the fruitful idea that a realist *attitude* in action and research (not reality itself!) is closely connected with seriousness, with close attention, with personal commitment, and thus with morality. As an element of morals, a realist attitude does not require metaphysical hypostasis ("each entity of the present theory exists in itself; each categorical statement of the present theory is true"). It only requires *normative rules* ("take seriously the entities and the categorical statements of a certain theory, until experimental evidence has undermined it and a new theory whose entities and categorical statements you can take seriously has been formulated"). This position corresponds quite closely to what we have called Schrödinger's quasi-realism in paragraph 2-1 of this essay, and it has obvious affinities with R. Harré's "policy realism"[2].

Another aspect of Schrödinger's association of an anti-realist metaphysical background and a realist attitude in science can be made clearer through a comparison with Husserl. Husserl's philosophy is sometimes interpreted as idealist, and sometimes as realist. If more careful time-discrimination is required, commentators usually say that Husserl was realist in his *logical investigations,* published in 1900, but that he defended an idealist position in his *Ideas,* published in 1913. However it has also been argued, on very serious grounds, that Husserl was *neither* an idealist *nor* a realist[3]. How can it be so? Husserl's phenomenological investigation is based on the method of *reduction* or *bracketing*. It aims at redirecting attention. Attention of any person caught in the "natural attitude" is directed towards the objects of the world; it is directed towards the entities which can be referred to in public discourse. After reduction or bracketing, which is typical of the philosophical method, attention is redirected towards the *noemata* which correspond to perceptual meanings. These perceptual meanings incorporate many elements such as memories, expectations, and the presently perceived profile. And they *also* incorporate directedness towards the perceived objects. Thus, in the philosophical attitude advocated by Husserl, directedness towards objects *only* appears as a component of a perceptual

[1] I. Murdoch, *Metaphysics as a guide to morals,* Penguin Books, 1993, p. 398

[2] R. Harré, *Varieties of realism,* op. cit.; A. Derksen (ed.), *The scientific realism of Rom Harré,* Tilburg University Press, 1994

[3] H. Hall, "Was Husserl a Realist or an Idealist?", in: H.L. Dreyfus (ed.), *Husserl, intentionality and cognitive science,* op. cit.

meaning. Any implicit belief in the existence of the purported object, which goes together with the "natural attitude", is suspended, and taken in turn as an object of enquiry. This is not to say that Husserl believes that intentionally aimed at objects do not exist, and only meanings exist. He has just *nothing* to tell us about the success or failure of reference, let alone about the existence or inexistence of objects. The philosophical attitude of *reduction* he adopts is only concerned with a reflective analysis of what is involved in the common actions and beliefs of someone who happens to adopt the "natural attitude"; not about his being right or wrong in adopting the "natural attitude".

One must therefore clearly separate the questions, according to whether they are asked from the standpoint of the "natural attitude" or from the standpoint of the phenomenological (of philosophical) attitude. "There is, first, a natural question about the existence of the real world. (...)". Even according to the "idealist thinker" Husserl, "There is nothing problematic about this question. It would be answered in the affirmative by anyone from the man in the street to the man of science - that is, by anyone from within the natural attitude". But besides this first question, "There is also a *philosophical* question concerning the existence of the real world. It asks for the meaning-contents of our natural acts which are aimed at this world (...)"[1]. There are thus two questions about the existence of the purported objects, not one. The question asked from within the "natural attitude" and the question asked from the standpoint of the philosopher who has performed the phenomenological bracketing.

But this remark is too weak. After all, emphasizing that one must not confuse the two kinds of questions still pertains to the philosophical attitude. From the standpoint of the "natural attitude", there is only *one* question and no confusion has thus to be feared. The true problem therefore does not bear on possible confusions, but about *choice* of attitude. Either one chooses to adopt the "natural attitude", and questions about the existence of the purported "things" in general must be given a positive answer. But this answer is trivial, not to say pointless. Or one chooses to adopt the philosophical attitude, and questions of existence become unanswerable to some extent because they can only be addressed at the reflective level of (implicit or explicit) beliefs, elements of justification, meanings of our acts, etc. They are completely cut off from what beliefs, justifications an meanings are about.

The true problem, in other words, is about *engagement (or commitment)*. When one is *engaged* in an activity of daily life or scientific research, this engagement manifests itself by actions and speech which presuppose a complete absence of doubt about their objects in general. This behaviour is characteristic of the "natural attitude". Conversely, expressing doubts, calling the presupposition a belief, asking for justifications, means that one is (partially or completely) disengaged from

[1]ibid. p. 185

the corresponding activity. The philosophical attitude can be viewed, accordingly, as systematic *disengagement*, although it often implies a different type of engagement or commitment (e.g. towards meanings considered as "essences"). Now engaged and disengaged behaviour are mutually exclusive. One can intermittently suspend a certain engaged behaviour in order to find better conditions for the efficiency of future engagement; yet nobody can *at the same time* practice an activity with a view to success and adopt a disengaged attitude towards it.

But this was exactly what Schrödinger tried to express when he alternated philosophical analysis and intentional directedness towards objects of science. According to him, "reality" is a concept which cannot be dispensed with *during the activities* of everyday life and of scientific research. Even though one may have good *philosophical* reasons to think that "things" of our environment and scientific entities are merely constructs, one has to forget everything about these reasons in circumstances where it is only required to *practice* life and science. True, it is sometimes useful to readopt the philosophical attitude during a short time in order to examine whether we could not free our view of the world from difficulties and paradoxes by changing the old entities into new ones. But as soon as the new entities are defined, one has to *engage* oneself in the practice which presupposes them and bracket again the philosophical doubts, just in the same way as the philosophical attitude brackets the tacit ontological commitment of the "natural attitude". This reciprocity of attitudes, this distinction between two levels of thought, was stated as clearly as possible by Schrödinger in "What is an elementary particle?" (just before the paragraph where he claimed that science is an extrapolation of everyday activities): "It does not necessarily follow (from the uncertainty relations) that we must give up speaking and thinking in terms of what is really going on in the physical world. It has become a convenient habit to picture it as a reality. In everyday life we all follow this habit, even those philosophers who opposed it theoretically, such as Bishop Berkeley. Such a theoretical controversy is on a different plane. Physics has nothing to do with it. Physics takes its start from everyday experience, which it continues by more subtle means".

Thus, in the eyes of Schrödinger as in those of Husserl, philosophical analysis on the one hand and the engaged behaviour which is characteristic of both everyday life and science on the other hand, are "on a different plane". One should not (and cannot) undermine the spontaneous realist attitude which is associated with the latter, by arguments which pertain to the former. In particular, according to Schrödinger, a scientific revolution should not be taken as an incentive to adopt the philosophical attitude on a permanent basis. It should rather be a transient occasion to determine (with the help of the philosophical method) which new set of entities scientists may focus on as soon as they revert to their characteristic way of extrapolating the "natural attitude".

At first, in the time of classical physics, scientists borrowed spontaneously the "natural attitude" from everyday life. And they also borrowed uncritically its purported objects. Later on, the quantum revolution put strong pressure on the traditional ontology of material bodies. Yet, even in this situation, Schrödinger thought that the "natural attitude" does not have to be abandoned; for it embodies serious commitment in research activities. It has only to be adapted, and transformed into self-conscious orientation towards the most appropriate entities.

5-10 *Form and individuality*

An important part of Schrödinger's discussion of the status of atoms and particles deals with individuality. In the course of his comparison between particles and "things" he had thus to address the question as to whether the marks of individuality of familiar "things" are also available for elementary particles.

So, what is it that determines the individuality of the material bodies of our environment? Schrödinger's answer is very clear: the individuality of "things" is usually determined by *form*. To illustrate this idea he takes the example of an "iron letter-weight in the shape of a Great Dane" that was bequeathed to him by his father, that he had to abandon in Austria in 1938, and that he finally recovered in 1947 while he was living in Ireland. He then goes on asking: "I am quite sure it is the same dog, the dog that I first saw more than fifty years ago on my father's desk. But why am I sure of it?"[1]. His straightforward answer is that "It is clearly the particular *form* or *shape* (of this dog) that raises the identity beyond doubt (...)"[2]. However, it has generally been recognized in the history of philosophy that identifying the principle of individuation to *form* raises difficult questions. A form can only define a kind, or a class: for instance the class of the iron Great Danes made in the same factory. And therefore the iron Great Dane which was recovered by Schrödinger could well be another copy of the same model. True, one might try to refine the analysis of form and find some elements of shape which apply only to one element of the iron Great Dane class; for instance a little scratch, or a special patina. But this is not satisfactory either. As Duns Scotus pointed out, these additional features of form may define a one-element class, but they do not define an individual as such. In other words, the fact that an object is individuated by some particular elements of form is contingent; it is due to the circumstance that no other existing individual happens to share these elements of form. There is nothing intrinsic in this way of individualizing objects. This is exactly the reason why Duns Scotus felt motivated to invent an ultimate principle of individuation, which cannot be reduced to form but which determines the form: the "*haecceitas*".

[1]E. Schrödinger, *Science and Humanism*, op. cit. p. 19
[2]ibid.

As for Schrödinger, one can easily guess why he did not retain either the solution which consists of adding more and more specific elements of form: this solution is irrelevant in microphysics, for no additional element of form can be invoked in order to distinguish a member of the class of protons from another member of the same class. If the discussion is to remain relevant to the question of individuality of elementary particles, one may perfectly ignore the kind of half-way solution to the problem of individuation which is represented by the concept of a distinctive element of form.

So, we have to consider other solutions, and to discuss their relevance in microphysics. Schrödinger's Great Dane could for instance be individuated by the *matter* which it is made of. But Schrödinger had many reasons to reject this idea. One of the reasons he gave is that purely material identity "(...) would be of very restricted interest"[1]; for if the iron which constituted the Great Dane had been melted and casted in another shape "I should probably not care very much about the identity or not of that mass of iron, and should declare that my souvenir has been destroyed"[2]. This reason to reject individuation by matter is purely interest-relative, and it is thus quite weak. However, it points indirectly towards another, and much better, reason a physicist could have to be quite reluctant to adopt matter as a criterion of individuality. Let us suppose we accept the matter which constitutes the letter-weight as the criterion of its individuality. We must then answer the question as to what makes the matter of the letter-weight different from any other matter. In the absence of such a second-level criterion, matter could hardly serve as a first-level criterion of individuality for the object which is made of it. Now the matter of the letter-weight might be different from any other matter because of the unique proportions of chemical elements in it, or because of the special configuration of its constituents (for instance the spatial distribution of atoms of iron and of carbon). But this would be tantamount to accepting individuation by form at the second level. The matter of the letter-weight might also be different from any other matter because *each of its constituent parts* are different from the constituent parts of any other object. But this raises in turn the question of the individuality of the constituent parts.

In a continuist framework of thought, this would lead to an infinite regress: the constituents might differ from each other either because of the configuration of *their* lower-order constituents (which would be tantamount to accepting individuation by form at the third level), or because these lower-order constituents (the constituents of the constituents) are different from each other. Then why do these lower-order constituents differ from one another? etc.

In an atomist paradigm the regress would stop at a certain level, because the atoms (or the particles) are supposed to be indivisible and thus

[1]ibid.
[2]ibid.

to have no constituent at all. From the standpoint of classical physics, this way of blocking the infinite regress would be taken as a major advantage of atomism. But certainly not from the standpoint of modern physics. For the whole burden of the criterion of individuality of ordinary objects has now been transferred on the elementary particles, whereas the inquiry was initiated because this criterion appeared *much more* problematic for elementary particles than for familiar objects of the mesoscopic scale.

There is nevertheless one more strategy for distinguishing the matter of one object from the matter of another object. This strategy starts from ostensive definition of a sample of matter, and accordingly from *location* of this sample of matter in space. It was used by St Thomas Aquinas under the name of "designated matter": "(...) the principle of individuation is not matter taken in just any way whatever but only designated matter"[1]. Designated matter is matter "considered under determined dimension". More generally, it is matter considered according to its spatial determinations, including location. It is matter construed as the spatio-temporally defined repository of form. The same form can be imposed on several samples of matter which were initially distinguished by their position. This gives rise to several individuals perfectly identical but numerically distinct[2]. Then, the identity of each individual at every instant is ascertained by the continuity of the history which goes from the initial location to the present one. Designated (or spatially determined) matter thus provides one with a *permanent* criterion of individuality. This account of individuality is closely akin to Kripke's theory of reference, where reference is traced back to an act of "initial baptism" through the intermediary of a continuous historical sequence.

As for Schrödinger, he also relied more or less implicitly on such a conception when he had to state precisely the reason why he believed that the iron letter-weight he received in Dublin in 1947 was *the same* as the one he had left in Austria nine years earlier: "It accompanied me to many places, until it stayed behind in Graz in 1938, when I had to leave in something of a hurry. But a friend of mine knew that I liked it so she took it and kept it for me. And three years ago, when my wife visited Austria, she brought it to me, and there it is again on my desk"[3]. One may wonder why Schrödinger did not make the spatio-temporal criterion of individuality more explicit in his text. The most likely explanation is that

[1]Aquinas, St Thomas, *On being and essence,* Translation and interpretation by J. Bobik, University of Notre-Dame Press, 1965, §23; see also §25

[2]This raises the Leibnizian question of the *identity of indiscernibles.* In 1673 (in his *Confessio Philosophi),* Leibniz still thought that it is possible to individualize a material body by extrinsic determinations such as location and time. But later on, Leibniz considered that locations are defined by reference to individual substances (or monads). Therefore, "In addition to the difference of time or of place there must always be an internal principle of distinction" (G.W. Leibniz, *New essays on human understanding,* Cambridge University Press, 1982, II, XXVII, §1).

[3]E. Schrödinger, *Science and Humanism,* op. cit. p. 19. Even Leibniz's "internal principle of distinction" can be considered as a mere internalization of the principle of continuous history. See e.g. *Discourse on metaphysics,* XIII: "(...)The concept of an individual substance includes once for all everything which can ever happen to it".

Schrödinger here anticipated the failure of this criterion in the case of quantum particles. He knew perfectly well that the possibility of ascertaining an uninterrupted historical sequence was just the paradigmatic criterion of permanent individuality for *classical* particles, and that an initial location followed by a continous trajectory was the form Boltzmann gave to it in his *First principles of mechanics* [1]. But he also realized the impossibility of applying this criterion to quantum particles (§4-1). He thus rather came back to *form,* insisting on its most attractive features. Even if form cannot be taken as a criterion of *individuation* in the most metaphysical sense of the word, delimitation of a one-member class by it has the advantage of being independent from any consideration about the identity of material substrata: "A man returns after twenty years of absence to the cottage where he spent his childhood. He is profoundly moved by finding the place unchanged. The same little stream flows through the same meadows (...) And so on. The shape and the organization of the whole place have remained the same, in spite of the entire 'change of material' in many of the items mentioned (...)"[2]. This obvious case of purely *formal* permanent "individuality" had a distinctive advantage in Schrödinger's eyes. It could be transposed without any difficulty to the situation of quantum mechanics. A situation where the attempts at reconstituting the trajectory of particles are doomed to fail, but where the theoretical entities referred to as *ψ-waves* can nevertheless be ascribed a purely formal permanent "individuality" (see paragraph 4-2, point (5)).

5-11 *Wholeness and individuality*

The question which has still to be addressed at this stage is the following. How are we to connect the familiar situation, where material objects are individuals, with the quantum situation, where the so-called elementary constituents of matter are non-individuals, and where the only "individuals" are the new formal entities called *ψ-waves?* "How does individuality arise at all in objects composed of non-individuals?"[3].

The clearest answer Schrödinger gave to this question is to be found in his notes for Dublin seminar of 1949[4]. There, he first points out that "The way in which particles constitute matter is (...) a very strange and novel one (...). Particles, having no individuality, constitute pieces of matter that have". In addition, he wonders "how is this transition from particles to individual, truly observable, pieces of matter (...) how is e.g. the pointer-hand of my instrument - or for that matter my pocket knife - thought of to be composed of particles". In order to address these questions,

[1]E. Schrödinger to H. Margenau, April 12, 1955, AHQP, microfilm 37, section 9

[2]E. Schrödinger, *Science and Humanism,* op. cit. p. 20. About the paradigmatic case of the permanence of a river, see Buridan's *Physics* (quoted by R. Paqué, *Le statut parisien des nominalistes,* P.U.F., 1985)

[3]E. Schrödinger, *Science and Humanism,* op. cit. p. 18

[4]in: E. Schrödinger, *The interpretation of quantum mechanics (Dublin seminars 1949-1955 and other unpublished texts),* op. cit.

Schrödinger makes a crucial shift in his vocabulary. The verbs "to compose" and "to constitute" usually connotate (even remotely) the idea of a spatial arrangement. The etymology of *compose* is "place together"; and the primary meaning of *constitute,* as given by the *Oxford English Dictionary* is "To set, place (in a specified state, situation, condition, etc.)". Using these verbs, together with some of their derivatives such as "component" or "composition", would then amount to making an implicit verbal concession to the idea that material bodies are made of spatial arrangements of smaller bodies called "particles" or "atoms".

Schrödinger therefore carefully avoids saying that particles are mere *components of* the macroscopic material objects, and speaks in a quite prudent way of the constitution of the latter objects. He rather emphasizes that "a number of (particles) *coalesce* to form a more extended gathering (...)". This verb "to coalesce" ("to cause to grow together, to unite, to combine" according to the *Oxford English Dictionary*) has the distinctive advantage of being much more evocative of the fusion of a multiplicity of observable states into a *single holistic* observable state than of the juxtaposition of a number of localized bodies.

Then, after the "coalescence" has taken place, the resulting whole corresponds to an observable which is so intricate that its form is distinct from any other form: "This coalescence gives rise to complicated observables - they *are* matter in the meaning of the philosopher". This sentence is obviously related to Schrödinger's general conception of individuality that he expressed in his 1949 seminar and in later texts as well. According to this conception, "(...) quantum mechanics, while depriving the particles of individuality, ascribes it to the observables, to the observable states. Matter in the meaning of the philosopher really consists of them - not of the particles"[1]. To summarize, Schrödinger's "matter in the meaning of the philosopher" does not consist of a *compound* individualized by the arrangement of its *components,* but of a *whole* individualized by the form of the global observable state which corresponds to it. *Form* is the dominant (non-metaphysical) criterion of individuality throughout the scale of spatial magnitudes. And not only form, but *pure* form; for in quantum mechanics, says Schrödinger, there is no point in thinking that form is the form *of* some material[2].

One could object that this conclusion conveys seriously damaging consequences for our conception of individual objects and classes. As we noticed previously, if the distinction we make between objects is only based on form, this does not yield any intrinsic criterion of individuality. At most, we are left with classes so narrowly defined that they only include one element. But this undermines the whole modern trend towards extensional definition of classes. For now, the elements can claim no priority over the classes. Several contemporary thinkers have come to the conclusion that these consequences are made unavoidable by quantum

[1]ibid. p. 98

[2]E. Schrödinger, *Science and Humanism,* op. cit. p. 21

physics. M.L. Dalla Chiara and G. Toraldo Di Francia[1] for instance argue that sets of particles are pure intensions; that they cannot be defined in extension. And in their recent work about quasi-sets, that they performed in parallel with D. Krause[2] and N. Da Costa, they tend to systematize this idea by altering the extensionality axiom of classical set theory. Far from being an isolated claim, Schrödinger's theory of individuality is thus likely to have anticipated some of the most promising directions of research of the end of this century.

[1]M.L. Dalla Chiara & G. Toraldo di Francia (1993), "Individuals, Kinds and names in physics", in: G. Corsi et al. (eds.) *Bridging the gap: philosophy, mathematics, and physics,* op. cit.

[2]D. Krause (1992), "On a quasi-set theory", Notre Dame Journal of Formal logic, 33, 402-411; See also N. Da Costa, D. Krause, & S. French, "The Schrödinger problem", in: M. Bitbol & O. Darrigol (eds.) *Erwin Schrödinger, Philosophy and the birth of quantum mechanics,* Editions Frontières 1993

CHAPTER 6

COMPLEMETARITY, REPRESENTATION AND FACTS

In this chapter, we shall discuss at length one of the most momentous features of Schrödinger's late interpretation of quantum mechanics. This feature has been referred to in paragraph 4-4 as the "structural parallelism" between the (ψ-wave) representation and the domain of facts. Let us recall that according to this conception, the wave-mechanical representation is never *factual,* for it does not describe observed facts and contains no immediate counterpart of these facts. Yet there is a strict *correspondence* between the sequences of facts and the end-product of the unitary evolution of wave-functions of composite systems, since these wave-functions provide us with information about the probability of facts and also about the mutual dependence of (or correlation between) the facts. As a consequence of this accepted dissociation between facts and representation, any motivation to modify the representation each time an experimental outcome is obtained disappears. No intrusion of fact-like discontinuities into the wave-mechanical representation is needed; nothing like collapse of the wave function is required. One has only to find the proper *translation* scheme, or empirical correspondence rules, between the representation and the (sequences of) facts.

Of course, as we noticed in paragraph 4-5, this strategy can be accused of *ignoring* the usual questions about actualisation (or about "objectification"), rather than addressing them. Thus, although the parallelist view does no harm in so far as the predictive content of the theory is concerned, it needs further philosophical justifications in order to become acceptable. At the very minimum, we should try to see whether this way of ignoring the question of actualisation and the measurement problem in general, cannot be associated to a careful "dissolution" of the problem in Wittgenstein's sense.

It is therefore the aim of this chapter to set the frame for a full philosophical appraisal of Schrödinger's "parallelism". But, as a preliminary, we must put this concept of *parallelism* in proper perspective, by comparing it to Bohr's concept of complementarity.

6-1 *Schrödinger's criticism of Bohr's complementarity*

In his classical book, W.T. Scott claims that Schrödinger made an extensive use of Bohr's concept of complementarity in his interpretation of quantum mechanics[1]. The examples Scott mentions are generally convincing. However, one may also quote many texts where Schrödinger expressed explicit rejection of complementarity. Most commentators have insisted on the latter texts, taking them as the clearest evidence of a radical conflict between Schrödinger's conception and Bohr's. So, things are quite

[1]W.T. Scott, *Erwin Schrödinger: an introduction to his writings,* University of Massachussetts Press, 1967, chapter IV, paragraph 5.

intricate. How are we to define Schrödinger's attitude towards the concept of complementarity? As we shall see later on, Schrödinger fully acknowledged the multiple *reasons* which prompted Bohr to adopt his views on complementarity, but he never accepted the synthetic conclusion. Bohr had inferred from them. Our approach will then consist in documenting Schrödinger's rejection of Bohr's views, before we analyze his way of dealing with the very constraints which led Bohr to formulate the concept of complementarity.

Very early in the history of quantum mechanics, in october 1926, a few days after his celebrated debate with Bohr in the latter's house in Copenhagen, Schrödinger wrote a letter to Bohr in which he rejected any attenuation of the difficulties by ideas which acted as a forerunner of complementarity: "(...) I cannot deduce that I am justified in continuing to operate with contradictory statements. One may weaken the statements, by saying e.g., that the collection of atoms 'in certain respects behaves as if ...' and 'in certain respects so as if ...', but this is so to speak merely a juridical expedient that cannot be converted into clear reasoning"[1]. Bohr had not yet developed a full-blown conception of complementarity at that time. According to Heisenberg, it was only a few months later, in February 1927, that the concept of complementarity became central to his thought[2]. But by the end of 1926 he had already claimed that the particle-conception and the wave-conception were both somehow essential for the quantum theory. He had also insisted that both pictures were only to be regarded symbolically, as Schrödinger himself mentioned in his letter. It was thus this early conception, referred to as a forerunner of complementarity, that Schrödinger rejected from the outset. Later on, even during the period of his strongest doubts (from 1929 to 1935), Schrödinger still resisted Bohr's conception of complementarity. He admitted in 1930 that both the atomistic and the continuous (wave-like) modes of thinking "(...) coincided in a formally mathematical way" in quantum mechanics. He also provisionally accepted to consider the wave description as a reflection of our knowledge, and as "(...) based on the fact that observations mutually disturb one another"[3]. But he did not synthetize these Copenhagen-like remarks into anything like Bohr's concept of complementarity.

Then, in the 1940's and 1950's, Schrödinger resumed his sharp criticisms of complementarity. In his Dublin seminars of 1949, he noticed gloomily: "Philosophical considerations about quantum mechanics have gone out of fashion. There is a widespread belief that they have become gratuitous, that everything is all right in this respect for we have been

[1]E. Schrödinger, Letter to N. Bohr (English translation), October 23, 1926, N. Bohr, *Collected works*, vol. 6, op. cit., p. 13

[2]W. Heisenberg, *Physics and beyond*, George Allen & Unwin, 1971, p.79; D. Murdoch, *Niels Bohr's philosophy of physics*, Cambridge University Press, 1987, p. 54

[3]E. Schrödinger, "The transformation of the physical concept of the world", Conference at the Deutsche Museum of Munich, May 1930, quoted by W. Moore, *Schrödinger, Life and Thought*, op. cit. p. 250

given the marvellously soothing word of *complementarity* (...)"[1]. To him, "Complementarity" was merely a *word* some leading physicists used in order to merge and smooth out the variety of interpretative problems which are associated to quantum mechanics. He developed this criticism in a letter to J. Synge written in 1959, where he denounced vigorously the "thoughtless slogan" of complementarity: "If I were not thoroughly convinced that (Bohr) is honest and really believes in the relevance of his - I do not say theory but - sounding word, I should call it intellectually wicked.

'For just when the concepts and logic are at the end of their tether
You are sure to hit on a word *to help you in your troubles'* (Faust I)"[2]

As a consequence, whenever Schrödinger happened to use the words "complementarity" or "complementary", he emphasized that he was not using them in the overarching (and in his eyes self-contradictory) sense which was promoted by Bohr: "By complementary I mean nothing mystical (...)"[3]. In Schrödinger's use of the word "complementarity", complementary features of a theoretical entity or of a phenomenon are merely *compatible aspects* of it. Whereas in Bohr's use of the word, complementary features are both mutually exclusive and jointly indispensible for a complete description[4].

6-2 *Bohr's complementarities*

We now have to dissociate the components of Bohr's concept of complementarity, in order to find their equivalents in Schrödinger's thought. This task is greatly facilitated by many recent and thorough studies of Bohr's philosophy of quantum mechanics[5]. To summarize, one may distinguish *three* applications of the concept of complementarity in quantum mechanics (we shall say nothing about Bohr's applications of the concept of complementarity outside the field of quantum mechanics). These three applications are the following:

[1]E. Schrödinger, *Notes for seminar 1949*, in: E. Schrödinger, *The interpretation of quantum mechanics (Dublin seminars 1949-1955 and other unpublished texts)*, op. cit. p. 103

[2]quoted by W. Moore, *Schrödinger, Life and Thought*, op. cit. p. 473

[3]E. Schrödinger, *July 1952 colloquium*, in: E. Schrödinger, *The interpretation of quantum mechanics (Dublin seminars 1949-1955 and other unpublished texts)*, op. cit. p. 20

[4]Bohr himself recognized in 1929 that the word "complementarity" (which was used in his Como lecture of 1927) connotates compatibility rather than mutual exclusiveness. He thus proposed to replace it by "reciprocity": "In order to emphasize that we are not concerned here with real contradictions, the author suggested in an earlier article the term 'complementarity'. In consideration of the above-mentioned reciprocal symmetry which occurs already in classical mechanics, perhaps the term 'reciprocity' is more suitable for expressing the state of affairs with which we are dealing". (N.Bohr, *Atomic Theory and the Description of Nature*, op. cit., 1987, p. 95). Later on, he came back to the word "complementarity", with an appropriate alteration of its meaning.

[5]H. Folse, *The philosophy of Niels Bohr: the framework of complementarity*, North-Holland, 1985; D. Murdoch, *Niels Bohr's philosophy of physics*, Cambridge University Press, 1987; J. Faye, *Niels Bohr: his heritage and legacy*, Kluwer, 1991; C. Held, "The meaning of complementarity", *Studies in History and Philosophy of Science*, 25, 871-893, 1994

(1) *The complementarity between causal relationship and description of phenomena in space-time.* This application was associated with the first explicit statement of complementarity in Bohr's Como lecture of 1927: "The very nature of quantum theory (...) forces us to regard the space-time coordination and the claim of causality, the union of which characterizes the classical theories, as complementary but exclusive features of the description, symbolizing the idealization of observation and definition respectively"[1],

(2) *The complementarity between the wave picture and the particle picture.* This paradigmatic version of complementarity appears in most early writings of Bohr, such as the 1927 Como lecture: "The individuality of the elementary electrical corpuscles is forced upon us by general evidence. Nevertheless, recent experience, above all the discovery of the selective reflection of electrons from metal crystals, requires the use of the wave theory superposition principles in accordance with the original ideas of L. de Broglie. Just as in the case of light, we have consequently in the question of the nature of the matter, so far as we adhere to classical concepts, to face an inevitable dilemma which has to be regarded as the very expression of experimental evidence. In fact, here again we are not dealing with contradictory but with complementary pictures of the phenomena which only together offer a natural generalization of the classical mode of description"[2]. Many commentators however consider that this is the most controversial version of complementarity, because it involves remnants of classical *representations* (rather than classical *variables,* classical *principles,* or classical constraints on the description of measuring apparatuses)[3]. In addition, these commentators notice that over the years (after 1935), Bohr "(...) tacitly abandons the idea of wave-particle complementarity"[4]. In Bohr's later writings, indeed, the concept of complementarity only allows indirect "clarification" of the dilemma of wave-particle dualism[5]; it does not operate directly on the corresponding couple of representations.

(3) *The complementarity between incompatible variables (the paradigmatic example being position and momentum).* "In quantum physics (...) evidence about atomic objects obtained by different experimental arrangements exhibits a novel kind of complementary relationship. Indeed, it must be recognized that such evidence which appears contradictory when combination into a single picture is attempted, exhaust all the conceivable knowledge about the object"[6]. Two difficulties associated with this version of complementarity have been widely discussed. The first one bears on the relation between the incompatibility

[1]N. Bohr, *Atomic Theory and the Description of Nature,* op. cit., p. 54
[2]ibid. p. 56
[3]D. Murdoch, *Niels Bohr's philosophy of physics,* op. cit. p. 79; C. Held, "The meaning of complementarity", loc. cit.
[4]C. Held, "The meaning of complementarity", loc. cit.
[5]N.Bohr, *Essays 1958-1962 on atomic physics and human knowledge,* Ox Bow Press 1987, p. 25
[6]ibid. p. 4

of experimental arrangements and the incompatibility of value ascription. From the incompatibility of experimental arrangements, one generally cannot derive the incompatibility of the corresponding value ascriptions to an object. For, after all, nothing precludes that it is only our experimental *knowledge* of position which is incompatible with our experimental *knowledge* of momentum, whereas the object "possesses" simultaneously both properties. This (purely epistemic) interpretation of incompatibility was ruled out by Bohr, when he endorsed the so-called "indefinability thesis". According to the indefinability thesis, the experimental arrangement is part of the *definition* of the variable; it is not a mere external intrument for reaching the alleged intrinsic value of the variable. Therefore, if an experimental setup cannot be used (because it is exclusive of another experimental setup which is currently being operated on the object), the value of the corresponding variable is not only unknown: there is no such value at all. Another difficulty of this version of complementarity bears on the sense in which one can say that the values of incompatible variables are *jointly* indispensible to exhaust knowledge about the object. If one accepts the indefinability thesis, this cannot mean that the object somehow *possesses simultaneously* values of the both incompatible variables. It has thus been proposed by D. Murdoch that joint indispensibility means that *successive* measurements of the said variables must be performed in order to exhaust the available knowledge about the object. But this solution is not satisfactory either, because the values can vary greatly according to the *order* of measurements and according to the *intermediate* values which serve as initial conditions for subsequent measurements. "Exhaustiveness" would then only be reached at the cost of an indefinite repetition of measurements, with various orders and various intermediate values; at any rate, under the successive measurement interpretation, joint completion does not involve only *two* measurements. Then, an alternative interpretation is needed. According to C. Held, who aims at providing such an alternative, exhaustiveness refers to the full list of *possibilities* of observation one has before any experimental arrangement has been chosen: "(...) it is particularly important to note that the jointly completing observables are a set of *possibilities,* as the arrangements defining them are possible ones before one is actually selected"[1]. It is only along this line of thought that one may reconcile mutual exclusion and completion without requiring a logical inconsistency: mutual exclusion pertains to *actual* experimental arrangements, whereas completion (or exhaustiveness) refers to *possible* measurements.

Actually, each application of the concept of complementarity to quantum mechanics comes up against much more fundamental difficulties than the ones we have just mentioned. Discussing the latter difficulties will

[1]C. Held, "The meaning of complementarity", loc. cit.

enable us to understand Schrödinger's approach. The easiest strategy is to take the three applications of complementarity in a different order.

To begin with, let us reconsider the complementarity between conjugate variables. It is quite easy to see why the feature of mutual exclusiveness which belongs to complementarity applies to conjugate variables. The incompatibility of the corresponding experimental arrangements, together with the indefinability thesis, entails the mutual exclusiveness of these variables. Even if one considers the recent studies on unsharp measurements, there is still a sense in which one may understand the idea of these variables being mutually exclusive (namely up to a certain precision). But what reasons do we have to believe that conjugate variables are *jointly indispensible?* From Bohr's writings, it is quite simple to figure out his reasons: according to him, conjugate variables are jointly indispensible because it is only when they are taken together that they exhaust the possible information *about* the objects under investigation. He insisted repeatedly on this *aboutness* of experimental information: "(...) evidence obtained under different experimental conditions (...) must be regarded as complementary in the sense that only the totality of the phenomena exhausts the possible information about the objects"[1]; "(...) together (these phenomena) exhaust all definable knowledge about the objects concerned"[2]; "(...) such phenomena together exhaust all definable information about the atomic objects"[3]. By saying so, Bohr tacitly assumes two things.

(i) On the one hand, he assumes that despite the limitations of experimental investigatability of the objects, it is still somehow possible to *refer to* them. But, as Schrödinger demonstrated convincingly, this possibility depends on the validity of the basic presupposition of the act of reference, namely reidentifiability; and he also showed that reidentification of individual atomic objects is generally impossible to ascertain. True, one might suspect that by using so often the plural of the word "object", Bohr was actually trying to refer to classes, not individuals. And classes (electrons, protons, etc.) are perfectly reidentifiable as such, by means of a set of characteristic values of superselective observables such as electric charge, spin, etc. However, this latter interpretation of Bohr's act of reference cannot be accepted. If a value of position at a certain time and a value of momentum at another time can be said to exhaust information about something, it can only be about a *particular* object, not about a class. Other particular objects belonging to the same class might perfectly give rise to different values of position and momentum. Bohr's enduring insistence on reference must thus concern individual objects. But in this case, it generally goes beyond the possibilities of any available experimental method.

[1]N. Bohr, *Essays 1933-1957 on atomic physics and human knowledge,* Ox Bow Press, 1987, p. 40
[2]ibid. p. 90
[3]ibid. p. 99

(ii) On the other hand, Bohr tacitly assumes that the atomic objects he refers to have the same general predicative structure as the objects of classical mechanics, even though they are not ascribed predicates as such. He considers that, just as the classical particles, the "atomic objects" can be construed as points of convergence of two families of conjugate characteristics; for instance as points of convergence of position *and* momentum. True, the indefinability thesis states that position and momentum are not *possessed simultaneously* by the atomic objects; the holistic conception of phenomena even precludes that each value of a variable is *possessed* by an object; but values of position and values of momentum are still both considered as possible information *about* one and the same object. They have *aboutness,* and this aboutness focus on one and the same object. It is only this way that it can be said that they are *jointly* indispensible. If one did *not* presuppose that experimental outcomes are *about* objects which have the same basic predicative structure as the moving material bodies of classical physics, one could perfectly accept that position and momentum are mutually exclusive without being *jointly* indispensible; a value of *one* of the two observables would only be *separately* indispensible in order to make the best possible probabilistic predictions allowed by quantum mechanics in a given experimental situation.

To summarize, Bohr implicitly accepts that the basic object-structure which underlies quantum mechanics is the same as the basic object-structure which underlies classical mechanics, even though he adds many qualifications to the latter in order to obtain the former. It is only in the framework of these last remnants of ontological conservatism that position and momentum have to be considered as *jointly* indispensible.

Let us come now to wave-particle complementarity. The major difficulty which undermines the idea of wave-particle complementarity is widely known among philosophers of physics[1]: there is *no* experimental arrangement in which one of these two pictures *alone* is sufficient to provide a satisfactory description of phenomena. On the one hand, in an interference experiment, where the wave aspects are predominant, the corpuscular aspects are not completely absent; for the isolated impacts on the detection screen are generally interpreted by using the corpuscularian picture. On the other hand, in experiments where the corpuscular aspects are predominant, namely those in which a good approximation of a *trajectory* can be obtained experimentally, the wave aspects are not completely absent either. They manifest themselves through phenomena which may be described as *diffraction* effects. Thus, those aspects of the wave picture and of the particle picture which intervene in quantum mechanics manifest themselves jointly in one and the same experimental configuration, although they never appear to their full extent. The conceptual revision which is required by quantum mechanics goes well

[1]See e.g. A. Grünbaum, "Complementarity in quantum physics and its philosophical generalization", *The Journal of philosophy,* 54, 717-735, 1957

beyond the alternate use of two classical pictures, be they construed as mere symbols.

Finally, let us reexamine complementarity between causality and the use of space-time concepts. The problem, here, is to clarify Bohr's ideas on this point. One of the most explicit statements of his position is to be found in his paper "The unity of human knowledge", published in 1960: "(...) this situation prevents the unlimited combination of space-time coordination and the conservation laws of momentum and energy on which the causal pictorial description of classical mechanics rests. (...) the use of any arrangement suited to study momentum and energy balance - decisive for the account of essential properties of atomic systems - implies a renunciation of detailed space-time coordination of their constituent particles"[1]. In the last sentence, Bohr suggests that complementarity between causality and space-time coordination mostly reduces to the complementarity between two couples of conjugate variables (momentum and energy on the one hand, and position and time on the other hand). In addition, his account of the complementarity between causality and space-time coordination is identical to his account of the complementarity between conjugate variables: in both cases, complementarity is related to incompatibility of experimental arrangements combined with the indefinability thesis. However, the first sentence of the quoted paragraph points towards a noticeably different direction. In this sentence, mutual exclusiveness bears on the pair (conservation *laws* and space-time coordination), not directly on the pair (momentum-energy coordination and space-time coordination). Several other texts of the earlier period[2], as well as Heisenberg's lucid interpretation of Bohr's position, favour this latter conception of the complementarity between causality and space-time coordination, namely a conception which does *not* tend to reduce it to the complementarity between conjugate variables. In his *Physical principles of the quantum theory,* Heisenberg insists on a conceptual asymmetry which seems to have disappeared in Bohr's texts of the late period. According to Heisenberg, the measurement of *any* variable, not only the measurement of space-time coordinates, involves spatio-temporal aspects. Thus, it is not only space-time coordination, but more generally a description of phenomena in space-time, which is incompatible with causality. The reason for this incompatibility, says Heisenberg, is that one

[1]N. Bohr, *Essays 1958-1962 on atomic physics and human knowledge,* op. cit. p. 11

[2]see e.g. N. Bohr, *Atomic Theory and the Description of Nature,* op. cit. p. 54: "(...) if in order to make observation possible we permit certain interactions with suitable agencies of measurement, not belonging to the system, an unambiguous definition of the state of the system is naturally no longer possible, and there can be no question of causality in the ordinary sense of the word. The very nature of the quantum theory thus forces us to regard the space-time co-ordination and the claim of causality, the union of which characterizes the classical theories, as complementary but exclusive features of the description, symbolizing the idealization of observation and definition respectively".

Here, "observation", occurring in space-time *irrespective of which variable is "observed",* is incompatible with definition of the state of the system. "Observation" is thus generally incompatible with application of a causal law to the state of the system. No distinction between "observation" of space-time variables and momentum-energy variables is relevant.

cannot observe *any* kind of spatio-temporally located phenomena without influencing them; and such an influence makes causal laws inapplicable to the observed phenomena. Causal relations pertain to "exact mathematical laws", not to spatio-temporally located phenomena. Spatio-temporally located phenomena obey the Uncertainty Relations; they are not ruled by a causal law. "(As) a geometric or kinematic description of a process implies observation, it follows that such a description of atomic processes necessarily precludes the exact validity of causality - and conversely"[1].

Heisenberg's interpretation of the complementarity between causality and space-time location has been adopted by many authors since then. According to C.F. Von Weizsäcker[2] and P. Mittelstaedt[3], this version of complementarity means a relation of mutual exclusiveness and joint completion between the abstract deterministic law of evolution of ψ-functions (i.e. the Schrödinger equation) and the measurement results observed in space-time. Using H. Wimmel's formulation (see paragraph 4-5), one could say that complementarity between causality and space-time location means complementarity between the ψ-model and the F-model, namely between the laws which rule the evolution of the ψ-function and the Facts. This type of complementarity, which concerns the relation between the symbolic predictive scheme and the phenomena, is clearly distinct from the complementarity between two couples of conjugate variables (namely position and time on one side, and momentum and energy on the other), which rather amounts to an internal relation between phenomena.

The problem is that whereas the latter authors attempted to promote a systematic distinction between various applications of complementarity, Bohr himself rather tended to merge these applications into a unified pattern. The ψ-F interpretation of complementarity between causality and space-time location was overtly rejected by Bohr in correspondence with Von Weizsäcker in 1956[4]. For, according to Bohr, complementarity can *only* hold between phenomena, not between the symbolic predictive scheme and the phenomena. Even though, as C. Held[5] points out, this late position of Bohr is not entirely faithful to his writings of the period 1927-1933, it partakes of a general trend towards reduction of the three major applications of complementarity to one all-comprehensive pattern, which was already at work during the early period. As we have seen, in his later texts, Bohr attempted to fit the three applications of complementarity into the model of incompatible variables complementarity. Complementarity between causality and space-time location was reduced to complementarity between momentum-energy variables and space-time variables. And

[1]W. Heisenberg, *The physical principles of the quantum theory,* op. cit. p. 63
[2]C. F. Von Weizsäcker, *Aufbau der physik,* Hanser, 1985
[3]P. Mittelstaedt, *Philosophical problems of modern physics,* in: *Boston Studies in the philosophy of science,* vol. 18, Reidel, 1976
[4]C. F. Von Weizsäcker, *Aufbau der physik,* op. cit., p. 293; see also C. Held, "The meaning of complementarity", loc. cit.
[5]C. Held, "The meaning of complementarity", loc. cit.

wave-particle complementarity was supposed to depend indirectly on the general model of incompatible variables complementarity through the indefinability thesis. But in earlier texts, other methods of mutual reduction of the various applications of complementarity were suggested instead. In his 1927 Como lecture, for instance, Bohr relates space-time description to the *wave-like* propagation of light, whereas the laws of conservation of energy and momentum find their expression in the *corpuscular* aspects of light. This dissociation implies "(...) the impossibility of a causal space-time description of the light phenomena"[1]. Here, the complementarity between space-time coordination and causality reduces to wave-particle complementarity. Yet, the system of relations on which this reduction is based is itself very questionable. It has been suggested that a reverted system of relations is at least as plausible as the direct one. After all, the Planck and de Broglie relations enable one to associate a frequency and wavelength (which are typical of a *wave-model)* to measured values of energy and momentum. And a precise measurement of position at a certain time can be construed as a measurement of the position *of a particle*. The latter remarks tend to support a relation between space-time description and the corpuscularian picture, and a relation between the energy-momentum conservation laws and the wave-picture, rather than the other way round. As D. Murdoch rightly points out, there is thus no univocal system of relations between the wave and particle pictures on one side, and time-position and energy-momentum measurement on the other side. The consequence is that "(...) wave-particle complementarity and kinematic-dynamic complementarity are logically independent notions"[2].

In all his reductive attempts, Bohr's key idea was to make a synthesis of the situations wherein two "features", two "phenomena", or two "pictures", are united in classical physics, but appear separated in quantum physics. Only in this way could he make sense of the correspondence principle which was so useful in the old theory of quanta, by proving that quantum mechanics is a "natural generalization of the classical physical theory"[3]. Unfortunately, the synthesis proved very difficult, not to say impossible; for, as we have emphasized, the various versions of complementarity displayed irreducibly specific features, and the only element able to unite these cases appeared to be little more than the *word* "complementarity" itself. The "natural generalization" of classical physics did not only involve the general scheme of pairwise separation and "reciprocity"; it also required recognition of the heterogeneity of the pairs, as well as a proper analysis of the conceptual root of this heterogeneity.

[1]N. Bohr, *Atomic Theory and the Description of Nature*, op. cit. p. 55-56
[2]D. Murdoch, *Niels Bohr's philosophy of physics*, op. cit. p. 67
[3]N. Bohr, *Atomic Theory and the Description of Nature*, op. cit. p. 19

6-3 *Schrödinger's "complementarities"*

The analysis of the divergent varieties of Bohr's complementarities was remarkably worked out by Schrödinger. Actually, he pushed his analysis far enough to be entitled to claim that there is no point in using a single word for all these cases. And he even commented at length the irreducibility of the cases to one another. In Schrödinger's writings one thus finds an exact equivalent of *each one* of Bohr's applications of complementarity. But the word "complementarity" is either missing, or used in its familiar sense of joint completion *without* mutual exclusiveness. Let us then discuss the list of three problems which correspond, in Schrödinger's framework of thought, to Bohr's three applications of complementarity.

We shall begin with Schrödinger's interpretation of those experimental features which were taken by Bohr as a manifestation of *wave-particle* complementarity. At first sight, Schrödinger appears to have rejected the idea of joint completion of the wave picture and of the particle picture throughout his career. In his early writings, the *factual and effective* ψ-waves were supposed to account for the corpuscular aspects of phenomena by means of the concept of wave packet. In his later writings, even though ψ-waves were no longer considered as factual, even though their relation with facts was restricted to providing *information* about the probability and the correlation of experimental outcomes, they remained the only warranted (continuous) "description". The atomistic aspects of phenomena were accounted for by the eigenstate scheme of second quantization, and this eigenstate scheme was itself underlied by a cosmological standing wave model (see end of paragraph 2-2). Schrödinger's conception of quantum mechanics thus remained pictorially and ontologically monistic from 1925 until his death.

In his Nobel lecture of 1933 and in his Seminars and articles of the 1950's, Schrödinger however pointed out that his unique wave-like description is able to accomodate a kind of structural dualism which is isomorphic to the traditional wave-particle dualism. The two features which are concerned by this structural dualism are (i) the "transversal structural interlacing" between events, which manifests itself by interference patterns, and (ii) the "longitudinal" structural interlacing, which manifests itself by tracks in cloud chambers. They are called "complementary" in the following sense: "(...) they depend on each other; that is: when and if, and to the same degree of approximation as, one of them exists, the other exists"[1]. Whereas the wave picture and the particle picture are logically exclusive of one another, the transversal and longitudinal "linkages" or "structural interlacings" are both implied by the ψ-wave model. Transversal linkage is expressed by wave- (or phase-) surfaces, while longitudinal linkage is expressed by wave-rays (or wave-

[1]E. Schrödinger, Dublin seminar, *July 1952 colloquium* (in: E. Schrödinger, *The interpretation of quantum mechanics (Dublin seminars 1949-1955 and other unpublished texts)*, op. cit. p. 20)

normals). "These two features, wave-surfaces and wave-normals, seem capable of representing the two observed features, namely the transversal structural interlacing and the longitudinal one, respectively. A particle phenomenon has not these two traits, only one : the paths or orbits or trajectories. They are capable, at most, of representing the longitudinal linkage; there is nothing in particles to correspond to the transversal one"[1]. This conception automatically entails a joint manifestation of some aspects of the wave picture and of the particle picture, although *not to their full extent*. The discontinuous events are evocative of particle impacts, and the *twofold* structural linkages between these point-like events can only be accounted for by a wave model. But the *purely* longitudinal linkage of the particle model is discarded, and the continuity which is typical of wave-like interference patterns *only* emerges from the accumulation of a very large number of discontinuous events of the usual type. In good agreement with our previous analysis of the interference experiments, no instrumental configuration endows any one of the two classical pictures (wave or particle) with an exclusive validity.

Let us then examine the various aspects of this lack of exclusive validity of each picture (wave or particle).

On the one hand, some elements of the two pictures are irretrievably intermingled. As Schrödinger noticed several years before Feynman, *if* something of the longitudinal linkage (or trajectory, or path) of the particle model is to be retained in quantum mechanics, it must be indissolubly associated to a wave-like transversal linkage between a multiplicity of (possible) paths. This idea appeared in his second paper of the seminal series "Quantization and proper values" of 1926: "(...) the whole manifold of paths of a system are bound together by equations, so that apparently a certain reciprocal action exists between the different paths"[2]. And it was expressed again in the Nobel lecture of 1933, though with some reservations which were dropped in the 1950's: "From the standpoint of wave-mechanics the innumerable multitude of possible particle paths would be only fictitious and no single one would have the special prerogative of being that actually travelled in the individual case"[3]; between these paths, "wave-fronts form a lateral connection"[4].

On the other hand, there are also some elements of the two pictures which are systematically missing in the experimental phenomena. What is missing in the phenomena is nothing less than continuity. Not only the transversal continuity of wave-fronts, but *also* the longitudinal trajectory of particle trajectories. Hence Schrödinger's insistance on confronting any classical picture, both the particle picture and the wave-picture, with the

[1]ibid.

[2]E. Schrödinger, "Quantization and proper values-II", in: *Collected papers on quantum mechanics*, op. cit., p. 26; see paragraph 1-7 of the present essay.

[3]E. Schrödinger, "The fundamental idea of quantum mechanics", in: E. Schrödinger, *Science and the human temperament*, op. cit. p. 152

[4]ibid. p. 153

raw material of isolated experimental events[1]. But once this extreme stage of analysis has been reached, one must recognize, he said, that the continuous wave mechanical model is able to afford adequate predictive information about *both* kinds of connections, longitudinal and transversal, particle-like and wave-like, between the isolated events. The so-called wave-particle complementarity is thus at the same time dissolved and accounted for by the wave mechanical model.

The second application of complementarity we have to reconsider from Schrödinger's standpoint is complementarity of conjugate variables. As we have noticed in the previous paragraph, the *mutual exclusiveness* of value ascription for conjugate variables can be construed as a consequence of (i) the incompatibility of the corresponding experimental arrangements and (ii) the "indefinability thesis". But Bohr's idea that these values are *jointly indispensible* arises from what we have called a remnant of ontological conservatism. It is only because Bohr's atomic objects have the same categorial form as the classical particles which possess a momentum and a position, that they are supposed to be points of convergence of both momentum-energy characterizations and space-time characterizations. One may thus easily avoid the idea of joint indispensibility of incompatible characterizations. This only requires a modification of the categorial form of the objects of physics. If the basic ontology is altered in such a way that its objects are not supposed to be characterized by couples of conjugate variables, but only by separate lists of compatible variables, then incompatible characterizations are no longer *jointly indispensible* to characterize the objects. Now this is exactly what Schrödinger proposed to do. Let us suppose, as he explicitly suggested, that the basic ontology of physics is an ontology of ψ-waves. By their very structure ψ-waves are such that one cannot focus upon them two incompatible characteristics. Ψ-waves are fully characterized by the values of a complete set of *commuting* observables. Information *about* such objects is exhausted without requiring that one goes beyond a list of *compatible* characterizations. Complementarity of conjugate variables is thus dissolved by an ontological shift.

Let us finally comment on Schrödinger's perception of what Bohr called the complementarity between causality and space-time coordination. As we have seen, one plausible interpretation of this application of complementarity is that it denotes the parallelism between the laws which rule the evolution of the ψ-function and the experimental facts; or more generally between the (continuous) wave mechanical model and the corpus of (discontinuous) experimental outcomes. This kind of parallelism was pointed out by Schrödinger quite early in the history of quantum mechanics, in 1929. He expressed it in a sentence that we have already quoted in paragraph 1-8, but that we must comment upon afresh at this

[1]See end of paragraph 4-1 of this essay

point: "(...) very likely, the wave-corpuscle contradiction is the manifestation of an important new fundamental principle: the non-identity of what is detailed in space-time, on the one hand, and of what is observable on the other"[1]. Here, Schrödinger's global claim is that the "wave-particle contradiction" is just a reflection of something more fundamental. This was also Bohr's position, especially in his later writings. But whereas Bohr tended to think that this more fundamental "something" is the incompatibility of the experimental arrangements which are involved in the definition of variables, Schrödinger identified it with what we have called the "post-modern distance" between the representation and the facts (i.e. the circumstance that the continuous wave-mechanical representation, which provides *information about* the facts, cannot be considered as a faithful *description of* the facts). True, in the sentence which we are commenting, Schrödinger expressed this distance in an unsatisfactory way, by putting "what is detailed in space-time" on one side and "what is observable" on the other side. But one must not forget that this mode of expression is typical of an early stage of his thought, where he still tended very strongly to identify the wave-mechanical model with a space-time description. Later on, this constraint was (partly) released, although the basic distinction between the description and the facts survived. Schrödinger insisted more and more upon the gap which exists between the continuous wave-mechanical model and the discontinuous events observed in the space-time framework of the laboratory, and less and less upon embedding the model itself in space-time (in spite of the circumstance that the second-quantization scheme offers such an opportunity). As we already know, in his texts of the 1950's, Schrödinger especially emphasized (i) that the list of the possible outcomes of an experiment in some given instrumental configuration (i.e. the list of possible "facts" in this configuration) *cannot* be derived from the wave-mechanical model; (ii) that it can only be determined by considering operators which are "loans from classical physics"[2], i.e. by calculating the eigenvalues of the so-called *observables;* and (iii) that the wave-mechanical model is only related indirectly to these *extrinsically* defined facts through Born's probabilistic empirical correspondence rules.

[1]E. Schrödinger, "Neue Wege in der Physik", *elektrotechnische Zeitschrift,* 50, 15-16, 1929

[2]E. Schrödinger, *Transformation and interpretation in quantum mechanics,* (Dublin seminar 1952), in: E. Schrödinger, *The interpretation of quantum mechanics (Dublin seminars 1949-1955 and other unpublished texts),* op. cit. p. 51; See § 4-4 of this essay (Point (8): *Preferred basis)*

BOHR		*SCHRÖDINGER*	
Causality / Space-Time Coordination	*Mutually exclusive & jointly exhaustive (complementary)*	Ψ-model / F-model	*Parallelism*
(P , Q)	*Mutually exclusive & jointly exhaustive (complementary)*	Ψ (P) /Ψ(Q)	*Transformation by a Fourier Integral*
Wave / Corpuscle	*Mutually exclusive & jointly exhaustive (complementary)*	Transversal Linkage / Longitudinal Linkage	*Compatible Features of the wave-mechanical model*

TABLE 1

To recapitulate (see Table 1), Bohr's wave-particle and incompatible variables complementarities were somehow dissolved by Schrödinger into the wider framework of wave mechanics. However, no dissolution occurred for complementarity between causality and space-time coordination. Instead, Schrödinger replaced it by a basic and irreducible parallelism between (a) the continuous wave-mechanical model and (b) the discontinuous Fact-model afforded by the eigenvalue scheme of the observables. Is it then possible to say that Schrödinger retained at least one application of Bohr's complementarity? Not exactly so. True, the wave-mechanical model and the Fact-model are *mutually exclusive,* in the same general sense as a continuous model and a discontinuous model are. But the two terms are not *jointly indispensible* to exhaust information about any third term. They rather point towards one another. The wave-mechanical model points towards the Fact-model, for it provides information *about facts.* And the Fact-model points towards the wave-mechanical model, for the correlations between facts are ruled by the laws of wave mechanics.

We can represent this momentous difference between Bohr's complementarity and Schrödinger's parallelism in the following diagram.

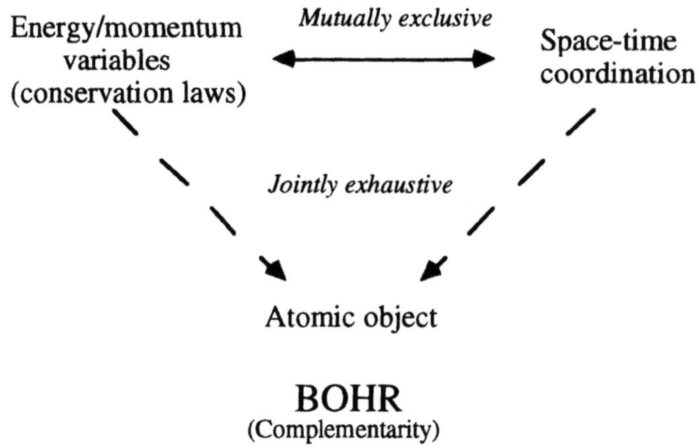

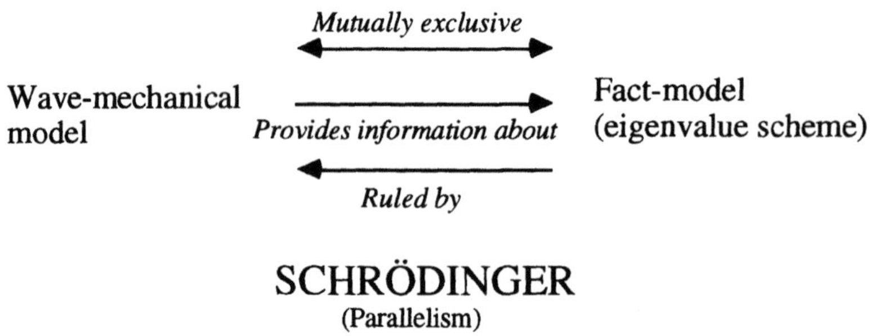

It is clear from this diagram that Bohr's concept of complementarity somehow calls for a remnant of the classical ontology (through the reference to the atomic object which supposedly focuses on it two families of canonically conjugate variables which, when taken together, constitute the classical state), whereas Schrödinger's concept of parallelism is completely independent of it. Schrödinger's parallelism is compatible with a purely instrumentalist conception of the ψ-model, as in Van Fraassen's version of the modal interpretation or in Wimmel's "critical interpretation" of quantum mechanics (see paragraph 4-5). But it can also pave the way for a new ontology of ψ-waves, as Schrödinger himself suggested insistently.

6-4 *Two parallelisms*

Schrödinger's twofold conviction was (i) that the Fact-model which is derived from the eigenvalue scheme of the observables can by no means be encompassed within the description provided by the wave-mechanical model, and (ii) that the wave-mechanical model has to be considered as something more than a symbolic instrument for predicting facts. He thus precluded in advance two opposite reductionist strategies. A strategy of reduction of fact-like appearances to complex features of the wave-mechanical representation, which the popular many-worlds interpretation of quantum mechanics and the strongest version of decoherence theories have tended to endorse; and, conversely, a strategy of reduction of the wave-mechanical model to a system of possible relations between facts which is typical of instrumentalism. According to Schrödinger, no cheap solution could dispense the physicist from accepting parallelism of the ψ-model and the F-model as a basic feature of quantum mechanics. But how are we to understand this alleged resistance of parallelism to any kind of reduction? And what is the meaning of the two parallel components?

In order to answer these questions it will prove useful to compare systematically the parallelism which is at stake in quantum mechanics, with another kind of parallelism which has already been a subject of intense philosophical discussions. This other kind of parallelism concerns human actions. Actions can be described in terms of intentions, beliefs, hopes, desires, and fears, which all involve *purpose* (namely a view on the *future);* and they can also be described, alternatively, in terms of *causes and effects,* behaviours, physiological events, neurological processes, which generally presuppose knowledge of a series of *past* events taken as conditions of the act. That the two accounts are not independent of each other, that there is a close correspondence between the elements of the intentional account of human action and the elements of the causal-physiological account, becomes more and more obvious the further neurological research progresses. But this does not make bare reductionism unproblematic. The meaning of the parallelism of the intentional account and the causal-physiological account of action is still very debatable.

Many questions remain open. Can the two accounts be projected (or not) on some dualist model of the mental and the physical? Are they reducible (or not) to one another? And if they are irreducible, why is it so? Which difficulties do the reductionist attempts meet? The interesting point, for us, is that these questions about the parallelism between the intentional and the causal accounts of action can be transposed almost exactly to the case of the parallelism between the Fact-model and the ψ-model in quantum mechanics. Moreover, the structure of the various answers which were proposed in both cases is remarkably similar.

6-5 *Being-in-a-body and being-in-the-world*

Before we undertake a systematic comparison between these two parallelisms (in paragraphs 6-6 to 6-8), it is useful to develop a picture which will help us understanding why this comparison involves more than a simple formal analogy.

The two distinct accounts of actions (causal and intentional) can be taken to correspond to the difference between a description of actions by an external observer, and a description of the same actions by someone who *is* or *could be* the agent. The causal account is typical of a situation where the person who describes the action is *facing* the body under study; and the intentional account is typical of a situation where the person who describes the action *is* or *could be* the person whose body is just the acting body. The enduring use of the intentional account, in spite of the continuous progress of neurophysiology, is thus likely to be due to the irreducible perspectival specificity of what it is like to be a *being-in-a-body*.

Now isn't it possible to find a similar difference between a description by an observer contemplating the world from an external standpoint and a description by someone who partakes of the world? This difference would be especially relevant in physics. The difficulty is that, in this case, the external point of view is not as unproblematically available as previously; reaching knowledge of the world "from outside" can at most be a project or a regulative idea. However, even if one considers that the external standpoint is not available in this case, even if no *outside* can meaningfully be opposed to the *inside,* there are still constraints which may be interpreted as preconditions for our being able to exist *in* this world, namely for our being (stable) integral parts of it. The conditions of possibility of our life, of our action, and also of the intellectual mastery of our activity, which happen to include the macroscopic scale and complexity of our bodies, are among these constraints. The enduring use of a description in terms of notions such as classical observables, "things", facts, etc., may then be related to the irreducible specificity of what it is like to be a *being-in-the-world*. Indeed, these notions are typical of the scale and complexity of our lives, and they are therefore relative to the interests of beings which are able to be stable inhabitants of the world. In the same way as the intentional model of actions is imposed on us by our being-in-a-body, we can say that the Fact/thing-model of physical processes is imposed on us by our being-in-the-world.

Further discussion is needed at this stage. After all, the concept of "being-in-the-world" is by no means new; it has not been developed in relation to quantum physics, not even exclusively in relation to physics. Let us then describe its bearing on everyday life and on classical physics before we come back to quantum physics.

The expression "being-in-the-world" has been coined by Heidegger in his *Being and Time* [1]. It aimed at pointing out that our relationship with our environment is not primarily one of objective knowledge, but also (and more fundamentally) one of practice, use, and tacit confidence in the permanent availability of the familiar elements of our surroundings. It was meant to emphasize that *knowing-how* has to be considered as the true basis for *knowing-that*. It also suggested that parts of our environment (at

[1]M. Heidegger, *Being and Time,* Harper & Row, 1962; H.L. Dreyfus, *Being-in-the-world (a commentary on Heidegger's Being and Time, division I),* M.I.T. Press, 1991; M. Grene, *A philosophical testament,* Open Court, 1995

least those parts that we call our *tools)* are tentatively treated by us as extensions of our body, as long as no unexpected difficulty arises. Thus being-in-the-world and being-in-the-body are not unrelated[1]. One may consider that being-in-the-world arises from an indefinite extrapolation of our being-in-a-body or, conversely and more likely, that being-in-the-body is a half-conventional and half-imposed limitation of our being-in-the-world.

Insistance on our position of beings-in-the-world, as opposed to some ideal all-encompassing position similar to God's viewpoint, was also considered by Cassirer[2] as typical of the conceptions of *classical physics in the late seventeenth century and during the eighteenth century*. After Descartes' attempt at identifying the "eternal truths" which are common to man and God, Newton proposed a more modest view of science. According to this view, the aim of science is not to break out through the boundaries of experience in order to give us access to the absolute essence of transcendent things. The aim of Newtonian science, in Cassirer's eyes, was only to provide us with unified rules of behaviour which may help our inhabiting comfortably the world of experience in any circumstance.

So, at first sight, it appears that there is nothing original or specifically quantum-mechanical in our former emphasis on being-in-the-world. Quantum mechanics can be better understood in the light of the concept of being-in-the-world, but isn't this also the case for Newtonian mechanics and even for the elementary structures of our everyday activity? Actually, there *are* noticeable differences between the case of quantum mechanics on the one hand and the case of both everyday life and Newtonian mechanics on the other hand. These differences account for the distinctive importance of the concept of being-in-the-world for the interpretation of quantum mechanics, and we thus have to comment on them.

To begin with, we shall stress a contrast between experiments in quantum physics and most other types of human activity. As we mentioned formerly, when the surrounding things of our environment which are "ready-at-hand" do not behave as we spontaneously expect, it is always possible to take them as objects of study in order to identify the reason for their departure from their familiar ways. This *objectification* presupposes that the surrounding things can, without harm, be made independent from the conditions of their ordinary use; that they can be put in very different conditions without losing the characteristic features which allowed their being thoughtlessly used as tools; that they are indifferent to particular contexts to such a large extent that they can be detached from *any* context within a certain range. This possibility of detachment is one of the circumstances which allow us to establish a clear-

[1]A systematic study of what it is like to be a *Being-in-a-body* can be found in: M. Merleau-Ponty, *Phenomenology of perception*, op. cit.

[2]E. Cassirer, *Die Philosophie der Aufklärung*, Mohr (Tübingen), 1932, Chapter 1

cut, though practical, difference between the tools of our environment and our body.

Consider the standard example of a stick used to orient oneself in the dark. "When (...) it is held firmly, we lose the sensation that it is a foreign body, and the impression of touch becomes immediately localized at the point where the stick is touching the body under investigation"[1]. At this stage, everything occurs exactly as if the stick was an integral part of our body: "If (...) someone asks me 'what are you now feeling in the fingers that hold the probe?' I might reply 'I don't know - I feel something hard and rough *over there*'"[2]. But the stick can also be held *loosely* or manipulated under various conditions. In this case, it is (and it *can* be) treated as if it were an independent entity, completely detached from the bodily conditions which allow one to use it as an extension of the fingers: "it appears to the sense of touch to be an object"[3].

Now in the experimental domain ruled by quantum mechanics, there is no such large context-independence, and thus no such possibility of detachment, except in very restricted circumstances. This characteristic feature of the quantum domain was forcefully brought out by Bohr. The so-called "individuality"[4] of quantum processes, by which he meant their "indivisibility"[5], soon became pivotal in his reflection on quantum mechanics. According to Bohr, the "essential wholeness" of each phenomenon cannot be broken in such a way that one may distinguish an object endowed with properties from the experimental context of its investigation. For, "(...) any attempt at its well-defined subdivision would require a change in the experimental arrangement incompatible with the appearance of the phenomenon itself"[6]. As a consequence, quantum physics has had an important effect on the common view of our position in the world: "(...) the new situation in physics has (...) forcibly reminded us of the old truth that we are both onlookers and actors in the great drama of existence"[7].

The idea of contextual dependence of quantum phenomena was also familiar to Schrödinger, and it is perfectly recognizable in his writings, although he expressed it in the framework of wave-mechanics rather than in general terms. Schrödinger's reaction to the Einstein-Podolsky-Rosen paper of 1935, for instance, has some interesting contextualistic elements in common with Bohr's. Let us recall that Einstein's argument prompted Bohr to insist less and less on direct disturbance of systems by the measuring agencies and more and more on the role of the experimental arrangement in the very *definition* of physical quantities. In this new conception, *projects* of measurements (in Bohr's terms, "(...) the very

[1]N. Bohr, *Atomic Theory and the Description of Nature*, op. cit. p. 99
[2]L. Wittgenstein, *Philosophical investigations*, Basil Blackwell, 1958, §626
[3]N. Bohr, *Atomic Theory and the Description of Nature*, op. cit. p. 99
[4]N. Bohr, *Essays 1933-1957 on Atomic physics and human knowledge*, op. cit. p.17
[5]ibid. p. 34
[6]ibid. p. 72
[7]N. Bohr, *Atomic theory and the description of nature*, op. cit. p. 119

conditions which define the possible types of predictions regarding the future behaviour of the system"[1]) became just as important as *effective* measurements. At the same time Schrödinger pointed out that, in the situation of "entanglement" of the wave-functions of two systems having interacted in the past, the list of possible final disentangled wave functions for one system is bound to depend on the *program* of experiments to be carried out on the other one. "It is necessary to envisage the dependence on the *programme"* [2]. True, in 1935, Schrödinger considered that this was a rather paradoxical feature of wave mechanics which might disappear in a future relativistic theory. But later on, in the 1950's, he tended to take the characteristics of wave mechanics more and more seriously. In his paper "What is an elementary particle?", he wrote, with no qualification whatsoever, that the possible states of a system "(...) are not absolutely defined; they depend on the arrangement of the - actual or imagined - experiment"[3].

In 1967, Kochen and Specker clarified the situation in a well-known paper[4]. For they demonstrated that contextuality is neither an idiosyncratic feature of Bohr's interpretation of quantum mechanics, nor a contingent characteristic of the wave mechanical formalism. Contextuality is a necessary aspect of any theory, including hidden variable theories, able to reproduce the same predictions as quantum mechanics. True, in hidden variable theories, detachment of properties with respect to any experimental context is maintained; but it is only at the cost of ascribing this detachment a purely formal status. As Bohm and Hiley recognize, "(...) the motion of the particles cannot properly be discussed in abstraction from the total experimental arrangement. This is reminiscent of Bohr's notion of wholeness, but it differs in that the entire process is open to our 'conceptual gaze' and can therefore be analyzed in thought, even if it cannot be divided in actuality without radically changing its nature"[5]. We could say that Bohm's theory is aimed at displaying a plausible reason for the impossibility of detachment, by means of a formal manipulation of detached representations.

Thus, in general, there is no possibility of detachment of the objects of quantum physics (except in a purely intellectual sense). The characteristics of the objects are indissolubly dependent on appropriate experimental contexts, even if they can always be distinguished *in abstracto,* within the framework of hidden variable theories, from the experimental conditions which determine them. Such a tight entanglement between parts of the world and the instruments of investigation is sufficient to understand why the concept of being-*in*-the-world (as against the traditional picture of a

[1]N. Bohr, "Can quantum-mechanical description of physical reality be considered complete?" in: J. A. Wheeler and W.K. Zurek, *Quantum mechanics and measurement,* op. cit.

[2]E. Schrödinger, "Discussion of probability relations between separated systems", loc. cit.

[3]E. Schrödinger, "What is an elementary particle?" loc. cit.

[4]S. Kochen & E. Specker, "The problem of hidden variables in quantum mechanics", *Journal of Mathematics and Mechanics,* 17, 59-87, 1967

[5]D. Bohm & B.J. Hiley, *The Undivided Universe,* op. cit. p. 38

being *facing* the world) appears to be so specifically relevant for the interpretation of quantum mechanics.

Let us notice incidentally, along this line of emphasis of the concept of being-*in*-the-world and of criticism of the picture of a being *facing* the world, that some connotations of the word "object" (etymologically: "thing thrown before") are quite inappropriate. We shall therefore replace the expressions "object" or "object of investigation" by the vaguer expression "domain of investigation" in the following paragraphs. This will avoid spurious connotations.

At this point, one may wonder whether the consequences of contextuality are not even stronger than the one we have just pointed out. As we mentioned previously, the possibility of detachment, the possibility of considering surrounding things as objects properly speaking, is one of the practical criteria which enable us to distinguish between the tools of our environment and our body. Since the domain of investigation of quantum physics lacks this sort of distinction, can't we say that it behaves somehow as an extension of our body? Can't we go beyond the comparative study of being-in-the-world and being-in-a-body we have undertaken, and merge the two concepts? Not at all, for at least two reasons.

Firstly, whereas the domain of investigation of quantum physics cannot be detached from the experimental conditions which define its determinations, the experimental devices themselves *can* be detached (in practice, though not in principle if quantum contextuality is considered universal), from the body of the experimenter. The status of the domain of quantum physics is thus closer to the status of something which cannot be made independent of the stick used for its exploration, than to the status of a stick which is too firmly held by a hand to be treated as an object separated from the body.

Secondly, a crucial feature of our body is that its parts have (*must* have) univocal control on one another. Our motor nerves control univocally the secretion of certain substances near the membranes of muscular cells, the concentration of these substances controls univocally the state of contraction of our muscles, the state of contraction of our muscles determines univocally the motion of our limbs, a certain level of excitation of sensory nerves determines univocally its electric response, etc. This chain of univocal control is a tacit precondition of both voluntary actions and perceptions. Without a reasonable amount of univocity of this control, which allows confidence and efficiency in everyday activities, random gesticulations would soon replace actions and scattered sensations would replace perceptions. But in quantum mechanics contextuality is closely associated to indeterminism; it is impossible, except in a few special cases, to control univocally the state of a system throughout its interactions, by means of some experimental device. The motor chain, and the sensory chain as well, breaks somewhere (say in the vicinity of the microscopic domain, for all practical purposes).

One must therefore emphasize that, even though the domain of investigation of quantum physics cannot be as completely detached from the instrumental extentions of our body as an *object* properly speaking should be, it is bound to manifest a certain amount of *independence* with respect to the law-like behaviour of the instruments. *This independence* is what makes definitely illegitimate, despite contextuality, any attempt at considering the domain of investigation of quantum physics as an extention of our body, or of this part of our instruments which we consider under our control. *This independence,* in other terms, is what constitutes the most important difference between being-in-a-body and being-in-the-world in this case.

To summarize, even in a situation where our environment cannot in principle be ascribed the status of a set of detached objects of study, being-in-the-world may still mean something distinct from being-in-a-body; but such a distinction arises only if some parts of this environment systematically elude our control, as it is the case in the domain ruled by quantum physics.

We must come now to further differences between the classical and the quantum ways of being-in-the-world. The major difference we shall point out is the following. Classical physics somehow incorporates many tacit presuppositions of everyday activity. By contrast, these presuppositions have no *direct* equivalent within quantum mechanics, in spite of the fact they still underlie the instrumental manipulations of the quantum physicists in their laboratories. In classical physics, the presuppositions which are needed in experimental investigations are already at work in the theories; they have at least a *counterpart* in the theories; they are so to speak intrinsic. But in quantum physics, these presuppositions are extrinsic; they can at most be justified as approximations when the scale of size and complexity of the instruments is such that the Planck constant becomes negligible. The presuppositions are paradigmatically justified in classical physics, whereas they are only practically justified, at our scale, in quantum physics.

This difference is especially important, since, as we shall show below, it accounts for the fact it is so much easier to forget our situation of *beings-in-the-world* in classical physics than in quantum physics. Let us then define more precisely what we mean by "presupposition" before we give a relevant example of paradigmatical and practical justification of a presupposition. To begin with, what is a *presupposition?* According to the usual account, a presupposition S of the proposition P is one of the propositions the speaker takes for granted when he states that P. The reductive semantic account of presuppositions even tends to eliminate the speaker from the picture by focusing attention on the set of propositions {S,P}: "P presupposes that S iff S must be true in order that P have a truth value at all". But, as Stalnaker[1] pointed out, this account clearly fails to grasp the sublety of the pragmatic concept of presupposition, for it is much too strong and rigid to do the job. A speaker can perfectly well carry through a certain presupposition S when he asserts P, even though he has no commitment to the truth of S. The "presupposition" S often does not go beyond the status of a useful tool, for the sake of communication with an audience which happens to share it. As Stalnaker rightly pointed out, in everyday dialogues a presupposition is sometimes much less than an assumption of truth, and even less than a belief. Just "a disposition to behave in one's use of language *as if* one had certain beliefs". Conversely, a presupposition can sometimes be much more than an assumption of truth or a simple belief: an artificially isolated part of a *form of life*. Wittgenstein pointed out that the word "presupposition" can be too weak to express the unconditioned commitment which is associated to a presupposition: "Doesn't a presupposition imply a doubt? And doubt may be entirely lacking. Doubting has an end"[2]. As for Searle[3], he insisted on the holistic features of what he calls the *background network:* one cannot always *individuate* presuppositions in it. The question as to whether a presupposition can be detached from (or individuated in) the overall background network thus has to be raised in each particular case; and the answer may depend not only on the type of presupposition which is at stake, but also of the type of background network from which it is to be isolated.

As we noticed previously, both our systems of practices and our theoretical paradigms may be said to imply certain presuppositions; some presuppositions implied by our systems of practices are taken over by certain theoretical paradigms, whereas they are *not* taken over by some other theoretical paradigms. Some presuppositions are taken over by classical physics, though not by quantum physics. Let us take an example which is especially relevant for the comparison of the two paradigms. This example concerns reference and identification. According to Searle[4], one may give a pragmatic account of reference as *speech act*. For, like any other action, reference can either succeed or fail. The ultimate criterion of success is unambiguous *identification*, by the speaker, of the object referred to, and communication of this identification to a hearer[5]. Conversely, impossibility of identification means a failure of reference. The difficulty is that, in most cases of everyday conversation, identification (which requires ostensive acts or descriptions) is not *actually* achieved. Searle thus makes a distinction between what he calls "fully consummated reference", wherein the object referred to has actually been identified, and simple successful reference which does not

[1]R.C. Stalnaker, "Pragmatic presuppositions" in: M.K. Munitz & P.K. Unger, eds. *Semantics and Philosophy*, New-York University Press, 1974

[2]L. Wittgenstein, *Philosophical investigations*, op. cit. , II, v

[3]J. Searle, *Intentionality*, Cambridge University Press, 1983, p. 142

[4]J. Searle, *Speech acts*, Cambridge University Press, 1969

[5]ibid. p. 82

convey any more than a *capacity* for identification. Successful reference, in this weaker acceptation, only requires that the speaker *could* identify the object he refers to on demand. The act of reference is tantamount to a *promise* of identification.

At this point, a more refined analysis is needed if we are to meet the requirements of the situation in physics. Following Searle's definition to its logical conclusion, reference cannot be unambiguously considered as a success or a failure until one has given an assessment of the crucial *capacity to identify*. To do this, we must make a further distinction between *practical* capacity to identify and *paradigmatical* capacity to identify. Having a *practical* capacity to identify means being able to pick out the object referred to by ostension or by partial descriptions, from among all the other objects which are present in the context of speech. Having a *paradigmatical* capacity to identify is quite a different matter. It means that the current theoretical paradigm allows one *in principle* to identify an object to which one is referring by making use of a theoretically acceptable individualizing description and/or spatio-temporal criteria of identification. Reference to a certain particle P is for instance paradigmatically successful in classical mechanics, for within this theory a particle can be individualized *in principle* by its position at a certain time and reidentified along a continuous trajectory. In other terms, we could say that reference to objects is paradigmatically successful in classical mechanics because the presupposition of permanence of objects is paradigmatically justified in the framework of this theory. But reference to a certain particle P is *not* paradigmatically successful in standard quantum mechanics, for the concept of continuous trajectory which enable reidentification has no equivalent in the formalism. True, there are equivalents of continuous trajectories in the formalism of Bohm's theory. Yet these equivalents admittedly elude *in principle* any experimental monitoring. They are thus unlikely to be interpreted as an actual or potential *capacity to identify*. One may still refer formally to particles in the quantum domain, as Bohm proposes; but, in so far as it is considered as a speech act equivalent to a promise of *effective* reidentification, reference is bound to remain paradigmatically unsuccessful.

The practical and paradigmatical justifications of presuppositions are definitely distinct in scope and extent, and none of them can *a priori* claim to afford an all-embracing criterion. It may happen that an act of reference is paradigmatically successful without being practically successful. Reference to a certain particle P, for instance, is *always* paradigmatically successful in classical mechanics, although no currently available experimental device enables one *in practice* to distinguish P from other particles above some threshold of tight packing.

Conversely, it may also happen that an act of reference is practically successful, although its success has no accepted paradigmatic counterpart. This latter situation often arises in the middle of scientific revolutions. During these periods, no paradigm is uncritically endorsed by scientists, and practical criteria of successful reference thus gain a renewed significance. A sign of this situation is provided by the predominance of phenomenalist and operationalist trends during scientific revolutions. Phenomenalist doctrines imply a widespread criticism of reference, and operationalist doctrines imply restriction of reference to the pieces of the experimental apparatuses. Phenomenalism is tantamount to denouncing *any usual kind* of reference as purely practical, and operationalism gives priority to practical modes of identification which operate in the laboratory work over any paradigmatical mode of identification in the domain of investigation. However this emphasis on practical modes of identification, and more generally on practical justification of any presupposition, is usually transient. Once the new paradigm is strongly rooted, paradigmatical justification of presuppositions prevail again over practical justification. The capacity of paradigmatical justification to give ground to every case of practical justification, its generality and rigidity, may even lead one to forget completely the former situation. Once this has happened, those presuppositions which are paradigmatically justified are often hypostasized (and can be so with no harm), until the next scientific revolution.

Let us consider the example of the Newtonian revolution. Newton's work was in some way construed as a reaction against Descartes's foundational conception of mechanics. A major component of Newton's philosophical attitude was his celebrated *"hypotheses non fingo"*, which may be translated: 'I do not conceive ambitious *a priori* systems of the world, as Descartes did; I rather try to find a unified framework for specific observational facts and specific laws'. During most part of the eighteenth century, this prudent attitude was highly praised. Voltaire, D'Alembert, and many Dutch physicists[1] took it as a model of what good natural philosophy should be. As for Kant, his attempt at going beyond pure empiricism did not imply a way back to any version of theological foundationalism; it only involved an analysis of the conditions which render (human) experience possible. However, the very success of classical mechanics prompted many scientists to forget the epistemological prudence of the revolutionary period which covers the end of the seventeenth century and part of the eighteenth century, and to project the presuppositions of this theory on "nature". Many scientists thoughtlessly identified the structure of the mature view of the world of classical physics with the structure of the world "as it is in itself". The incorporation of many presuppositions of the "natural attitude" into the background network of classical physics made especially appealing their being taken as features of the world "out there".

Now it is precisely this kind of forgetfulness which has been made much more difficult by quantum mechanics. In quantum physics, the light thrown on the practical status of our presuppositions is not a transient

[1]See E. Cassirer, *Die Philosophie der Aufklärung*, op. cit., chapter 2

feature of the revolutionary period. There are some presuppositions, such as reidentifiability of material bodies, which are indispensible in the daily experimental work of the laboratory but which have *no* direct counterpart in the theoretical framework. Their practical status is thus a permanent feature of the present situation in physics. For instance, practically successful reference to pieces of experimental equipment has no proper paradigmatic basis in a theory where the constituents of matter have no experimentally available trajectory, and it must thus be considered as *irreducibly* practical. At most, one can ascribe them a basis within *the former* paradigm, that is within the classical paradigm; but since the classical paradigm incorporates most basic presuppositions of everyday life and speech, ascribing them a classical basis is just a sophisticated way of aknowledging their practical status. One can also show that these presuppositions are *compatible* with the quantum paradigm, provided the size and complexity of the processes are such that the Planck constant is negligible. But a statement of reasonable quantitative compatibility between a presupposition and a paradigm in a certain (macroscopic) range, is by no means equivalent to the statement that the presupposition underlies the paradigm. Showing that a presupposition needed in daily experimental work may only arise, in this paradigm, as an *approximation* valid in the vicinity of *our* scale, is rather one more way of emphasizing its purely practical character.

The consequences of this situation are very important. As we noticed previously, if the set of those presuppositions which are needed by any scientific inquiry is somehow incorporated in the background network of a theory, and if moreover the success of this theory is considered as a sign that it provides a faithful description of the world, one may be tempted to hypostasize them. One may try to justify the presuppositions by saying that they just reflect intrinsic features of the world. But if, among the presuppositions which underlie the experimental work, there are some which are definitely foreign to the new theoretical paradigm, and which just happen to be approximately compatible with it at the scale of magnitude and complexity of laboratory activities where the classical paradigm is still (approximately) adequate, this kind of justification is no longer acceptable for them. Justification of these presuppositions by their bearing a likeness to constituents *of the world* must be replaced by justification of the presuppositions by their being appropriate, for all practical purposes, to *our position in the world*. Bohr's insistence on the perennial value of the correspondence principle with classical physics for modern quantum mechanics, was thus another way to express the impossibility to eliminate any consideration about our position in the world from the interpretation of quantum mechanics. And Schrödinger's claim that the eigenvalue scheme of the observables (i.e. the *Fact-model)* is a "loan from classical physics"[1] which can by no means be derived from

[1]E. Schrödinger, *Transformation and interpretation in quantum mechanics,* in: E. Schrödinger, *The interpretation of quantum mechanics (Dublin seminars 1949-1955 and other unpublished texts),* op. cit. p.

pure wave-mechanics (i.e. from the *ψ-model),* yields exactly the same conclusion. One more argument pointing in the same direction has been forced upon us quite recently; it is the impossibility of working out decoherence theories without using (mentalistic, evolutionist or practical) meta-theoretical assumptions[1].

We can now see more clearly what makes quantum mechanics so different from previous physical theories. In previous physical theories, it was possible to conceal the role played by our position *in* the world by incorporating it to a structure which could be interpreted as a description *of* the world. In quantum mechanics, no such method of concealment is available, except in the purely formal sense of Bohm's theory. Thus, before quantum mechanics (or without explicit reference to quantum mechanics), the emphasis put by some philosophers on "being-in-the-world" could just be an isolated attempt at awaking the western culture from the objectivist slumber favoured by naïve scienticism. But since the advent of quantum mechanics, the irreducibility of our being-in-the-world to any disengaged description wherein we are only supposed to be *facing* the world, is progressively being recognized as a pivotal interpretative element of one of our most basic scientific theories.

6-6 *The body, the world, and dualism*

We can now undertake a systematic comparison of two parallelisms. Parallelism between intentional expressions and causal accounts of action on the one hand, and parallelism between Fact-model and ψ-model on the other hand. Parallelism between what it it is like to be a being-in-a-body and the external description of the body's behaviour in the first case, and parallelism between what it it is like to be a being-in-the-world and the external description of the world in the second case.

One of the most widespread ways of dealing with the first kind of parallelism is *dualism.* We could define dualism as an attempt at aknowledging the momentous difference between the two parallel descriptions, while continuing to *project* them on the *same* plane. Descartes' dualism of *res extensa* and *res cogitans* is the most typical example: the specificity of intentional and causal descriptions is fully retained, but both are projected on the plane of objectified *substances.* In *Mind and Matter,* Schrödinger also provided us with a short but very sharp account of this strategy. The act of *"objectivation",* he says, means that "We step with our own person back into the part of an onlooker who does not belong to the world, which by this very procedure becomes an objective world"[2]. But my body, and the bodies of other people as well, are *in* the world. Since "(...) there is no distinction between myself and

51

[1]See paragraph 4-4, point (8) of the present essay

[2]E. Schrödinger, *Mind and matter,* op. cit. p. 127

the others, but on the contrary full symmetry for all intents and purposes (...)", I have no reason to doubt that the other's bodies "(...) are, as it were, seats of spheres of consciousness" of the same type as mine. I am therefore tempted to project their "spheres of consciousness", and mine as well, in the objective world. My "sentient, percipient and thinking ego" is thus reintegrated within the very domain (the objective world) which had been formerly created by its withdrawal. "I so to speak put my own sentient self (...) back into it - with the pandemonium of disastrous logical consequences that flow from the aforesaid chain of faulty conclusions"[1].

What are these disastrous logical consequences? There are two such embarassing consequences, which both arise from the difficulty of bridging two heterogeneous elements in the same objective world. The first consequence is that nobody can tell exactly *where* (and *when)* the bridging takes place; and the second consequence is that nobody can tell *how* it operates.

Descartes proposed a tentative answer to these questions. He located the human soul in the pineal gland[2], and he used a wrong law of conservation of momentum (i.e. a law which states the conservation of the quantity, not the direction, of momentum) in order to account for the mutual action of the soul and the body. But he also recognized, in a letter to the Princess Elizabeth, the paradoxical aspects of his dualism: "It does not seem to me that the human mind is capable of conceiving quite distinctly and at the same time both the distinction between mind and body and their union; because to do so, it is necessary to conceive them as a single thing, and at the same time to conceive them as two things, which is self-contradictory"[3].

As for Schrödinger, he considered that the above questions about where, when, and how the briging between mind and body takes place, are just unanswerable because their background assumptions are flawed. The question(s) about the *mode* of mutual interaction of mind and body, to begin with: "How does matter act on mind, to produce in it the sensory qualities - and also how does mind act on matter, to move it at will? These questions cannot, so I believe, be answered in this form (...)"[4]. And then, the question about the *place* where this interaction occurs: "(...) the conscious mind itself remains a stranger within that construct (i.e. within the objective world), it has no living space in it, you can spot it nowhere in space"[5].

[1]ibid. p. 128

[2] See A. Bitbol-Hespériès, *Le principe de vie chez Descartes*, Vrin, 1990

[3]R. Descartes, *Oeuvres* III, p. 693, C. Adam & P. Tannery (eds.), Vrin, 1964-75. Quoted and translated by M. Grene, *Descartes*, Harvester Press, 1985, p. 19

[4]E. Schrödinger, *Science, philosophy and the sensates*, (chapter 3 of a series of lectures which were intended to be given in Harvard, as William James Lectures, in 1954; in: E. Schrödinger, *The interpretation of quantum mechanics (Dublin seminars 1949-1955 and other unpublished texts)* op. cit. p. 145

[5]E. Schrödinger, *Mind and matter*, op. cit. p. 132

These two problems, bearing on the mode of interaction of the two entities and the location of their interaction, also arise in other versions of dualism. One version of dualism is especially interesting because it is less metaphysically loaded than the cartesian version, and because its formal characteristics can be transposed almost immediately to a similar structure which arises in the interpretation of quantum mechanics. It manifests itself in the so-called "practical inferences" or "practical syllogisms" which often tend to project the reasons (for acting) and the causes (of events) [or the intentions (of the agent) and the objects (of the world)] on the same line of deduction. An example of such inference (extrapolated from G.H. Von Wright[1]) is the following:

(i) Agent A intends to bring about X at time t *if* Y occurs before time t' (t'<t)

(ii) Y has occurred before time t'

(iii) Therefore, A sets himself to bring about X at t, unless he forgets about time or is prevented.

Point (i) involves an intention bearing on an event (X); point (ii) involves an event (Y) which has actually occurred at a certain time; and point (iii) involves an action "prompted by" the combination of the intention and the event Y. Now the problem of the mode of interaction of the two levels of speech has been raised by some philosophers; and it has been solved provisionally by adding a dubious "quasi-law" which makes the action a necessary consequence of the combination of the intention and the event. Something like: "For any agent A, events X and Y, and times t and t', If A intends to bring about X at t *if* Y occurs before t', and if Y occurs before t', then, *under normal conditions,* A will bring about X at t' "[2]. With this quasi-law, the practical inference can apparently be ascribed the status of a Deductive-Nomological explanation wherein intentions occur on a par with events. In the framework of this explanation, intentions and reasons are considered as causes. But, as R. Tuomela[3] pointed out in his critique of Von Wright's conception, the former inference is a Deductive-Nomological explanation only in a "formal and pickwikian sense". For the so-called "quasi-law" is a formulation of a *normative* standard of behaviour, not an empirically falsifiable law.

The attempt at projecting intentions and causes on the same deductive line meets another difficulty. It concerns the place (and the time) of the transition between intentional and causal description. Each outcome of a practical syllogism is an action which starts a causal chain in the world[4]. But where does the action end, and where does the causal chain start? Let

[1]G.H. Von Wright, *Explanation and understanding,* Routledge & Kegan Paul, 1971, p. 107

[2]See a critical discussion of a similar "quasi-law" in: K.O. Apel, "Causal explanation, motivational explanation, and hermeneutical understanding", in: G. Ryle (ed.), *Contemporary aspects of philosophy,* Oriel Press, 1976. This critical discussion is developed in paragraph 6-8 below.

[3]quoted by K.O. Apel, "Causal explanation, motivational explanation, and hermeneutical understanding", loc. cit.

[4]see P. Ricoeur, *Soi-même comme un autre,* Seuil, 1991, p.134

us come back to the Bohrian example of a stick used to orient oneself in the dark. If the stick is held firmly, one will tend to say that the action encompasses the stick: in this case, one acts *by means of* the stick, not *on* the stick. However, if the stick is thrown away, the usual modes of expression are less univocal. It can both be said that one has acted *on* the stick, if attention is to be called on the balistic step of the action, or that one has acted *by means of* the stick, if attention is to be called on the final result of the action.

A slightly different and well-known example is G.E.M. Anscombe's man who "(...) is pumping (poisoned) water into the cistern which supplies the drinking water of a house"[1]. Here, the basis for a distinction between intentional and causal description is provided by the different answers which are expected after the questions "why?" (asked to the agent *qua* agent) and "what?" (asked from the standpoint of an external observer) are asked. More explicitly, the questions are the following: 'why are you moving your arm up and down?' and 'what happens during and after this process?'. The problem, here again, is that there is no clear boundary between the domain where the question "why?" is relevant and the domain where the question "what?" is to be asked. To the question "why are you moving your arm up and down?", the agent may just answer "because I am pumping water". This answer pushes all the further consequences of pumping water towards the domain of the question "what?": "what is happening next (after water have been pumped)?"; answer: "the cistern is filled, the people in the house drink water, they are poisoned, etc.". But, when asked the question "why?", the agent could also have answered directly "I am moving my arm up and down because I want to poison the people up there". And in this case, the domain of relevance of the question "what?" would have been restricted to what happens after the poisoning has occurred (or failed).

Another problem of boundary arises upstream from the action properly speaking. Usually, we do not ask somebody "why are you moving such and such muscles (with latin names)?" or "why are you generating those substances in your nerve fibers?". As a rule, the question "why?" bear on the reasons one has of acting, or on his intentions, not on the bodily tools he uses thoughtlessly when he acts. These bodily tools rather fall into the domain of a certain kind of question "what?" asked by an external observer called a *physiologist*. But here again, the boundary between the two domains may fluctuate. If the agent is perfectly aware of his own anatomy and of the muscles he is using during his work, the question "why?" directed towards his use of a certain set muscles is no longer awkward; it just belongs to a quite specialized language game.

Hence, the dualist mode of description (intentional and causal) raises at least one, perhaps two, problems of boundaries. Descartes was already aware of this difficulty. After he had hypostasized the dualist mode of

[1]G.E.M. Anscombe, *Intention*, Basil Blackwell, 1957, §23

description into a dualism of substances, he tried to solve the two problems of boundaries at once by pushing both boundaries as close as possible to the (problematic) interface between the soul and the body. According to Descartes, the soul *is* a true (may be the *only* true) agent: "The ones I call its actions are all of our volitions, because we find by experience that they come directly from our soul (...)"[1]. Beyond these volitions of the soul, which modify the direction but not the *quantity* of motion of the animal spirits, the only question which can rightly be asked is the question "what?": what does happen to the "machine of our body" after the volition; what are the muscles which contract when the animal spirits have flowed into them? The question "why?" is (or should be) restricted to the soul. However, the decision of pushing the boundaries in the vicinity of the soul does not solve all the difficulties. The boundaries still have to be located precisely somewhere in (or close to) the hypothetical seat of the soul, namely the pineal gland. True, the imprecision has become microscopic, but it is still there.

Now we meet exactly the same kind of difficulties, and the same types of tentative solutions in the interpretation of quantum mechanics. The similarity of the problems is even so striking that it has led many authors to *merge* them into one.

In quantum mechanics, we start with the parallelism of the Fact-model and the ψ-model. And we use a typical dualist strategy when we attempt to project them both on the *same* plane of objective description without trying to reduce one to the other one. Here, the two relevant questions, namely the two questions which provide a basis for the distinction between a partial description according to the F-model and a partial description according to the ψ-model, are the following: (i) "which facts are possible in given experimental circumstances and which fact has effectively occurred?" and (ii) "what is the (probabilistic) link between the facts?". Indeed, the first question clearly pertains to the Fact-model: the possible experimental facts are the eigenvalues of an observable, and the effective experimental fact is one of these eigenvalues. As for the second question, it can only be answered, at least within the framework of *standard* quantum mechanics, by specifying an appropriate ψ-function, calculating its evolution by means of the Schrödinger equation, and using Born's empirical correspondence rules at the final stage. But if we accept to consider that facts and inter-factual evolution of the ψ-functions are elements of a single time-series, if in other terms we project them on the *same* plane of objective description, then the following questions arise. *How* does the transition between the ψ-model and the facts occur in this time-series; *where* and *when* does it occur? How, where and when does the so-called "wave-packet collapse" occur? These questions are exactly isomorphic to those which arise in the case of mind-body (or intentionality-causality) dualism; and they are just as embarassing.

[1]R. Descartes, *The passions of the soul,* (Translated by S. Voss) Hackett publishing company, 1989, art. 17

The original form of dualism in the interpretation of quantum mechanics is Bohr's "cut" between a classical and a quantum domain. In addition to his insistance on wholeness, or on the indivisibility of the measurement chain, Bohr pointed out that a "cut" has to be introduced somewhere in the middle of the measurement chain: "The essentially new feature in the analysis of quantum phenomena is, however, the introduction of a *fundamental distinction between the measuring apparatus and the objects under investigation*. This is a direct consequence of the necessity of accounting for the functions of the measuring instruments in purely classical terms (...)"[1]. This "fundamental distinction" does not contradict the statement of indivisibility. Indivisibility reflects an essential impossibility of analysing the interactions between the parts of the measurement chain, whereas the necessity of a "cut" is imposed on us by pragmatical or epistemological constraints. The pragmatical constraint is that "(...) only with the help of classical ideas is it possible to ascribe an unambiguous meaning to the results of observation"[2]. And the epistemological constraint is that "(...) the distinction between object and agency of measurement (is) inherent in our very idea of observation"[3]. But here again the traditional questions of dualism arise, although they are attenuated by the admittedly pragmatic status of the "cut". True, the modalities of the interaction between the classical and the quantum side of the chain are not really a problem, for the cut has not been given any "substantial" meaning. However, the question as to where the cut takes place is slightly more embarassing, because one *has* to locate it somewhere in each case, even though no univocal location can be given. According to Bohr, the location of the cut is *a priori* arbitrary: "(...) the place within each measuring procedure where this discrimination is made is (...) largely a matter of convenience"[4]. Nevertheless, some practical considerations lead one to define a certain range within which the cut can be located without harm: "(...) we have only a free choice of this place within a region where the quantum-mechanical description of the process concerned is effectively equivalent with the classical description"[5].

In these circumstances, one may try to use the (Cartesian) strategy of pushing the boundary between the two domains as far as possible; that is, so far that *almost nothing* is left in the domain of the Fact-model, whereas *almost everything* is ruled by the ψ-model. Schrödinger sketched half-seriously this kind of solution when he evoked the role of the "living subject" or of a "mental act" as a possible end of the measuring process which falls entirely within the domain of the ψ-model (see paragraphs 4-3 and 4-4). But the true emblematic figures of this strategy were J. Von Neumann, F. London & E. Bauer, and E.P. Wigner[6]. At its most extreme

[1]N. Bohr, *Essays 1958-1962 on atomic physics and human knowledge*, op. cit. p. 3-4
[2]N. Bohr, *Atomic theory and the description of nature*, op. cit. p. 17
[3]ibid. p. 68
[4]N. Bohr, "Can quantum-mechanical description of reality be considered complete?", loc. cit.
[5]ibid.
[6]J. Von Neumann, *Mathematical foundations of quantum mechanics*, Princeton University Press, 1955;

point of development, their approach led to confine the occurrence of a fact to the "I" or the "abstract self". Yet, this limiting case was soon felt unsatisfactory, for it led to solipsism. Thus, in order to avoid solipsism, which contradicts a basic methodological presupposition of experimental research, both London & Bauer and Wigner proposed to locate the ψ / Fact transition in the vicinity of *any "consciousness" whatsoever,* not only "mine". Unfortunately, instead of providing a satisfactory solution to the difficulties raised by the quantum version of dualism, this kind of proposal was tantamount to translating them in terms of the difficulties raised by the mind-body version of dualism. Merging the two kinds of dualism into one, mixing up the "Cartesian cut" and the "Bohr-Heisenberg cut"[1], did not provide any additional clue.

There is a final possibility of rescuing dualism. It is to push the "cut(s)" one step further, beyond the personal "ego", rather than stepping back to the popular concept of "consciousness". This means replacing ordinary dualism by what we have called "functional dualism" in paragraph 4-4 point (9). In this abstract kind of dualism, "Mind" is pure subject; it has no content, no location, no time (According to Schrödinger it is "(...) always *now"* [2]), but it still retains a *function:* it is the '*to whom* ' physical events appear. In the quantum case, one would also say that it is the '*for whom* ' facts can be said to occur. The problem of solipsism is thus avoided, because "Mind" or the pure "knowing subject" in its more abstract, transcendental, universal, sense, does not pertain to any particular person. And the problem of location of the "cut(s)" does not arise, because the pure knowing subject is nothing spatial or temporal; it is just the '*for whom*' there are spatio-temporally located happenings, and its acts always occur *here and now.* To use Wittgenstein's Tractarian concepts, we could say that in this conception both "cuts" (the Cartesian cut and the Bohr-Heisenberg cut) are identical to the *limit of the world - not a part of it* [3].

True, functional dualism does not help to solve the problem of the mode of interaction of the "subject" and the physical world; the transition between happenings in the world and appearances for the subject is just as mysterious in functional dualism as it is in any other version of dualism. But it suggests an original way of sorting out the difficulty of the *place and time* of the transition. The transition coincides with the awareness of a result by the pure "knowing subject"; it occurs "here and now", as any other act of this subject.

E.P. Wigner, "Remarks on the mind-body question"; F. London & E. Bauer, "The theory of observation in quantum mechanics"; both reprinted, (in English translation for London and Bauer), in: J. A. Wheeler and W.K. Zurek, *Quantum mechanics and measurement,* op. cit.

[1]H. Atmanspacher, "Is the ontic/epistemic distinction sufficient to describe quantum systems exhaustively?", loc. cit.

[2]E. Schrödinger, *Mind and Matter,* op. cit. p. 145

[3]L. Wittgenstein, *Tractatus logico-philosophicus,* op. cit. § 5.641. This expression is applied by Wittgenstein to the knowing subject itself, not to the *cut* between the subject and the world. But here, obviously, there is no way of distinguishing the subject from the cut.

Functional dualism also gives a metaphorical expression to an idea which will be developed (in paragraph 6-8) in the framework of Wittgenstein's late philosophy rather than in the framework of the*Tractatus:* elements of the intentional description can be *located nowhere* in the causal description; and, similarly, elements of the Fact-model can be *located nowhere* in the ψ-model. Intentions can be connected to the chain of causes in a certain context of speech; and, similarly, facts can be connected to the time-development of wave functions in a certain context of (human) interest.

6-7 *The body, the world, and monism*

With all the paradoxical features of dualism, the most natural alternative is monism. And the most straighforward version of monism arises from reductionism. When applied to the parallel between intentional expressions and causal expressions, reductionism could mean either reducing causal account to intentional account or reducing intentional account to causal account. The first reduction would yield an idealistic view. The second reduction is by far the most usual, for it fits better with the enduring project of an all-encompassing objective description; it fits better, in other terms, with the dream of generalizing the disengaged account (or God's eye view) to the detriment of the engaged account. This kind of reduction has assumed two forms in the early and late mid-century: behaviourism and mind-brain identity theory.

But in recent years, the epistemological aspects of this kind of reductionism have been found quite difficult to maintain, and monism has taken a more subtle form: it has been based on a flexible thesis, referred to as "*supervenience*" after G.E. Moore. Davidson called this position "anomalous monism" (that is, *lawless* monism), and he described it in the following terms: "Anomalous monism resembles materialism in its claim that all events are physical, but reject the thesis, usually considered essential to materialism, that mental phenomena can be given purely physical explanations. (...) Although the position I describe denies there are psychophysical laws, it is consistent with the view that mental characteristics are in some sense dependent, or supervenient, on physical characteristics"[1]. To sum up, anomalous monism is clearly distinct from dualism; it has so many features in common with materialist reductionism that one can consider it as a version of materialist monism; but it also maintains certain elements of parallelism (between the intentional and the causal modes of expression). It appears as a distorted form of parallelism with a strong bias in favour of the causal mode of expression.

It is quite easy to find formal equivalents to these versions of materialist monism among the available interpretations of quantum mechanics[2]. Just as in materialist monism intentional descriptions were

[1] D. Davidson, "Mental events", in: *Essays on actions and events*, Oxford University Press, 1982, p. 214

[2] A formal equivalent of idealist monism can also be found among the available interpretations of quantum

reduced to, or considered as supervenient on, causal descriptions, there exist interpretations of quantum mechanics where the elements of the Fact-model are reduced to, or supervenient on, some characteristics of the ψ-model. Let us begin with reductionism. The first interpretation of quantum mechanics which was based on reduction of the Fact-model to the ψ-model was Schrödinger's original wave-interpretation. In the terminology we introduced in paragraph 2-1, Schrödinger assumed in 1926 that the ψ-functions were both effective and factual. He thought that certain configurations of waves (that is, wave-packets) were able to account for particle-like series of facts. But many difficulties, especially those linked to his straightforward interpretation of the concept of wave-packet, prevented him very soon from maintaining his original conception of quantum mechanics. His major move in the 1950's, as we saw in the two first chapters of this book, was to recognize the irreducibility of the Fact-model to the ψ-model and to advocate a kind of parallelism between them. Two other, more recent, reductionist attempts, are De Witt's popular many-worlds interpretation, and the strongest version of the decoherence program. Both strategies differ from Schrödinger's original proposal of 1926 on one crucial point: they tend to encompass the *observers,* not only the putative objects, within the ψ-model construed as a disengaged description of the world. They both try to accomplish what A. Shimony[1] calls "closing the circle" of the knower and the known, by reducing the knower to a fraction of the known, and by eliminating any first-person account in favour of an all-encompassing third-person account. In the many-worlds interpretation each fact corresponds to an event which occurs in the particular world which is associated to the corresponding eigen-component of the overall wave-function; and in the strongest version of the decoherence program, namely Gell-Mann's and Hartle's decoherent histories[2], each fact corresponds to a particular state of the "Information Gathering and Utilizing Systems (IGUS)".

Both interpretations aimed at restoring what H. Putnam calls the "God's-Eye View conception of physics"[3], by considering that what is under God's eye is the whole universe as described by an all-encompassing wave-function, *including the observer.* However, both interpretations also met insuperable difficulties of the same type. Let us just recall a relevant example of these difficulties. Both the many-worlds interpretation and the decoherent histories interpretation stumbled on the preferred basis problem. The many-worlds interpretation could not provide a solution of the preferred basis problem within its own

mechanics. It is pure instrumentalism: the elements of the Fact-model are given absolute priority, and the ψ-model is considered as a mere symbolic scheme allowing probabilistic prediction of (experimental) facts.

[1]A. Shimony, "Reality, causality and closing the circle", in: *Search for a naturalistic world view I,* op. cit., p. 44

[2]M. Gell-Mann & J.B. Hartle, "Classical equations for quantum systems", loc. cit.; M. Gell-Mann, *The Quark and the Jaguar,* Princeton University Press, 1994

[3]H. Putnam, *Realism with a human face,* Harvard University Press, 1990, p. 9

framework, and the decoherent history interpretation could not account for its decoherent basis without making use of some meta-theoretical characteristics of its IGUS (say practical, epistemological, or adaptative characteristics)[1].

Everett's relative state interpretation, which acted in many respects as an ancestor of the many-worlds interpretation, is less ambitious. On the one hand, it may be construed as a *monist* interpretation because it assumes that the ψ-model is sufficient to describe the universe as a whole. Yet, on the other hand, it uses special symbols clearly distinguished from the ψ-symbols, in order to express *facts*. This fact-like series of symbols is the set of the memory brackets and their contents. Everett himself used the concept of "psycho-physical parallelism" when he had to explain the relation between the memory brackets and the wave-function. But in his formalism, the link between the increasing content of the memory brackets and the time-development of the wave function is certainly looser than what the expression "psycho-physical parallelism" implies. In Everett's theoretical frame, for instance, a given memory content *could* in practice be associated with a wide range of observer's ψ-states, just in the same way as a value of a thermodynamical variable can be associated with a wide range of microscopic states. And moreover, a single global ψ-state is generally associated with a multiplicity of memory contents (this multiplicity is precisely what corresponds to Everett's "many branches"). The link between ψ-states and memory contents is thus likely to manifest itself through weaker constraints. In Everett's theory, these constraints are the following: (i) there cannot be two observer's ψ-states alike in all respects but differing in the *set* of their associated memory brackets; and (ii) a *set* of memory brackets cannot be altered if its associated ψ-state has not been altered. Now, this fits in almost exactly with Davidson's definition of supervenience: "(...) supervenience might be taken to mean that there cannot be two events alike in all physical respects but differing in some mental respect, or that an object cannot alter in some mental respect without altering in some physical respect"[2]. Everett's version of wave-mechanical monism can therefore be taken as the quantum equivalent of the "anomalous (or lawless) monism" advocated by Davidson in the mind-body controversy. Memory contents about *facts* supervene on the ψ-model; they do not identify themselves to features of the wave-functions, and they are not rigidly associated to them.

As a result, it is quite easy, in the frame of Everett's original interpretation, to overcome the difficulties which plague strictly reductionist conceptions such as the popular many-worlds interpretation. Let us take again the preferred basis problem as an example. In the reductionist interpretations, the preferred basis should arise from special features of the ψ-model, because the Fact-model has no specificity and no autonomous feature whatsoever. The preferred basis should, in other

[1]See paragraph 4-4 point (8)

[2]D. Davidson, "Mental events", in: *Essays on actions and events*, op. cit., p. 214; see p. 253

terms, *emerge* from a ψ-background. In this context of thought, the acknowledged necessity of imposing anthropocentric meta-conditions bearing on the fact-structure, in order to derive a preferred basis, can only be considered as a spurious irruption of something alien from the unique ψ-model. As against this, in an anomalous monist perspective, it is not unnatural to make the preferred basis arise from constraints which are exerted on the Fact-model; for, here, the Fact-model has enough specificity to be considered apart and to be imposed conditions of its own. As Ben-Dov[1] cogently pointed out, it is perfectly acceptable, in Everett's original interpretation, to impose that the decomposition of the global ψ-function of the composite system (system + apparatus + observer) be such that mutually *exclusive* facts are recorded *permanently* in the memory bracket of the observer. This decomposition does not represent a mode of split of the whole universe, but only a method for selecting a set of special states relevant for any viable observer.

However, everything is not solved at this stage. The anomalous monist interpretation of quantum mechanics makes it *acceptable* to impose conditions and constraints to the Fact-model rather than directly and exclusively to the ψ-model, but, in its framework, the reason why one *should* focus on these constraints remains mysterious. True, the occurrence of facts supervenes on the evolution of the wave-functions; it does not reduce to it, and this makes possible a certain amount of independence between Fact-conditions and ψ-conditions. Yet, in an anomalous monist conception, as in any monist conception, the ψ-model is admittedly the basic level of description. We are thus given no real clue as to why one *must* consider the supervenient level of description when the preferred basis problem is at stake, rather than restricting attention on the characteristics of the underlying level. We are given no reason to think that imposing conditions on the Fact-level is more than a provisional makeshift. Further investigations into the characteristics of anomalous monist doctrines are needed in order to clarify this point. These investigations will prompt us to substitute what we shall call "anomalous parallelism" for anomalous monism.

6-8 *The body, the world, and anomalous parallelism*

How are we to understand *non-dualistically* the need of an intentional description of actions beside the causal description, and of a Fact-model beside the ψ-model? That is, how are we to understand their joint necessity without projecting them onto a single plane of objective description? In order to answer these questions, we shall start from recent philosophical criticisms directed towards both monism and standard dualism. These criticisms have been formulated by P. Ricoeur and K.O.

[1]Y. Ben-Dov, "Everett's theory and the 'many-worlds' interpretation", Am. J. Phys., 58, 829-832, 1990

Apel, two philosophers who defend the specificity of the hermeneutic method against objectivist reductionism.

To begin with, Ricoeur criticizes Davidson's attempt at reducing actions to events, and he especially criticizes Davidson's claim that "A primary reason for an action is its cause"[1]. He notices that this reduction is only made possible by a systematic bias in favour of retrospective thinking as against teleological considerations. Reasons of acting, according to Davidson in his earlier essay "Actions, Reasons, and causes", are *a posteriori* rationalizations, or rational reconstructions. For all that, in the same essay, Davidson gave absolute priority to the retrospective aspect of the concept of intention ("having acted with a certain intention") over the prospective aspect ("indending to act"). Ricoeur gives Davidson credit for having recognized in his later writings that he was wrong[2] to think that *intending to act* would be easy to understand in terms of *acting with an intention* and *acting intentionally*. But this early mistake, according to Ricoeur, was only a superficial sign of a more systematic and more fundamental neglect which is a permanent feature of Davidson's thought and of analytic philosophy in general: neglect of the agent in favour of action and its effects; neglect of the pragmatics of action in favour of a pure semantics of action. Even what is supposed to concern most specifically *what it is like to be an agent,* namely teleological explanation of actions by reasons, tends persistently to be included by Davidson in the frame of causal explanation[3]. Identification of "primitive reasons" to mental events is a momentous step in this direction[4].

Ricoeur also develops another example in order to dissociate himself from the objectivist reductionism which he considers as a characteristic of analytical philosophy: he points out that Anscombe's analysis of intention is characterized by the same type of neglect as Davidson's. Long developments are devoted, in Anscombe's *Intention,* to the difference of scope of the questions "what?" and "why?", but very little is to be found about the question "who?": *who* is the actor? *who* is responsible for this state of affairs?[5]. Yet, says Ricoeur, when the question "who?" is raised, the special features of agency, as against simple series of events, are immediately manifest. Many intricacies arising when one wonders which events count as actions are dissolved at once by consideration of the agent. In order to show this, let us consider the peculiar relationship between the agent and his actions, and let us contrast it with the relationship between causes and effects. Whereas the inquiry about the causes of an event has no true limit, the distinctive feature of an inquiry about the agent is that it stops somewhere. We can always find a cause upstream from any causes, but if we are asked "*who* did this?", the answer we give is bound to be

[1]D. Davidson, "Actions, Reasons, and causes", in: *Essays on actions and events,* op. cit., p. 12; P. Ricoeur, *Soi-même comme un autre,* op. cit. p. 100
[2]D. Davidson, *Essays on actions and events,* op. cit., p. XIII
[3]P. Ricoeur, *Soi-même comme un autre,* op. cit. p. 107
[4]ibid. p. 104-105
[5]ibid. p. 87

univocal and finite. An action could then be *defined* as that for which there is an individually and socially acceptable answer to the question "who?". Of course, a clear-cut answer to the question "who?" is *not* supposed to *interrupt* by itself the unbounded series of causes. When I claim that I am the author of some action, this means that I (*qua* person and agent) accept to *endorse the responsibility* of the action; this does *not* mean that I reject the possibility of an uninterrupted causal description of the associated gesticulations of my limbs and of their remote consequences in the world. The claim of responsibility and the causal description operate on two different planes and with two different attitudes: the plane of mutual understanding between subjects acting in the world and the plane of objective explanation; the attitude of commitment and the attitude of attempted detachment.

These remarks had a considerable importance in the history of philosophy. The opposition between first beginning *in media res* (from an agent) and causation *in infinitum* was the basic remark from which Kant's third antinomy of pure reason stemmed. And the double level of description of actions, from the standpoint of a community of agents or in the perspective of objective knowledge, was one of the main motivations which prompted Kant to write the *Critique of practical reason* (which deals with freedom) after the *Critique of pure reason* (which deals with causality according the law of nature)[1].

A very similar kind of criticism was formulated by K.O. Apel[2] against G.H. Von Wright's "new dualist" analysis of *practical syllogisms*. As any other variety of dualism, Von Wright's "new dualism" tends to project two levels of description on the same plane of objective knowledge (describing them along a single time-series, or according to a single linear deductive sequence). According to Von Wright, the difference between understanding and explanation simply reflects the difference between motives and causes taken respectively as intentional and non-intentional *objects* [3]. Understanding and explanation are thus distinguished, but they both point towards elements of the same objectified series. The basis of this distinction is not strong enough to prevent one from encompassing intentional and non-intentional objects within a single theoretical *explanatory* scheme. No wonder that "quasi-naturalistic social sciences" drop these last remnants of the traditional distinction between explanation and understanding, and make use of "quasi-causal and even quasi-nomological motivational explanations"[4].

In order to make sense of the *necessity* of understanding besides explanation, one has to withdraw attention from the (scientific) task of describing objects or objectified data within a nomological frame, and to

[1]See e.g. L. W. Beck, *A commentary on Kant's critique of practical reason*, University of Chicago Press, 1960; L. W. Beck, *The actor and the spectator*, Yale University Press, 1975

[2]K.O. Apel, "Causal explanation, motivational explanation, and hermeneutical understanding", loc. cit.

[3]G.H. Von Wright, *Explanation and understanding*, op. cit. p. 135

[4]K.O. Apel, "Causal explanation, motivational explanation, and hermeneutical understanding", loc. cit.

concentrate on the (hermeneutic) task of integrating oneself into the subject-cosubject relationship of intersubjective communication[1]. The necessity of a language-game of motivations, reasons, beliefs, etc. does not rely on the emergent features of a certain class of non-intentional *objects;* nor does it rely on the role that motivations, reasons, and beliefs can play, as intentional objects, in the *objectified* series of practical inferences. It relies on their being *interest-relative,* namely relative to the position of a subject belonging to a community of intercommunicating subjects.

It is interesting to notice that, when he wants to characterize the relations between the language game of explanation of objectified data and the language game of hermeneutic understanding, K.O. Apel finds no better tool than Bohr's concept of *complementarity*. For, he says, the two language games are both mutually exclusive and jointly indispensible. They are mutually exclusive "(...) in as far as it is not possible to *objectify scientifically* the *behaviour* of a human being in a strict sense so long as one maintains with regard to this human being the relationship of *communicative understanding between co-subjects"* [2]. And they are jointly indispensible (Apel says they "complement each other" in the traditional sense of the verb "to complement"), because explanation of objectified data *presupposes* communicative understanding. As it stands, however, this kind of relation between the language games of understanding and explanation does not correspond very accurately to Bohr's concept of "complementarity". For there is a kind of asymmetry in Apel's account of the two language games. Apel claims that explanation (of objectified data) presupposes understanding (between co-subjects), but nowhere does he say that understanding presupposes explanation and/or objectification. Yet, this converse relation holds as well, because it is also true that mutual understanding between co-subjects is usually *about* something. Mutual understanding arises about an *object* of possible agreement. But in this case, can one still speak of mutual exclusiveness? Can one still argue that objective description is sometimes *incompatible* with communication? The only case of mutual exclusiveness arises when *every* potential co-subject of communication is not recognized as such, but is taken as an object of discourse. For instance, if X refused to speak to Y and Z, and rather described every element of Y's and Z's present behaviour *qua* behaviour, then communication with Y and Z would be impossible. However, it is also conceivable that X describes Y's present behaviour to Z, so that some sort of communication and mutual understanding, here between X and Z *about* Y, would still go on. Retrospectively, the caricatural situation where X contents himself with describing Y's and Z's behaviour can be interpreted as one in which there is communication between X and a *potential* co-subject S *about* Y and Z, not as one with no communication whatsoever. Even more important, one may conceive a partial overlapping of the co-subject of communication

[1]ibid.
[2]ibid.

and the object of discourse: nothing prevents X and Y from communicating *about* Y's past behaviour taken as an object of discourse. Therefore, Apel's claim that it is not possible to *objectify* the *behaviour* of a human being so long as one maintains with regard to him the relationship of *communicative understanding* must be qualified. It is not indispensible that the object of discourse and the co-subject of communication be *physically* distinct (as human beings); one must only act and speak in such a way that they are *pragmatically* distinguished in every relevant circumstance (with respect to their *roles* in conversation).

To summarize, the language games of objective description and mutual understanding are not incompatible *in abstracto*. As a rule, they *are* used in combination in ordinary speech. One must only be careful, in order to maintain their peaceful coexistence, to establish in each context of speech a reasonable *functional* distinction between the object of description and the sought co-subject of communication.

In view of these remarks, we shall drop Apel's Bohrian vocabulary and adopt another instead. We shall say that Apel's ideas about understanding and explanation, or about intentional and non-intentional language-games, reflects an *"anomalous parallelist"* doctrine, rather than a doctrine of "complementarity". The word "parallelist" is aimed at distinguishing Apel's position from both dualism (with its projection of two language-games on the same plane) and any kind of monism (with its bias in favour of one of the two language-games). As for the qualification "anomalous", it indicates (as in Davidson's "anomalous monism") that there is here no strict correlation law between the two "parallel" language games.

It is interesting to notice, in view of further developments on quantum mechanics, that Schrödinger's own position on the explanation-understanding issue was strikingly close to this *anomalous parallelist* conception. In a paper of 1935[1], he emphasized that:

(i) "Natural sciences do not rely exclusively on experiments, but also on a certain fundamental hypothesis; a hypothesis which is very, very, obvious, and which is accepted by everyone (...)". This hypothesis states that other human beings have similar perceptions to mine. It can be called, in short, an "anti-solipsistic hypothesis". It is a necessary presupposition for communication between scientists and it thus pertains to pragmatics and hermeneutics.

(ii) Yet, this hypothesis *cannot* be justified scientifically. It does not belong to the domain of science: "Science is not self-sufficient; it needs a fundamental axiom, an axiom coming from outside". Even worse, a natural scientist (say a physiologist) who would like to use the anti-solipsistic hypothesis as an element of his explanations would be accused of betraying his own science: "(...) do you believe a physiologist (*qua* physiologist) should answer that the man shouts because he feels pain? Not

[1]E. Schrödinger, "Quelques remarques au sujet des bases de la connaissance scientifique", Scientia, LVII, 181-191, 1935

at all, for by answering thus he would close his eyes on the true scientific problem". The anti-solipsist hypothesis cannot be used in the solution of each particular scientific problem "(...) *in spite,* and may be because, of the fact science as a whole relies on this hypothesis". *Its necessity does not arise from any feature within the field of scientific explanation, but from the circumstance that* we *(men and scientists) belong to a community of intercommunicating subjects.*

One year after, in 1936, R. Carnap wrote an answer to Schrödinger's paper and published it in the same journal[1]. His conclusion, opposite to Schrödinger's, was that "(...) no premiss of science eludes empirical control". The detail of the arguments shows that, in this controversy, the position of Schrödinger with respect to Carnap was quite similar to the position of Ricoeur and Apel with respect to the philosophers of the analytic tradition[2].

At this point, we can carry on with our systematic comparison between the explanation / understanding controversy, and the continuous ψ–model / discontinuous Fact-model problem. Just as, according to Ricoeur, Anscombe's analysis of intention focused too exclusively on the couple of questions "why?" and "what?", leaving almost completely aside the question "who?", one may notice that many modern interpretations of quantum mechanics focus too exclusively on the couple of questions "which?" (which fact *can* occur in a certain experimental context, and which fact *does in fact* occur in an experiment?) and "what?" (what kind of [causal or stochastic, ontological or formal] link is there between one fact and a subsequent fact?). Bohr's insistance on the questions "where from?", or "for whom?" (*from which* scale of size and complexity, does the idea that discontinuous facts occur makes sense; or *for whom* are the concepts of unambiguous experimental results and classical variables indispensible?) is only beginning to attract renewed attention. These questions obviously correspond to the question "who?" in the analysis of intention, for they also tend to point reflexively towards the subject of agency and knowledge. The only difference between Bohr's subject and Ricoeur's subject is that Bohr's subject is a *generical* "being-in-the-world", whereas Ricoeur's subject is an *individual* "being-in-a-body".

Now, as soon as the questions "where from?" and "for whom?" are raised, the measurement problem, that is the problem of the relations between the continuous ψ–model and the discontinuous Fact-model, can be seen in a very different light. To show this, and to display at the same time an analogy with the explanation / understanding controversy, we

[1]R. Carnap, "Existe-t-il des prémisses de la science qui soient incontrôlables?", Scientia, LX, 129-135, 1936

[2]See M. Bitbol, "L'alter-ego et les sciences de la nature, *Autour d'un débat entre Schrödinger et Carnap* ", in: A. Soulez & A. Sebestik, *Science et philosophie en Autriche et en France entre 1880 et 1930,* (to be published); see also F. Nef, "A propos d'une controverse entre Carnap et Schrödinger", in: M. Bitbol & O. Darrigol (eds.), *Erwin Schrödinger, Philosophy and the birth of quantum mechanics,* Editions Frontières, 1992

shall follow as closely as possible the general pattern of our exposition of Ricoeur's criticism of Anscombe's position.

Whereas the description of an experiment by pure unitary quantum mechanics (i.e. by the ψ–model) can go on for ever, from one subsystem to another, the distinctive feature of an inquiry among members of the scientific community about some experiment is that it generally stops when somebody claims that such-and-such a fact has occurred. One can always introduce more and more steps in the pure wave-mechanical description of a measurement chain, but if we ask some member of the human community to which we belong a question about the occurrence of some experimental fact, the unambiguous and finite answer we get puts a socially acceptable end to the investigation. A fact might even be defined, conversely, as that which promotes individual and social agreement about where to stop the chain of questions concerning the outcome of an experiment.

The essential point to notice is that in the context of this pragmatic conception of facts, a clear-cut answer, by a member of our community, to a question bearing on the occurrence of some fact, is *not* supposed to *interrupt* the development of the overall wave function of the measurement chain including himself. A socially acceptable end is not tantamout to a descriptive end. When I state that some experimental fact has occurred, this means that I accept, in the name of the scientific community to which I belong, and from the position *we* (myself and the community) hold in the world, to endorse the responsibility of the claim; this does *not* mean that I have any reason to contend that something has interrupted the wave mechanical description of the whole experimental process in which I am myself involved. The claim of scientific responsibility about "facts", and the wave-mechanical description, operate on two completely different planes and with two different attitudes: the plane of mutual understanding between subjects making experimental manipulations in the world on the one hand, and the plane of theoretical description on the other hand; the attitude of (social) commitment and the attitude of (epistemological) detachment.

This being granted, the standard account of measurements in terms of (i) truncated episodes of continuous evolution of a wave function according to the Schrödinger equation, and (ii) sudden "collapses", appears as a typical dualist attempt at projecting two heterogeneous levels of discourse on one and the same plane.

The latter dualist strategy includes some well-known and some less well-known loopholes. It has sometimes been said that the concept of "wave-packet collapse" makes perfect sense, as long as one sticks to a purely instrumentalist attitude according to which the ψ-model is merely a mathematical tool for calculating probabilities for the occurrence of experimental facts. But if one takes seriously the instrumentalist position and develop it to its ultimate consequences, the concept of wave-packet collapse becomes merely pointless. A proper instrumentalist tackling of quantum mechanics only involves three steps: (i) associating a wave-function to a certain experimental preparation, (ii) letting it change according to the Schrödinger equation, and finally (iii) calculating a list of probabilities by means of the Born rule. There is nowhere that wave-packet collapse has to be introduced, for the ternary scheme (preparation, evolution, probabilities) is universally applicable. It is also applicable to the paradigmatic situation where "collapse" is supposed to occur, i.e. when a *second* experiment is performed on a system after a preliminary experiment has given a certain result on the "same" system. When one has to predict the probability of the results of this second experiment, one determines a *new* wave function which corresponds to the combined preparation (initial preparation + first experiment having given a certain result). Saying that this is tantamount to "collapsing" a former wave function sounds very artificial, for this former wave function is no longer relevant.

Expressing the *substitution* of wave function in terms of (discontinuous) *change* of the former wave function then appears as a contrived attempt at maintaining at any cost some kind of temporal continuity between two definitely distinct predictive calculations. Now this is exactly what Schrödinger pointed out in his 1935 cat's paper (see §4-3) when he insisted that, in the instrumentalist framework where the wave function is construed as a "catalogue" of expectations, one should not say that it "changes" during a measurement, but that it *disappears,* and is *created anew* in order to make further predictions. This is also what Margenau pointed out when he insisted on the momentous conceptual difference between preparation and measurement.

The concept of wave-packet collapse actually points towards a more ambitious conception of the wave-function; one where the wave function is not considered as a mere mathematical instrument that may be dropped without harm as soon as it has provided the sought probabilities of a certain set of outcomes after a certain preparation, but rather as a *descriptive* element which retains some identity (or which refers to the state of the *same* system), in spite of its discontinuous changes. However, if the wave-function is ascribed a descriptive role, and if no other level of thought than this description is retained, the necessity of a discourse in terms of facts, and the very occurrence of wave-packet collapses, become very mysterious; for, without any additional *ad hoc* element such as Ghirardi's, Rimini's and Weber's *spontaneous collapse* terms[1], no feature of the standard unitary wave-mechanical description is able to accomodate by itself a counterpart of the discontinous occurrence of facts, namely the discontinuous wave-function collapses.

Is it then impossible to make sense of *facts* when (i) a descriptive role is ascribed to wave-functions, and (ii) the standard unitary model of quantum mechanics *(without spontaneous collapse terms)* is retained? Not

[1]See G.C. Ghirardi & A. Rimini, "Old and new ideas in the theory of quantum measurement", A.I. Miller (ed.), *Sixty-two years of uncertainty,* NATO-ASI series B226, Plenum Press, 1990

at all. However, in this case, one has to accept the idea that the Fact-model is *irreducible* to any feature of the ψ-model; that both models operate *on a completely different plane*. The *necessity* of the Fact-model besides the ψ-model can only be understood in this context if one provisionnally withdraws attention from the (epistemic) task of describing wave-mechanical entities within a nomological frame and concentrates on the (hermeneutic) task of integrating oneself into intersubjective communication with other members of the scientific community. This being accepted, the Fact-model does not rely on some emergent features of the ψ-model; nor does it rely on some sudden spontaneous projections of the ψ-functions. The Fact-model is made necessary, quite independently from the disengaged wave-mechanical description, by its being *interest-relative,* namely relative to *our* position. Let us add a precision to this remark. In the framework of a quasi-realist attitude, wherein the symbols of the most advanced physical theories are taken at face value, one could consider (with no metaphysical implication, but rather with an intentional and semantical connotation) that wave machanics is a description *of* the world; and in this case, the Fact-model would be made necessary relative to our situation *in* this world.

The position we have just developed can be called an *anomalous parallelist* conception of quantum mechanics. The term "anomalous" (or "lawless") expresses the circumstance that the series of experimental facts is not supposed to be linked by a strict law to the time development of some overall wave function, but only through Born's probabilistic correspondence rules. As for the term "parallelism", it contrasts with both dualism and monism. The anomalous parallelist conception is clearly distinct from any version of classical / quantum dualism or discontinuous collapse / continuous evolution dualism, for in it the language-game of facts is *not* projected on the same plane as the language-game of wave-mechanical description; the two language-games develop in parallel on two different planes. Anomalous parallelism is also distinct from standard monist conceptions of quantum mechanics, for in it the language-game of facts is not supposed to express a mere epiphenomenon of the language-game of wave-mechanical description. Last but not least, anomalous parallelism has many features in common with Everett's anomalous monist conception, but it adds an important piece of information which was lacking (or was merely implicit) in Everett's interpretation: in the anomalous parallelist conception, the introduction of a Fact-model besides the ψ-model finds a clear *justification,* quite apart from any questionable position in the mind-body problem debate. In the anomalous parallelist conception, the need of a Fact-model in addition to the ψ-model is justified by its pertaining to a pragmatic-hermeneutic attitude as against an objectifying attitude.

Table 2 summarizes the principal elements of our analogy between what it is like to be in a body (mind-body problem) and what it is like to be in the world (measurement problem of quantum mechanics).

	Being-in-a-body	***Being-in-the-world*** (Quantum mechanics)
Descriptions	Intentional / Causal	Fact-model / ψ - model
Dualism	Mind / Body (cartesian cut)	-Classical apparatus/quantum object (Bohr-Heisenberg cut) -Discontinuous collapse/continuous evolution (Von Neumann cut)
Reductionist Monism	-Only Mind (idealism) -Only Body (materialism)	-Only F, Ψ symbolic (instrumentalism) -Only Ψ, F emergent (Popular Many-worlds; Decoherence; Schrödinger 1926)
Anomalous monism	Supervenience of mental on physical (Davidson)	Supervenience of F-model on ψ -model (Everett's original relative state interpretation)
Anomalous parallelism	Two language-games and two attitudes: hermeneutic (intentions) and objectifying (causes)	Two attitudes: participation to intersubjective intercourse (F-model) and objectifying detachment (ψ model) (Schrödinger 1950 [with qualifications])

TABLE 2

We can now connect our analysis of the anomalous parallelist conception of quantum mechanics with Schrödinger's views (and justify the last line of Table 2). As we have shown repeatedly (paragraphs 4-4 point (6), 4-5, and 6-4), Schrödinger's approach to the interpretation of quantum mechanics in the 1950's was predominantly parallelist. We can now add that his approach fits almost exactly what we have called *anomalous parallelism*.

In Schrödinger's conception, there was on the one hand an *uninterrupted* continuous evolution of the wave functions, and on the other hand a completely autonomous eigenvalue scheme. This is what one can call, after Wimmel, a statement of *parallelism* between the ψ-model and the Fact-model. Schrödinger then emphasized (in contradistinction with Van Fraassen or Wimmel) that according to him the ψ-model must be ascribed a descriptive value, not only a predictive role. And finally, he recognized that the connection between the two models is essentially probabilistic. The models are connected by means of the Born rules, or, even better, by means of an expression of the expectation values of observables associated with the correspondence principle. It is this purely probabilistic link which accounts for the epithet "anomalous" before the word "parallelism". Therefore, Schrödinger's approach in the 1950's had clearly every ingredient of an *anomalous parallelist* interpretation of quantum mechanics.

True, such a strategy does not *solve* the measurement problem, and Schrödinger sometimes manifested signs of discomfort about this lack of solution. Yet, it tends to make the problem pointless by removing every motivation for its very formulation. Let us recall that the measurement problem is essentially a statement of the absence of any apparent fact-like structure in pure unitary quantum mechanics. In d'Espagnat's concise wordings "Within standard quantum mechanics, (there are) no 'really existing' facts"[1]. A solution of the measurement problem would then consist in displaying a fact-like structure by manipulating somehow pure unitary quantum mechanics (or complementing it by additional formalism). However, if one accepts to consider that pure unitary quantum mechanics is completely separated from the Fact-model, that the two models develop in parallel, and that this parallelism is not strict but "anomalous" (since the two models only connect *via* a probabilistic empirical correspondence rule), then the problem of finding a fact-like structure within the ψ-model does not even arise. In a Wittgensteinian sense, the measurement problem has been *dissolved* by the anomalous parallelist strategy.

But could Schrödinger appreciate the underlying *motivation* of anomalous parallelism, namely the distinction between a pragmatic-hermeneutic attitude and an objectifying attitude; a distinction between participation to a community of intercommunicating subjects and pure intentional directedness? Could he realize that the necessity of a Fact-model independent from the ψ-model is the mark, in quantum mechanics, of the impossibility of ignoring the peculiarity of our position *in* the world whenever we try to formulate a theory *of* the world? There are indications in his writings that he was in possession of the intellectual tools which could have enabled him to reach these conclusions. The idea that science includes, broadly speaking, *position-relative* or *interest-relative* elements, can be found in many of his papers and books. It constitutes the main topic of two papers, published respectively in 1932 and 1948[2]. In 1939 he even advocated explicitly a relativist conception of science that he

[1]B. d'Espagnat, "Towards an empirical separable reality?", *Found. Phys.* 20, 1147-1172, 1990

[2]E. Schrödinger, "Ist Naturwissenschaft milieubedingt?", in: E. Schrödinger, *Über Indeterminismus in der Physik*, Barth, Leipzig, 1932; English Tr. in: E. *Schrödinger, Science and the human temperament,* op. cit. E. Schrödinger, "Die Besonderheit des Weltbilds der Naturwissenschaft", *Acta Physica Austriaca,* 1, 201-245, 1948

called "practical subjectivism"[1]; according to this view, many features of science are determined by the cultural, social and historical background of the scientific community, through the selection of relevant methods and aims. His book *Nature and the Greeks* contains several examples of "special features" that the scientific undertaking owes to our cultural inheritance. To him, as to the Hellenist T. Gompertz, "(Science) is (...) something special, something that has grown historically over many centuries, not the general, the only possible way of thinking about nature"[2]. In addition, he pointed out in *Mind and Matter* that our cognitive capacities and their general orientation are to be traced back to our phylogenetic origins[3]. In short, Schrödinger considered that many elements of our way of dealing with the surrounding world, and of our view *of* the world, are to be traced back to our (phylogenetical, historical, and cultural) position *in* the world. Generalizing these ideas would have been sufficient for him to reach Bohr's conclusion that the concepts of phenomenon, of experimental fact, and of classical variable, stem from certain characteristics of our position in the world, i.e. from our particular scale and complexity, and from the norms of intersubjective communication. Schrödinger's reference to the "assumption of correspondence"[4] (by which he meant a modern version of Bohr's principle of correspondence), and to the circumstance that operators associated to measurement procedures are a "loan from classical physics" was a first step in this direction. The second step would have consisted in recognizing, as Bohr did, that part of classical physics is a loan from our special position both in the world and in a community of speakers. This would have been enough to make clear that the Fact-model is position- and interest-relative.

The reason why Schrödinger did not go as far as to make explicit each step leading to the conclusion that the Fact-model is interest-relative, is almost certainly that he feared the kind of flat instrumentalist consequences one could quite naturally draw from them. If the concept of fact and the associated eigenvalue scheme are relative to our position in the world, can't one extend this conclusion to the ψ-model and say, with Bohr, that "Strictly speaking, the mathematical formalism of quantum mechanics and electrodynamics merely offers rules of calculation for the deduction of expectations obtained under well-defined experimental conditions specified by classical physical concepts"[5]? Isn't the ψ-model *only* a tool for linking the elements of the Fact-model? This is exactly what Schrödinger could not accept, because he thought that making

[1]E. Schrödinger, Unpublished notes about A. Eddington's *Philosophy of physical science*, Cambridge University Press, 1939 (E. Schrödinger's Library, Alpbach, Austria)

[2]E. Schrödinger, *Nature and the Greeks*, op. cit. p. 88

[3]E. Schrödinger, Mind and Matter, op. cit. chapters 1 and 2

[4]E. Schrödinger, *Transformation and interpretation in quantum mechanics*, (Dublin seminar 1952), in: E. Schrödinger, *The interpretation of quantum mechanics (Dublin seminars 1949-1955 and other unpublished texts)*, op. cit. p. 51

[5]N. Bohr, *Essays 1958-1962 on atomic physics and human knowledge*, op. cit. p. 60

pictures, and maintaining some intentional directedness towards our theoretical entities, is *also* part of our cultural inheritance. Pictures and directedness are also parts of our most ingrained traditions because they partake of our epistemological commitment (knowledge points *towards* something) and of our mode of communication (we communicate *about* something). The growing influence of the instrumentalist tendency in the interpretation of quantum mechanics then prompted Schrödinger to react as follows: "Physical science, which aims not only at devising fascinating new experiments, but at obtaining a rational understanding of the results of observations, incurs at present, so I believe, the grave danger of getting severed from its historical background"[1].

Now, anomalous parallelism is perfectly able to reconcile these two apparently diverging elements of the anthropological background. One should even say that anomalous parallelism makes sense *only* in the perspective of their reconciliation. Indeed, the two indissociable components of the anomalous parallelist strategy are: acceptance of the autonomy of the interest-relative Fact-model with respect to the wave mechanical model, *and* acceptance of the autonomy of pure unitary evolution of wave-functions with respect to any element of the Fact-model. The first component arises from recognition of the role our particular position in the world must play in the formulation of physical theories, and the second component manifests the persistence of our *seriousness* (whose paradigm is the "natural attitude" of everyday life) about theoretical entities and nomological rules. This being granted, one must realize that Schrödinger's methodological realism, or *quasi-realism,* about theoretical entities like wave-functions fits much more easily the anomalous parallelist interpretation of quantum mechanics than, say, Wimmel's instrumentalist conception of wave-functions. For, if wave-functions were to be considered as mere instruments of calculation of the probabilities of future facts, why should one insist so much on the autonomy of their law of evolution with respect to any element of the Fact-model? Why should one avoid so carefully using a certain wave-function between the preparation and the measurement and then drop it as soon as it has been used to calculate the relevant probabilities, like an ordinary instrumentalist? In a strictly instrumentalist perspective, the wave-packet reduction strategy which consists in choosing a new appropriate wave-function for each new predictive situation, is just as good (and just as devoid of significance) as the parallelist strategy which consists in following the unitary evolution of an ever-increasingly entangled wave-function, and using it in order to calculate the probability of certain *sequences* of facts by means of Born's rule. But in a quasi-realist perspective, the intentionally aimed at entities cannot be treated so light-heartedly, and the parallelist strategy must thus be preferred.

[1]E. Schrödinger, "Are there quantum jumps?", op. cit.

To summarize, *anomalous parallelism* was not only one of the dominant components of Schrödinger's late interpretation of quantum mechanics; it also agreed more closely with the general trends of his philosophical outlook than with most alternative conceptions.

CHAPTER 7

CONCLUSION

Quantum theories are unanimously considered as the result of a collective work. Each physicist, within a list including Planck, Einstein, Bohr, de Broglie, Heisenberg, Schrödinger, Dirac, Pauli, Born, Jordan, etc. is ascribed a role in the edification of the formalism and the empirical correspondence rules of quantum mechanics. By contrast, the *interpretation* of quantum mechanics is usually regarded as a subject of conflict or an instrument of sociological leadership, rather than a field of cooperation. This view is supported by the hot debates which took place between the founders of quantum mechanics, and which still take place today, about the meaning and scope of this theory. Yet, the further the debate progresses, the clearer it appears that parts of the conceptions which were once considered conflicting are likely to be incorporated together into an acceptable mature attitude towards quantum mechanics.

Let us then see how this could possibly happen, by stating briefly the main contribution of each author to the collective reflection on the interpretation of quantum mechanics, as well as the limits of his conceptions. Bohr's major discovery in this field is certainly his holistic conception of phenomena, which plays a role in many modern interpretations of quantum mechanics, and which, according to the Kochen and Specker theorem *must* have a counterpart in every acceptable theory able to reproduce the predictions of quantum mechanics. As we mentioned previously, even Schrödinger accepted that wave functions are only definable *relatively*, in relation to programs of observation; and, in his ontological interpretation of quantum mechanics, Bohm gave Bohr's holism the form of contextualism. But Bohr did not stop at this very strong point. He also tried to associate his holism (and the associated "indefinability thesis") with some remnants of ontological conservatism, through his concept of "complementarity" (see §6-2). This concept remains one of the most controversial aspects of his thought. Even Heisenberg, who shared so many ideas with Bohr, took complementarity as only *one* of the possible answers to the dilemma of the inapplicability of ordinary language structure and classical pictures to the quantum domain[1].

But Heisenberg's most specific contribution to the debate on the interpretation of quantum mechanics is not his discussion of complementarity; it is rather his insistance on the role of *dispositions*, through his ascription of the status of Aristotelian *potentia* (and thus of an intermediate "kind of reality") to wave-functions. After many transformations, this idea has given rise to Popper's concept of propensity, and it has been revived in recent discussions about quantum field theory and about the concept of vacuum[2]. But Heisenberg was quite

[1]W. Heisenberg, *Physics and Philosophy*, op. cit. chapter X

[2]R. Harré, "Tracks and affordances: the sources of a physical ontology", International studies in the

hesitant concerning the universal applicability of his dispositional view of quantum mechanics. He continued speaking in terms of atoms and particles, though *very* critically and with *many* qualifications, instead of focusing exclusively the attention on a more primitive layer of dispositions such as the energy (that he himself considered as some sort of *materia prima* in the sense of Aristotle), or, even better, the "hohlraum oscillator" of quantum field theory.

Schrödinger was more daring in this respect. He was the physicist who showed that, on the one hand, one can manage without *any* remnant of classical ontology, and that, on the other hand, one has good reasons to take very seriously the new entities of quantum mechanics. The concept of vacuum which arises from an analysis of states in Fock space was particularly crucial in his own conception of these new entities.

As we have seen in previous chapters, Schrödinger grounded his approach on a completely non-metaphysical, or *quasi-realist,* definition of ontologies; and he applied it to a dissolution of the measurement problem, by means of an "anomalous parallelist" conception of the relations between facts and wave-functions. The intentional gaze of the physicist, which is an integral part of the efficiency of his investigations, could thus be redirected towards new sets of entities such as state-vectors, instead of being either prohibited or confined to a "symbolic" use of some aspects of the old entities in restricted experimental conditions. But these remarkable advances were hidden to the eyes of Schrödinger's contemporaries, due to his persistent use of modes of expression which are usually associated to uncritical versions of realism, and to classical wave representations.

Today, the three trends respectively favoured by Bohr, Heisenberg and Schrödinger can be perfectly reconciled in a synthetic interpretation of modern quantum theories, provided they are dissociated from the conflicting original *weltanschauung* of their champions. As we have seen, these three constitutive trends are (i) *contextuality,* (ii) *dispositionalism, and* (iii) *quasi-realism about the theoretical entities, associated with anomalous parallelism between these entities and the experimental data.*

But what about hidden variable theories? Even though they aim at going beyond quantum mechanics, they too contribute, by contrast, to a proper appraisal of its significance. They do so by displaying directly and convincingly that the interpretation of quantum mechanics, and the worldview associated with it, is *underdetermined* by the constraints which are embodied in its predictive formalism[1]. The fact that some outstanding physicists have formulated hidden variable theories, and the arguments

philosophy of science, 4, 149-158, 1990; S. Saunders & H.R. Brown, *The philosophy of vacuum,* Oxford University Press, 1991

[1] C. Chevalley, «Le conflit de 1926 entre Bohr et Schrödinger: un exemple de sous-détermination des théories», in: M. Bitbol & O. Darrigol (eds.), *Erwin Schrödinger, Philosophy and the birth of quantum mechanics,* op. cit.

some other outstanding physicists use in order to advocate them, show that the choice between the different worldviews which are allowed by the predictions of quantum mechanics is a matter of values. A quantum physicist *can* decide without harm, in the name of some explicit or implicit values, to put strong emphasis on one of the regulative principles which are used as criteria for the elaboration of scientific theories, to the detriment of the others. Emphasis on certain regulative principles will lead to adopt suitable hidden variable theories, and emphasis on other regulative principles will rather lead to take seriously the worldview(s) suggested by the formalism of quantum field theory.

But what are these regulative principles? According to W. James they are: "usefulness for prediction, (...) conservation of past doctrine, simplicity, and coherence"[1]. In the case of the most popular hidden variable theories, the predominant criterion is obviously conservation of past classes of objects (essentially particles), to the detriment of simplicity and complete conceptual coherence[2]. By contrast, Schrödinger tended to put emphasis on simplicity and coherence even if it means dropping any remnant of the traditional ontology of material bodies. He thus anticipated Kuhn's idea that the emergence of a new scientific paradigm means the emergence of a new world, by proposing nothing less than a complete change of the ontological furniture of the world in addition to the change of its laws.

[1]H. Putnam, *Pragmatism*, Basil Blackwell, 1995, p. 10

[2]Bohm's hidden variable theory initially consisted in introducing one kind of "beable": corpuscle-like beables. It nowadays introduces *two* different sets of "beables" in order to derive the empirical consequences of relativistic quantum theories: corpuscle-like beables for fermions and field-like beables for bosons. Yet, as Bohm and Hiley notice in their recent book, a field theory *alone* would be perfectly able to account for corpuscle-like behaviour as well. The only reason they give for using *two* models and *two* different sets of beables instead of one (thus for losing some simplicity and conceptual coherence) is that a fermion field theory relies on anti-commutation relations which have no proper *classical limit*. We can then identify very clearly one of their dominant values: not historical *conservatism* (for their beables have a very un-classical behaviour), but historical *continuism*.

BIBLIOGRAPHY

Agostino, S. d', "Continuity and completeness in physical theory: Schrödinger's return to the wave conception of quantum mechanics in the 1950's", in: M. Bitbol and O. Darrigol (eds.), *Erwin Schrödinger, philosophy and the birth of quantum mechanics,* Editions Frontières, 1993

Agostino, S. d', "Boltzmann and Hertz on the Bild-conception of physical theory", History of Science, 28, 380-398, 1990

Aharonov, Y., Anandan, J., & Vaidman, L., "Meaning of the wave function", *Phys. Rev.* A47, 4616-4626, 1993

Albert, D.Z., & Loewer, B., "Interpreting the many-worlds interpretation", Synthese, 77, 195-213, 1988

Albert, D. Z., *Quantum mechanics and experience,* Harvard University Press, 1993

Alberti, L.B., *De Pictura* , Macula Dédale, 1992

Anscombe, G.E.M., *Intention,* Basil Blackwell, 1957

Apel, K.O., "Causal explanation, motivational explanation, and hermeneutical understanding", in: G. Ryle (ed.), *Contemporary aspects of philosophy,* Oriel Press, 1976

Aquinas, St Thomas ,*On being and essence,* Translation and interpretation by J. Bobik, University of Notre-Dame Press, 1965

Aris, R., Davis, H.T., and Stuewer, R.H., (eds.), *Springs of scientific creativity,* University of Minnesota Press, 1983

Atmanspacher, H., "Is the ontic/epistemic distinction sufficient to describe quantum systems exhaustively?", in: K.V. Laurikainen, C. Montonen, & K. Sunnaborg, *Symposium on the foundations of modern physics 94,* Editions Frontières 1994

Auyang, S. Y., *How is quantum field theory possible?,* Oxford University Press, 1995

Barrow, J.D., and Tipler, F., *The anthropic cosmological principle,* Oxford University Press, 1986

Beck, L. W., *A commentary on Kant's critique of practical reason,* University of Chicago Press, 1960

Beck, L. W.,*The actor and the spectator,* Yale University Press, 1975

Beller, M., "Born's probabilistic interpretation: a case study of 'concepts in flux'", Stud. Hist. Phil. Sci., 21, 563-588, 1990

Beller, M., "Schrödinger's dialogue with Göttingen-Copenhagen physicists", in: M. Bitbol and O. Darrigol, *Erwin Schrödinger, Philosophy and the birth of quantum mechanics,* Editions Frontières, 1993

Ben-Dov, Y., "Everett's theory and the 'many-worlds' interpretation", Am. J. Phys., 58, 829-832, 1990

Ben-Menahem, Y., "Struggling with causality: Schrödinger's case", Stud. Hist. Phil. Sci., 20, 307-334, 1989

Ben-Menahem, Y., "Struggling with realism: Schrödinger's case", in: M. Bitbol and O. Darrigol (eds.), *Erwin Schrödinger, Philosophy and the birth of quantum mechanics,* Editions Frontières, 1993

Berlin, I., *Concepts and Categories,* Oxford University Press, 1980

Bertotti, B., & Curi, U., (eds.), *Erwin Schrödinger scienziato e filosofo,* Il poligrafo, 1994

Bitbol, M., *L'élision,* in: E. Schrödinger, *L'esprit et la matière,* Editions du Seuil, 1990

Bitbol, M., "Perspectival realism and quantum mechanics", in: P. Lahti and P. Mittelstaedt, (eds.), *Proceedings of the symposium on the foundations of modern physics 1990,* World Scientific, 1991

Bitbol, M., *La Clôture de la représentation,* in: E. Schrödinger, *La nature et les grecs,* Editions du Seuil, 1992

Bitbol, M., and Darrigol, O., (eds.), *Erwin Schrödinger, Philosophy and the birth of quantum mechanics,* Editions Frontières, 1993

Bitbol, M., "Esquisses, forme, et totalité", in: M. Bitbol and O. Darrigol (eds.), *Erwin Schrödinger, Philosophy and the birth of quantum mechanics,* Editions Frontières, 1993

Bitbol, M., & Ruhnau, E., (eds.), *Now, time and quantum mechanics,* Editions Frontières, 1994

Bitbol, M., *Mécanique quantique: une introduction philosophique,* Flammarion 1996

Bitbol, M., *De l'intérieur du monde,* (in preparation)

Bitbol-Hespériès, A., *Le principe de vie chez Descartes,* Vrin, 1990

Blackburn, S., *Essays in Quasi-realism,* Oxford University Press, 1993

Block, I., (ed.), *Perspectives on the philosophy of Wittgenstein,* Basil Blackwell, 1981

Bohm, D., & Hiley, B.J., *The undivided universe,* Routledge, 1993

Bohr, N., Kramers, H.A., and Slater, J.C., "The quantum theory of radiation", Phil. Mag. 47, 785-802, 1924

Bohr, N., "The quantum postulate and the recent development of atomic theory", Nature, 121, 580-590

Bohr, N., *Atomic theory and the description of nature,* Cambridge University Press, 1934 (Reprint Ox Bow Press, 1987)

Bohr, N., "Can quantum-mechanical description of physical reality be considered complete?", Phys. Rev. 48, 696-702, 1935

Bohr, N., *Collected works,* E. Rüdinger (gen. ed.), vol. 6, J. Kalckar (ed.), North-Holland, 1985

Bohr, N., *Essays 1933-1957 on atomic physics and human knowledge,* Ox Bow Press, 1987

Bohr, N., *Essays 1958-1962 on atomic physics and human knowledge,* Ox Bow Press 1987

Boltzmann, L., *Theoretical physics and philosophical problems,* (B. Mac Guinness, ed.), Reidel, 1974

Born, M., "Zur Quantenmechanik der Stossvorgänge" Zeitschrift für physik, 37, 863-867, 1926; Tr. "On the quantum mechanics of collisions" in: J.A. Wheeler and W.H. Zurek, *Quantum theory and measurements,* Princeton University Press, 1983

Born, M., "Quantenmechanik der Stossvorgänge", Zeitschrift für physik, 38, 803-827, 1926

Born, M., "Das Adiabatenprinzip in der Quantenmechanik", Z. Phys. 40, 167-192, 1926.

Born, M., *Natural philosophy of cause and chance,* Oxford University Press, 1949

Born, M., "The interpretation of quantum mechanics", Brit. J. Phil. Sci., 4, 95-106, 1953

Born, M., "Physical reality", Phil. Quart., 3, 139-49, 1953

Born, M., *Physics in my generation,* Pergamon Press, 1956

Born, M., (ed.), *The Born-Einstein letters,* Mac Millan, 1971

Brown, H.R., Dewdney, C., & Horton, G., "Bohm particles and their detection in the light of neutron interferometry", *Foundations of physics,* **25**, 329-347, 1995

Carnap, R., *The logical structure of the world,* Routledge & Kegan Paul, 1967

Carnap, R., "Existe-t-il des prémisses de la science qui soient incontrôlables?", Scientia, LX, 129-135, 1936

Carnap, R., *Meaning and Necessity,* The University of Chicago Press, 1956

Cassirer, E., *Die Philosophie der Aufklärung,* Mohr (Tübingen), 1932

Cassirer, E., *Determinismus und Inderterminismus in der modernen Physik,* Götheborgs Högskolas Arsskrift 42, 1937; Tr. *Determinism and indeterminism in modern physics,* Yale University Press, 1956

Chevalley, C., "Le conflit de 1926 entre Bohr et Schrödinger: un exemple de sous-détermination des théories", in: M. Bitbol & O. Darrigol (eds.) *Erwin Schrödinger, Philosophy and the birth of quantum mechanics,* Editions Frontières, 1993

Cini, M., and Lévy-Leblond, J.M., (eds.), *Quantum theory without reduction,* Adam Hilger, 1990

Da Costa, N., Krause, D. , & French, S., "The Schrödinger problem", in: M. Bitbol & O. Darrigol (eds.) *Erwin Schrödinger, Philosophy and the birth of quantum mechanics,* Editions Frontières, 1993

Dalla Chiara, M.L., and Toraldo di Francia, G., "Individual, Kinds and Names in Physics", in: Corsi et al. (eds.) *Bridging the Gap: Philosophy, mathematics and physics,* Kluwer, 1993

Darrigol, O., "The origin of quantized matter waves", H.S.P.S., 16, 197-239, 1986

Darrigol, O., *From c-numbers to q-numbers,* University of California Press, 1992

Darrigol, O., "Schrödinger's statistical physics and some related themes", in: M. Bitbol & O. Darrigol (eds.) *Erwin Schrödinger, Philosophy and the birth of quantum mechanics,* Editions Frontières, 1993

Davidson, D., *Essays on actions and events,* Oxford University Press, 1982

Dennett, D., *Consciousness explained,* Penguin, 1993

Derksen, A., (ed.), *The scientific realism of Rom Harré,* Tilburg University Press, 1994

Descartes, R., *Oeuvres* III, C. Adam & P. Tannery (eds.), Vrin, 1964-75

Descartes, R.,*The passions of the soul,* (Translated by S. Voss) Hackett publishing company, 1989

Deutsch, D., "Quantum theory as a universal physical theory", Int. J. Theor. Phys., 24, 1-41, 1985

De Witt, B.S., and Graham, N., *The many-worlds interpretation of quantum mechanics,* Princeton University Press, 1973

Dieks, D., "Resolution of the measurement problem through decoherence of the quantum state", Phys. Lett. A-142, 439-446, 1989

Dieks, D., "Modal interpretation of quantum mechanics, measurements, and macroscopic behavior", Phys. Rev. A49, 2290-2300, 1994

Dickson, M., "An empirical reply to empiricism: protective measurement opens the door for quantum realism", *Philosophy of science,* 62, 122-140, 1995

Dorling, J., "Schrödinger's original interpretation of the Schrödinger equation: a rescue attempt", in: C.W. Kilmister (ed.), *Schrödinger, centenary celebration of a polymath,* Cambridge University Press, 1987

Dreyfus, H.L., (ed.), *Husserl, intentionality and cognitive science,* MIT Press, 1982

Dreyfus, H.L., *Being-in-the-world (a commentary on Heidegger's Being and Time, division I),* M.I.T. Press, 1991;

Eddington, A., *Relativity theory of protons and electrons,* Cambridge University Press, 1936

Einstein, A.,*The meaning of relativity,* Princeton University Press, 1974

Einstein, A., "Quantentheorie des einatomigen idealen Gases", Berlin Akademie der Wissenschaft Sitzungsberichte, 1925, 3-14

Einstein, A., Podolsky, B., and Rosen, N., "Can quantum-mechanical description of physical reality be considered complete?" Phys. Rev. 47, 777-780, 1935

Einstein, A., *The meaning of relativity,* (1921), Princeton University Press, 1974

Einstein, A., "A generalization of the relativistic theory of gravitation", Ann. Math. Princeton, 46, 578; 47, 146,731, 1945

Einstein, A., *Oeuvres choisies, I, Quanta,* Editions du Seuil, 1989

Electrons et photons, 5th Solvay conference (1927), Paris 1928

Espagnat, B. d', *Conceptual foundations of quantum mechanics,* W.A. Benjamin, 1976

Espagnat, B. d', "Towards an empirical separable reality?" Found. Phys. 20, 1147-1172, 1990

Espagnat, B. d', "Appearance of a local world", Phys. lett., A 171, 17-20, 1992

Espagnat, B. d', *Veiled reality,* Addison-Wesley, 1995

Everett, H., "'Relative state' formulation of quantum mechanics", Rev. Mod. Phys., 29, 454-462, 1957

Everett, H., "Theory of the universal wave function" in: B.S. De Witt and N. Graham, *The many-worlds interpretation of quantum mechanics,* Princeton University Press, 1973

Faye, J., *Niels Bohr: his heritage and legacy,* Kluwer, 1991

Fine, A., *The shaky game,* University of Chicago Press, 1986

Folse, H., *The philosophy of Niels Bohr: the framework of complementarity,* North-Holland, 1985

Foster, S., & Brown, H., "On a recent attempt to define the interpretation basis in the many-worlds interpretation of quantum mechanics", Int. J. Theor. Phys., 27, 1507-1531, 1988

Gell-Mann, M., & Hartle, J.B., "Classical equations for quantum systems", Phys. Rev., D47, 3345-3382, 1993

Gell-Mann, M.*The Quark and the Jaguar,* Princeton University Press, 1994

George, A., (ed.), *Louis de Broglie physicien et penseur,* Albin Michel, 1953

Goodman, N., "The significance of Der logische Aufbau der Welt", in: P. A. Schilpp (ed.), *The philosophy of Rudolf Carnap,* Open court, 1963

Goodman, N., *Ways of worldmaking,* Hackett Publishing Company, 1978

Grene, M., *Descartes,* Harvester Press, 1985

Grene, M., *A philosophical testament,* Open Court, 1995

Grünbaum, A., "Complementarity in quantum physics and its philosophical generalization", The Journal of philosophy, 54, 717-735, 1957

Hanle, P.A., *Erwin Schrödinger's statistical mechanics (1912-1925),* Ph. D. dissertation, Yale University, 1975

Harré, R., *Varieties of realism,* Basil Blakwell, 1986

Harré, R., "Tracks and affordances: the sources of a physical ontology", International studies in the philosophy of science, 4, 149-158, 1990

Healey, R.,*The philosophy of quantum mechanics,* Cambridge University Press, 1989

Heelan, P., *Quantum mechanics and objectivity,* Martinus Nijhoff, 1965

Heelan, P., *Space perception and the philosophy of science,* University of California Press, 1983

Heidegger, M., *Being and Time,* Harper & Row, 1962

Heisenberg, W., "The physical content of quantum kinematics and mechanics" in: J.A. Wheeler & W.H. Zurek, *Quantum theory and measurement,* Princeton University Press, 1983

Heisenberg, W., "Schwankungerscheinungen und Quantenmechanik", Z. Phys., 40, 501-506, 1927

Heisenberg, W., *The physical principles of the quantum theory,* The University of Chicago Press, 1930

Heisenberg, W., *Physics and Philosophy,* Harper and Brothers, 1958

Heisenberg, W., *Physics and Beyond, encounters and conversations,* George Allen & Unwin, 1971

Hegel, G.W.F., *The phenomenology of Mind,* Trad. J.B. Baillie, Mac Millan, 1910

Held, C., "The meaning of complementarity", *Studies in History and Philosophy of Science,* 25, 871-893, 1994

Hertz, H.,*The principles of mechanics,* Mac Millan, 1899

Hume, D., *A Treatise of Human Nature,* (ed. L.A. Selby-Bigge), Oxford University Press, 1960

Husserl, E., (1913), *Ideas (general introduction to pure phenomenology),* G. Allen & Unwin, 1931

Husserl, E., *Grundlegende Untersuchungen zum phänomenologischen Ursprung der Räumlichkeit der Natur,* in: M. Farber, *Philosophical essays in memory of E. Husserl,* Harvard University Press, 1940

Jammer, M., *The conceptual development of quantum mechanics,* Mc Graw Hill, 1966

Kidd, R., Ardini, J., & Anton, A., "Compton effect as a double Doppler shift", Am. J. Phys., 53, 641-644, 1985

Kilmister, C.W., (ed.), *Schrödinger, centenary celebration of a polymath,* Cambridge University Press, 1987

Klein, M.J., "Einstein and the wave-particle duality", The natural philosopher, 3, 3-49, 1964

Kochen, S., & Specker, E., "The problem of hidden variables in quantum mechanics", *Journal of Mathematics and Mechanics,* 17, 59-87, 1967

Kochen, S., "A new interpretation of quantum mechanics", in: P. Lahti & P. Mittelstaedt, (eds.) *Symposium on the foundations of modern physics, World Scientific, 1985*

Krause, D., (1992), "On a quasi-set theory", Notre Dame Journal of Formal logic, 33, 402-411

Krikpe, S., "Wittgenstein on rules and private language", in: I. Block (ed.), *Perspectives on the philosophy of Wittgenstein,* Basil Blackwell, 1981

Kuhn, H., ("The phenomenological concept of 'horizon'", in: D. Farber, *Philosophical essays in memory of E. Husserl,* Harvard University Press, 1940

Laurikainen, K.V., *Beyond the atom,* Springer-Verlag, 1988

Laurikainen, K.V., Montonen, C., & Sunnaborg, K., *Symposium on the foundations of modern physics 94,* Editions Frontières 1994

Leibniz, G.W., *New essays on human understanding,* Cambridge University Press, 1982

Lentzen, V., "The interaction between subject and object in observation", Erkenntnis, 6, 326-333, 1936

Lockwood, M., *Mind, Brain, and the Quantum,* B. Blackwell, 1989

Lockwood, M., "What Schrödinger should have learnt from his cat", in: M. Bitbol & O. Darrigol (eds.) *Erwin Schrödinger, Philosophy and the birth of quantum mechanics,* Editions Frontières, 1993

Lockwood, M., "'Many-minds' interpretations of quantum mechanics", Brit. J. Philos. Sci. (To be published)

Mach, E.,*The analysis of sensations,* Dover, 1959

Mac Kinnon, E.M., *Scientific explanation and atomic physics,* The University of Chicago Press, 1982

Margenau, H., "Critical points in modern physical theory", Philos. sci., 4, 337-370, 1937

Margenau, H., "Measurements in quantum mechanics", Annals of physics, 23, 469-485, 1963

Mehra, J., and Rechenberg, H., *The historical development of quantum mechanics,* 5-2, Springer-Verlag, 1987

Melgar, M.F., "The philosophy of Erwin Schrödinger: a diachronic view of Schrödinger's thought", Found. Phys., 18, 357-371, 1988

Merleau-Ponty, M., *Phenomenology of perception,* Routledge & Kegan Paul, 1962

Mill, J.S., *An examination of Sir W. Hamilton's philosophy,* Longman & Green, 1865

Miller, A.I., (ed.), *Sixty-two years of uncertainty,* NATO-ASI series B226, Plenum Press, 1990

Mittelstaedt, P., *Philosophical problems of modern physics,* in: *Boston Studies in the philosophy of science,* vol. 18, Reidel, 1976

Moore, W., *Schrödinger, Life and thought,* Cambridge University Press, 1989

Mott, N.F., "The wave mechanics of α-ray tracks", Proc. Roy. Soc. Lond., A126, 79-84, 1929

Munitz, M.K., & Unger, P.K., eds. *Semantics and Philosophy,* New-York University Press, 1974

Murdoch, D., *Niels Bohr's philosophy of physics,* Cambridge University Press, 1987

Murdoch, I., *Metaphysics as a guide to morals,* Penguin Books, 1993

Nef, F., "A propos d'une controverse entre Carnap et Schrödinger", in: M. Bitbol and O. Darrigol, *Erwin Schrödinger, Philosophy and the birth of quantum mechanics,* Editions Frontières, 1993

Omnès, R., *The interpretation of quantum mechanics,* Princeton University Press, 1994

Paqué, R., *Le statut parisien des nominalistes,* P.U.F., 1985

Paz, J.P., & Zurek, W.H., "Environment-induced decoherence, classicality, and consistency of quantum histories", Phys. Rev. D48, 2728-2737, 1993

Pippard, A.B., "The interpretation of quantum mechanics", Eur. J. Phys., 43-48, 1986

Popper, K., *The logic of scientific discovery,* Hutchinson, 1968

Popper, K., *Unended quest,* Routledge, 1974

Popper, K., *Quantum theory and the schism in physics,* Hutchinson, 1982

Primas, H., "A propos de la mécanique quantique des systèmes macroscopiques", in: M. Bitbol and O. Darrigol (eds.), *Erwin Schrödinger, Philosophy and the birth of quantum mechanics,* Editions Frontières, 1993

Przibram, K., *Letters on wave mechanics,* (Tr. M.J. Klein), Philosophical Library, 1967

Putnam, H., *Realism with a human face,* Harvard University Press, 1990

Putnam, H., *Pragmatism,* Basil Blackwell, 1995

Quine, W.V., *The roots of reference,* Open Court, 1974

Raman, V.V., & Forman, P., "Why was it Schrödinger who developed de Broglie's ideas?", HSPS, 1, 291-314, 1969

Redhead, M., & Teller, P., "Particle labels and the theory of indistinguishable particles in quantum mechanics", Brit. J. Phil. Sci., 43, 201-218, 1992

Ricoeur, P., *Soi-même comme un autre,* Editions du Seuil, 1991

Rimini, G.C., & Rimini, A. , "Old and new ideas in the theory of quantum measurement", A.I. Miller (ed.), *Sixty-two years of uncertainty,* NATO-ASI series B226, Plenum Press, 1990

Rosenfeld, L., Commentary on Born's probabilistic interpretation, in: J.A. Wheeler and W.H. Zurek, *Quantum theory and measurement,* Princeton University Press, 1983

Rüger, A., "Atomism from cosmology: Erwin Schrödinger work on wave mechanics and space-time structure", HSPS, 18, 378-401, 1988

Russell, B., *Our knowledge of the external world,* Open Court, 1914

Russell, B., *The analysis of Mind,* G. Allen & Unwin, 1971

Russell, B., *Mysticism and logic,* Unwin Paperbacks, 1986

Ryle, G., *The concept of Mind,* Hutchinson, 1949

Ryle, G., (ed.), *Contemporary aspects of philosophy,* Oriel Press, 1976

Saunders, S., & Brown, H.R., *The philosophy of vacuum,* Oxford University Press, 1991

Saunders, S., "Decoherence, Relative states, and Evolutionary adaptation", Found. Phys. 23, 1553-1587, 1993

Saunders, S., "Time and quantum mechanics", in: M. Bitbol & E. Ruhnau (eds.), *Now, time and quantum mechanics,* Editions Frontières, 1994

Schmidt, E., "Zur Theorie der linearen und nicht linearen Integral Gleichungen (I)", Math. Annalen, 63, 433-476, 1907

Schrödinger, E., "Bohrs neue Strahlungshypothese und der Energiesatz", Die Naturwissenschaften, 12, 720-724, 1924

Schrödinger, E., "Bemerkungen über die statistische Entropie-definition beim idealen Gas", Berlin Akademie der Wissenschaft Sitzungsberichte, 1925, 434-441

Schrödinger, E., "Zur Einsteinschen Gastheorie", Phys. Z., 27, 95-101, 1926

Schrödinger, E., "Quantization as a problem of proper values" (I-IV), in: *Collected papers on wave mechanics,* Blackie and son, 1928.

Schrödinger, E., "The exchange of energy according to wave mechanics", in: *Collected papers on wave mechanics,* Blackie & son, 1928

Schrödinger, E., "The continuous transition from micro- to macro-mechanics", in: *Collected papers on wave mechanics,* Blackie & son, 1928

Schrödinger, E., "On the relation between the quantum mechanics of Heisenberg, Born and Jordan, and that of Schrödinger", in: *Collected papers on quantum mechanics,* Blackie & son, 1928

Schrödinger, E., "The Compton effect" in: *Collected papers on wave mechanics,* Blackie and son, 1928

Schrödinger, E., *Collected papers on wave mechanics,* Blackie and son, 1928

Schrödinger, E., *Four lectures on wave mechanics*, Blackie & son, 1928

Schrödinger, E., "Neue Wege in der Physik", *elektrotechnische Zeitschrift,* 50, 15-16, 1929

Schrödinger, E., "Die Erfassung der Quantengesetze durch kontinuierliche Funktionen", Naturwissenschaften, 26, 486-489, 1929

Schrödinger, E., "Ist Naturwissenschaft milieubedingt?", in: *Über Indeterminismus in der Physik,* Barth, Leipzig, 1932

Schrödinger, E., *Über Indeterminismus in der Physik,* Barth, Leipzig, 1932

Schrödinger, E., "Sur la theorie relativiste de l'électron et l'interprétation de la mécanique quantique", Ann. Inst. H. Poincaré, 2, 269-310, 1932

Schrödinger, E., "Über die Unanwendbarkeit der geometrie im Kleinen", Naturwissenschaften, 22, 518-520, 1934

Schrödinger, E., "Quelques remarques au sujet des bases de la connaissance scientifique", Scientia, LVII, 181-191, 1935

Schrödinger, E., "Die gegenwärtige Situation in der Quantenmechanik", Naturwissenschaften, 23, 807-812, 823-828, 844-849, 1935

Schrödinger, E., "Discussion of probability relations between separated systems", Proc. Cambridge Phil. Soc., 31, 555-563, 1935

Schrödinger, E., *Science and the Human temperament,* G. Allen & Unwin, 1935

Schrödinger, E., "Probability relations between separated systems", Proc. Camb. Phil. Soc., 32, 446-452, 1936

Schrödinger, E., "World structure", Nature, 140, 742-744, 1937

Schrödinger, E., "Sur la théorie du monde d'Eddington", 15, Nuovo cimento, 246-254, 1938

Schrödinger, E., "The proper vibrations of the expanding universe", Physica, 6, 899-912, 1939

Schrödinger, E., *Statistical Thermodynamics,* Cambridge University Press, 1944

Schrödinger, E., "Der Geist der Naturwissenschaft", Eranos Jahrbuch, 14, 491-520, 1946

Schrödinger, E., "The final affine laws", Proc. Roy. Irish Acad. 51A, 163-171, 1947

Schrödinger, E., "The foundations of the theory of probability", Proc. R.I.A., 51A, 51-66, 141-146, 1947

Schrödinger, E., "Die Besonderheit des Weltbilds der Naturwissenschaft", Acta Physica Austriaca, 1, 201-245, 1948; Trad. "On the peculiarity of the scientific world-view", in:*What is life? and other essays,* Doubleday anchor, 1957

Schrödinger, E., "What is an elementary particle?", Endeavour, 9, 109-116, 1950

Schrödinger, E., *Space-time structure,* Cambridge University Press, 1950

Schrödinger, E., *Science and Humanism,* Cambridge University Press, 1951

Schrödinger, E., "Are there quantum jumps?", Brit. J. Phil. Sci., 3, 109-123, 233-242, 1952

Schrödinger, E., "The meaning of wave mechanics", in: A. George (ed.), *Louis de Broglie physicien et penseur,* Albin Michel, 1953

Schrödinger, E., *Nature and the Greeks,* Cambridge University Press, 1954

Schrödinger, E.,*What is life? and other essays,* Doubleday anchor, 1957

Schrödinger, E., "Might perhaps energy be merely a statistical concept?", Nuovo cimento, 9, 162-170, 1958

Schrödinger, E., *My view of the world,* Cambridge University Press, 1964

Schrödinger, E.,*What is life?* and *Mind and matter,* Cambridge University Press, 1967

Schrödinger, E., *Gesammelte abhandlungen,* Verlag der österreichischen Akademie der Wissenschaften, Friedrich Wievweg & Sohn, 1984

Schrödinger, E., *The interpretation of quantum mechanics (Dublin seminars 1949-1955 and other unpublished texts),* edited and with introduction by M. Bitbol, Ox Bow Press, 1995

Scott, W.T., *Erwin Schrödinger: an introduction to his writings,* University of Massachussetts Press, 1967

Searle, J., *Speech acts,* Cambridge University Press, 1969

Searle, J., *Intentionality,* Cambridge University Press, 1983

Searle, J., *The Construction of Social Reality,* Allen Lane, The Penguin Press, 1995

Sellars, W., *Science perception and reality,* Routledge & Kegan Paul, 1963

Shimony, A., *Search for a naturalistic world view I,* Cambridge University Press, 1993

Squires, E.J., "How to test for Cartesian dualism by quantum experiments", in: P. Lahti and P. Mittelstaedt, (eds.), *Proceedings of the symposium on the foundations of modern physics 1990,* World Scientific, 1991

Stalnaker, R.C., "Pragmatic presuppositions" in: M.K. Munitz & P.K. Unger, eds. *Semantics and Philosophy,* New-York University Press, 1974

Steiner, F., "Schrödinger's discovery of coherent states", Physica B 151, 323-326, 1988

Strnad, J., "The Compton effect: Schrödinger's treatment", Eur. J. Phys., 7, 217-221, 1986

Treder, H.J., and von Borzeszkowski, H.H., "Interference and interaction in Schrödinger's wave mechanics", Found. Phys., 18, 77-93, 1988

Van der Waerden, B.L (ed.), *Sources of quantum mechanics,* North Holland, 1967

Van Fraassen, B. C., "The Einstein-Podolsky-Rosen paradox", Synthese, 29, 291-309, 1974

Van Fraassen, B. C., *Quantum mechanics, an empiricist view,* Oxford University Press, 1991

Von Neumann, J., *Mathematical foundations of quantum mechanics,* Princeton University Press, 1955

Von Weizsäcker, C. F., *Aufbau der physik,* Hanser, 1985,

Von Wright, G.W., *Explanation and understanding,* Routledge & Kegan Paul, 1971

Vujicic, M., and Herbut, F., "Distant steering: Schrödinger's version of non-separability", J. Phys A (Math. Gen.) 21, 2931-2939, 1988

Wentzel, G., "Zur Theorie des photoelektrischen Effekts", Z. Phys., 40, 574-589, 1926

Wessels, L., *Schrödinger's interpretations of wave mechanics,* Ph. D. dissertation, Indiana University, 1975

Wessels, L., "Erwin Schrödinger and the descriptive tradition", in: R. Aris, H.T. Davis, and R.H. Stuewer (eds.), *Springs of scientific creativity,* University of Minnesota Press, 1983

Wheeler, J. A., and Zurek, W.K., *Quantum mechanics and measurement,* Princeton University Press, 1983

Wimmel, H., *Quantum physics and observed reality,* World Scientific, 1992

Wittgenstein, L., *Tractatus logico-philosophicus,* Routledge & Kegan Paul, 1974

Wittgenstein, L., *Philosophical investigations,* Basil Blackwell, 1958

Wittgenstein, L., *On certainty,* Basil Blackwell, 1974

Zambrini, J.C., "Probability in quantum mechanics according to Schrödinger", Physica B 151, 327-331, 1988

Zurek, W.H., "Environment-induced superselection rules", Phys. Rev. D26, 1862-1880, 1982

INDEX

—A—

—B—

—C—

—D—

—E—

—J—

—K—

—L—

—M—

—N—

—O—

—S—

—T—

—U—

—V—

—W—

—Y—

—Z—

60. T. Nickles (ed.): *Scientific Discovery.* Case Studies. 1980 ISBN 90-277-1092-9; Pb 90-277-1093-7
61. M.A. Finocchiaro: *Galileo and the Art of Reasoning.* Rhetorical Foundation of Logic and Scientific Method. 1980 ISBN 90-277-1094-5; Pb 90-277-1095-3
62. W.A. Wallace: *Prelude to Galileo.* Essays on Medieval and 16th-Century Sources of Galileo's Thought. 1981 ISBN 90-277-1215-8; Pb 90-277-1216-6
63. F. Rapp: *Analytical Philosophy of Technology.* Translated from German. 1981 ISBN 90-277-1221-2; Pb 90-277-1222-0
64. R.S. Cohen and M.W. Wartofsky (eds.): *Hegel and the Sciences.* 1984 ISBN 90-277-0726-X
65. J. Agassi: *Science and Society.* Studies in the Sociology of Science. 1981 ISBN 90-277-1244-1; Pb 90-277-1245-X
66. L. Tondl: *Problems of Semantics.* A Contribution to the Analysis of the Language of Science. Translated from Czech. 1981 ISBN 90-277-0148-2; Pb 90-277-0316-7
67. J. Agassi and R.S. Cohen (eds.): *Scientific Philosophy Today.* Essays in Honor of Mario Bunge. 1982 ISBN 90-277-1262-X; Pb 90-277-1263-8
68. W. Krajewski (ed.): *Polish Essays in the Philosophy of the Natural Sciences.* Translated from Polish and edited by R.S. Cohen and C.R. Fawcett. 1982 ISBN 90-277-1286-7; Pb 90-277-1287-5
69. J.H. Fetzer: *Scientific Knowledge.* Causation, Explanation and Corroboration. 1981 ISBN 90-277-1335-9; Pb 90-277-1336-7
70. S. Grossberg: *Studies of Mind and Brain.* Neural Principles of Learning, Perception, Development, Cognition, and Motor Control. 1982 ISBN 90-277-1359-6; Pb 90-277-1360-X
71. R.S. Cohen and M.W. Wartofsky (eds.): *Epistemology, Methodology, and the Social Sciences.* 1983. ISBN 90-277-1454-1
72. K. Berka: *Measurement.* Its Concepts, Theories and Problems. Translated from Czech. 1983 ISBN 90-277-1416-9
73. G.L. Pandit: *The Structure and Growth of Scientific Knowledge.* A Study in the Methodology of Epistemic Appraisal. 1983 ISBN 90-277-1434-7
74. A.A. Zinov'ev: *Logical Physics.* Translated from Russian. Edited by R.S. Cohen. 1983 [*see also* Volume 9] ISBN 90-277-0734-0
75. G-G. Granger: *Formal Thought and the Sciences of Man.* Translated from French. With and Introduction by A. Rosenberg. 1983 ISBN 90-277-1524-6
76. R.S. Cohen and L. Laudan (eds.): *Physics, Philosophy and Psychoanalysis.* Essays in Honor of Adolf Grünbaum. 1983 ISBN 90-277-1533-5
77. G. Böhme, W. van den Daele, R. Hohlfeld, W. Krohn and W. Schäfer: *Finalization in Science.* The Social Orientation of Scientific Progress. Translated from German. Edited by W. Schäfer. 1983 ISBN 90-277-1549-1
78. D. Shapere: *Reason and the Search for Knowledge.* Investigations in the Philosophy of Science. 1984 ISBN 90-277-1551-3; Pb 90-277-1641-2
79. G. Andersson (ed.): *Rationality in Science and Politics.* Translated from German. 1984 ISBN 90-277-1575-0; Pb 90-277-1953-5
80. P.T. Durbin and F. Rapp (eds.): *Philosophy and Technology.* [*Also* Philosophy and Technology Series, Vol. 1] 1983 ISBN 90-277-1576-9
81. M. Marković: *Dialectical Theory of Meaning.* Translated from Serbo-Croat. 1984 ISBN 90-277-1596-3

Boston Studies in the Philosophy of Science

82. R.S. Cohen and M.W. Wartofsky (eds.): *Physical Sciences and History of Physics.* 1984. ISBN 90-277-1615-3
83. É. Meyerson: *The Relativistic Deduction.* Epistemological Implications of the Theory of Relativity. Translated from French. With a Review by Albert Einstein and an Introduction by Milić Čapek. 1985 ISBN 90-277-1699-4
84. R.S. Cohen and M.W. Wartofsky (eds.): *Methodology, Metaphysics and the History of Science.* In Memory of Benjamin Nelson. 1984 ISBN 90-277-1711-7
85. G. Tamás: *The Logic of Categories.* Translated from Hungarian. Edited by R.S. Cohen. 1986 ISBN 90-277-1742-7
86. S.L. de C. Fernandes: *Foundations of Objective Knowledge.* The Relations of Popper's Theory of Knowledge to That of Kant. 1985 ISBN 90-277-1809-1
87. R.S. Cohen and T. Schnelle (eds.): *Cognition and Fact.* Materials on Ludwik Fleck. 1986 ISBN 90-277-1902-0
88. G. Freudenthal: *Atom and Individual in the Age of Newton.* On the Genesis of the Mechanistic World View. Translated from German. 1986 ISBN 90-277-1905-5
89. A. Donagan, A.N. Perovich Jr and M.V. Wedin (eds.): *Human Nature and Natural Knowledge.* Essays presented to Marjorie Grene on the Occasion of Her 75th Birthday. 1986 ISBN 90-277-1974-8
90. C. Mitcham and A. Hunning (eds.): *Philosophy and Technology II.* Information Technology and Computers in Theory and Practice. [*Also* Philosophy and Technology Series, Vol. 2] 1986 ISBN 90-277-1975-6
91. M. Grene and D. Nails (eds.): *Spinoza and the Sciences.* 1986 ISBN 90-277-1976-4
92. S.P. Turner: *The Search for a Methodology of Social Science.* Durkheim, Weber, and the 19th-Century Problem of Cause, Probability, and Action. 1986. ISBN 90-277-2067-3
93. I.C. Jarvie: *Thinking about Society.* Theory and Practice. 1986 ISBN 90-277-2068-1
94. E. Ullmann-Margalit (ed.): *The Kaleidoscope of Science.* The Israel Colloquium: Studies in History, Philosophy, and Sociology of Science, Vol. 1. 1986 ISBN 90-277-2158-0; Pb 90-277-2159-9
95. E. Ullmann-Margalit (ed.): *The Prism of Science.* The Israel Colloquium: Studies in History, Philosophy, and Sociology of Science, Vol. 2. 1986 ISBN 90-277-2160-2; Pb 90-277-2161-0
96. G. Márkus: *Language and Production.* A Critique of the Paradigms. Translated from French. 1986 ISBN 90-277-2169-6
97. F. Amrine, F.J. Zucker and H. Wheeler (eds.): *Goethe and the Sciences: A Reappraisal.* 1987 ISBN 90-277-2265-X; Pb 90-277-2400-8
98. J.C. Pitt and M. Pera (eds.): *Rational Changes in Science.* Essays on Scientific Reasoning. Translated from Italian. 1987 ISBN 90-277-2417-2
99. O. Costa de Beauregard: *Time, the Physical Magnitude.* 1987 ISBN 90-277-2444-X
100. A. Shimony and D. Nails (eds.): *Naturalistic Epistemology.* A Symposium of Two Decades. 1987 ISBN 90-277-2337-0
101. N. Rotenstreich: *Time and Meaning in History.* 1987 ISBN 90-277-2467-9
102. D.B. Zilberman: *The Birth of Meaning in Hindu Thought.* Edited by R.S. Cohen. 1988 ISBN 90-277-2497-0
103. T.F. Glick (ed.): *The Comparative Reception of Relativity.* 1987 ISBN 90-277-2498-9
104. Z. Harris, M. Gottfried, T. Ryckman, P. Mattick Jr, A. Daladier, T.N. Harris and S. Harris: *The Form of Information in Science.* Analysis of an Immunology Sublanguage. With a Preface by Hilary Putnam. 1989 ISBN 90-277-2516-0

Boston Studies in the Philosophy of Science

105. F. Burwick (ed.): *Approaches to Organic Form.* Permutations in Science and Culture. 1987
ISBN 90-277-2541-1
106. M. Almási: *The Philosophy of Appearances.* Translated from Hungarian. 1989
ISBN 90-277-2150-5
107. S. Hook, W.L. O'Neill and R. O'Toole (eds.): *Philosophy, History and Social Action.* Essays in Honor of Lewis Feuer. With an Autobiographical Essay by L. Feuer. 1988
ISBN 90-277-2644-2
108. I. Hronszky, M. Fehér and B. Dajka: *Scientific Knowledge Socialized.* Selected Proceedings of the 5th Joint International Conference on the History and Philosophy of Science organized by the IUHPS (Veszprém, Hungary, 1984). 1988 ISBN 90-277-2284-6
109. P. Tillers and E.D. Green (eds.): *Probability and Inference in the Law of Evidence.* The Uses and Limits of Bayesianism. 1988 ISBN 90-277-2689-2
110. E. Ullmann-Margalit (ed.): *Science in Reflection.* The Israel Colloquium: Studies in History, Philosophy, and Sociology of Science, Vol. 3. 1988
ISBN 90-277-2712-0; Pb 90-277-2713-9
111. K. Gavroglu, Y. Goudaroulis and P. Nicolacopoulos (eds.): *Imre Lakatos and Theories of Scientific Change.* 1989 ISBN 90-277-2766-X
112. B. Glassner and J.D. Moreno (eds.): *The Qualitative-Quantitative Distinction in the Social Sciences.* 1989 ISBN 90-277-2829-1
113. K. Arens: *Structures of Knowing.* Psychologies of the 19th Century. 1989
ISBN 0-7923-0009-2
114. A. Janik: *Style, Politics and the Future of Philosophy.* 1989 ISBN 0-7923-0056-4
115. F. Amrine (ed.): *Literature and Science as Modes of Expression.* With an Introduction by S. Weininger. 1989 ISBN 0-7923-0133-1
116. J.R. Brown and J. Mittelstrass (eds.): *An Intimate Relation.* Studies in the History and Philosophy of Science. Presented to Robert E. Butts on His 60th Birthday. 1989
ISBN 0-7923-0169-2
117. F. D'Agostino and I.C. Jarvie (eds.): *Freedom and Rationality.* Essays in Honor of John Watkins. 1989 ISBN 0-7923-0264-8
118. D. Zolo: *Reflexive Epistemology.* The Philosophical Legacy of Otto Neurath. 1989
ISBN 0-7923-0320-2
119. M. Kearn, B.S. Philips and R.S. Cohen (eds.): *Georg Simmel and Contemporary Sociology.* 1989 ISBN 0-7923-0407-1
120. T.H. Levere and W.R. Shea (eds.): *Nature, Experiment and the Science.* Essays on Galileo and the Nature of Science. In Honour of Stillman Drake. 1989 ISBN 0-7923-0420-9
121. P. Nicolacopoulos (ed.): *Greek Studies in the Philosophy and History of Science.* 1990
ISBN 0-7923-0717-8
122. R. Cooke and D. Costantini (eds.): *Statistics in Science.* The Foundations of Statistical Methods in Biology, Physics and Economics. 1990 ISBN 0-7923-0797-6
123. P. Duhem: *The Origins of Statics.* Translated from French by G.F. Leneaux, V.N. Vagliente and G.H. Wagner. With an Introduction by S.L. Jaki. 1991 ISBN 0-7923-0898-0
124. H. Kamerlingh Onnes: *Through Measurement to Knowledge.* The Selected Papers, 1853-1926. Edited and with an Introduction by K. Gavroglu and Y. Goudaroulis. 1991
ISBN 0-7923-0825-5
125. M. Čapek: *The New Aspects of Time: Its Continuity and Novelties.* Selected Papers in the Philosophy of Science. 1991 ISBN 0-7923-0911-1

Boston Studies in the Philosophy of Science

126. S. Unguru (ed.): *Physics, Cosmology and Astronomy, 1300-1700.* Tension and Accommodation. 1991 ISBN 0-7923-1022-5
127. Z. Bechler: *Newton's Physics on the Conceptual Structure of the Scientific Revolution.* 1991 ISBN 0-7923-1054-3
128. É. Meyerson: *Explanation in the Sciences.* Translated from French by M-A. Siple and D.A. Siple. 1991 ISBN 0-7923-1129-9
129. A.I. Tauber (ed.): *Organism and the Origins of Self.* 1991 ISBN 0-7923-1185-X
130. F.J. Varela and J-P. Dupuy (eds.): *Understanding Origins.* Contemporary Views on the Origin of Life, Mind and Society. 1992 ISBN 0-7923-1251-1
131. G.L. Pandit: *Methodological Variance.* Essays in Epistemological Ontology and the Methodology of Science. 1991 ISBN 0-7923-1263-5
132. G. Munévar (ed.): *Beyond Reason.* Essays on the Philosophy of Paul Feyerabend. 1991 ISBN 0-7923-1272-4
133. T.E. Uebel (ed.): *Rediscovering the Forgotten Vienna Circle.* Austrian Studies on Otto Neurath and the Vienna Circle. Partly translated from German. 1991 ISBN 0-7923-1276-7
134. W.R. Woodward and R.S. Cohen (eds.): *World Views and Scientific Discipline Formation.* Science Studies in the [former] German Democratic Republic. Partly translated from German by W.R. Woodward. 1991 ISBN 0-7923-1286-4
135. P. Zambelli: *The Speculum Astronomiae and Its Enigma.* Astrology, Theology and Science in Albertus Magnus and His Contemporaries. 1992 ISBN 0-7923-1380-1
136. P. Petitjean, C. Jami and A.M. Moulin (eds.): *Science and Empires.* Historical Studies about Scientific Development and European Expansion. ISBN 0-7923-1518-9
137. W.A. Wallace: *Galileo's Logic of Discovery and Proof.* The Background, Content, and Use of His Appropriated Treatises on Aristotle's *Posterior Analytics.* 1992 ISBN 0-7923-1577-4
138. W.A. Wallace: *Galileo's Logical Treatises.* A Translation, with Notes and Commentary, of His Appropriated Latin Questions on Aristotle's *Posterior Analytics.* 1992 ISBN 0-7923-1578-2
     Set (137 + 138) ISBN 0-7923-1579-0
139. M.J. Nye, J.L. Richards and R.H. Stuewer (eds.): *The Invention of Physical Science.* Intersections of Mathematics, Theology and Natural Philosophy since the Seventeenth Century. Essays in Honor of Erwin N. Hiebert. 1992 ISBN 0-7923-1753-X
140. G. Corsi, M.L. dalla Chiara and G.C. Ghirardi (eds.): *Bridging the Gap: Philosophy, Mathematics and Physics.* Lectures on the Foundations of Science. 1992 ISBN 0-7923-1761-0
141. C.-H. Lin and D. Fu (eds.): *Philosophy and Conceptual History of Science in Taiwan.* 1992 ISBN 0-7923-1766-1
142. S. Sarkar (ed.): *The Founders of Evolutionary Genetics.* A Centenary Reappraisal. 1992 ISBN 0-7923-1777-7
143. J. Blackmore (ed.): *Ernst Mach – A Deeper Look.* Documents and New Perspectives. 1992 ISBN 0-7923-1853-6
144. P. Kroes and M. Bakker (eds.): *Technological Development and Science in the Industrial Age.* New Perspectives on the Science–Technology Relationship. 1992 ISBN 0-7923-1898-6
145. S. Amsterdamski: *Between History and Method.* Disputes about the Rationality of Science. 1992 ISBN 0-7923-1941-9
146. E. Ullmann-Margalit (ed.): *The Scientific Enterprise.* The Bar-Hillel Colloquium: Studies in History, Philosophy, and Sociology of Science, Volume 4. 1992 ISBN 0-7923-1992-3

Boston Studies in the Philosophy of Science

147. L. Embree (ed.): *Metaarchaeology.* Reflections by Archaeologists and Philosophers. 1992 ISBN 0-7923-2023-9
148. S. French and H. Kamminga (eds.): *Correspondence, Invariance and Heuristics.* Essays in Honour of Heinz Post. 1993 ISBN 0-7923-2085-9
149. M. Bunzl: *The Context of Explanation.* 1993 ISBN 0-7923-2153-7
150. I.B. Cohen (ed.): *The Natural Sciences and the Social Sciences.* Some Critical and Historical Perspectives. 1994 ISBN 0-7923-2223-1
151. K. Gavroglu, Y. Christianidis and E. Nicolaidis (eds.): *Trends in the Historiography of Science.* 1994 ISBN 0-7923-2255-X
152. S. Poggi and M. Bossi (eds.): *Romanticism in Science.* Science in Europe, 1790–1840. 1994 ISBN 0-7923-2336-X
153. J. Faye and H.J. Folse (eds.): *Niels Bohr and Contemporary Philosophy.* 1994 ISBN 0-7923-2378-5
154. C.C. Gould and R.S. Cohen (eds.): *Artifacts, Representations, and Social Practice.* Essays for Marx W. Wartofsky. 1994 ISBN 0-7923-2481-1
155. R.E. Butts: *Historical Pragmatics.* Philosophical Essays. 1993 ISBN 0-7923-2498-6
156. R. Rashed: *The Development of Arabic Mathematics: Between Arithmetic and Algebra.* Translated from French by A.F.W. Armstrong. 1994 ISBN 0-7923-2565-6
157. I. Szumilewicz-Lachman (ed.): *Zygmunt Zawirski: His Life and Work.* With Selected Writings on Time, Logic and the Methodology of Science. Translations by Feliks Lachman. Ed. by R.S. Cohen, with the assistance of B. Bergo. 1994 ISBN 0-7923-2566-4
158. S.N. Haq: *Names, Natures and Things.* The Alchemist Jābir ibn Ḥayyān and His *Kitāb al-Aḥjār* (Book of Stones). 1994 ISBN 0-7923-2587-7
159. P. Plaass: *Kant's Theory of Natural Science.* Translation, Analytic Introduction and Commentary by Alfred E. and Maria G. Miller. 1994 ISBN 0-7923-2750-0
160. J. Misiek (ed.): *The Problem of Rationality in Science and its Philosophy.* On Popper vs. Polanyi. The Polish Conferences 1988–89. 1995 ISBN 0-7923-2925-2
161. I.C. Jarvie and N. Laor (eds.): *Critical Rationalism, Metaphysics and Science.* Essays for Joseph Agassi, Volume I. 1995 ISBN 0-7923-2960-0
162. I.C. Jarvie and N. Laor (eds.): *Critical Rationalism, the Social Sciences and the Humanities.* Essays for Joseph Agassi, Volume II. 1995 ISBN 0-7923-2961-9 Set (161–162) ISBN 0-7923-2962-7
163. K. Gavroglu, J. Stachel and M.W. Wartofsky (eds.): *Physics, Philosophy, and the Scientific Community.* Essays in the Philosophy and History of the Natural Sciences and Mathematics. In Honor of Robert S. Cohen. 1995 ISBN 0-7923-2988-0
164. K. Gavroglu, J. Stachel and M.W. Wartofsky (eds.): *Science, Politics and Social Practice.* Essays on Marxism and Science, Philosophy of Culture and the Social Sciences. In Honor of Robert S. Cohen. 1995 ISBN 0-7923-2989-9
165. K. Gavroglu, J. Stachel and M.W. Wartofsky (eds.): *Science, Mind and Art.* Essays on Science and the Humanistic Understanding in Art, Epistemology, Religion and Ethics. Essays in Honor of Robert S. Cohen. 1995 ISBN 0-7923-2990-2 Set (163–165) ISBN 0-7923-2991-0
166. K.H. Wolff: *Transformation in the Writing.* A Case of Surrender-and-Catch. 1995 ISBN 0-7923-3178-8
167. A.J. Kox and D.M. Siegel (eds.): *No Truth Except in the Details.* Essays in Honor of Martin J. Klein. 1995 ISBN 0-7923-3195-8

Boston Studies in the Philosophy of Science

168. J. Blackmore: *Ludwig Boltzmann, His Later Life and Philosophy, 1900–1906.* Book One: A Documentary History. 1995 ISBN 0-7923-3231-8.
169. R.S. Cohen, R. Hilpinen and Q. Renzong (eds.): *Realism and Anti-Realism in the Philosophy of Science.* Beijing International Conference, 1992. 1995 ISBN 0-7923-3233-4
170. I. Kuçuradi and R.S. Cohen (eds.): *The Concept of Knowledge.* The Ankara Seminar. 1995 ISBN 0-7923-3241-5
171. M.A. Grodin (ed.): *Meta Medical Ethics*: The Philosophical Foundations of Bioethics. 1995 ISBN 0-7923-3344-6
172. S. Ramirez and R.S. Cohen (eds.): *Mexican Studies in the History and Philosophy of Science.* 1995 ISBN 0-7923-3462-0
173. C. Dilworth: *The Metaphysics of Science.* An Account of Modern Science in Terms of Principles, Laws and Theories. 1995 ISBN 0-7923-3693-3
174. J. Blackmore: *Ludwig Boltzmann, His Later Life and Philosophy, 1900–1906* Book Two: The Philosopher. 1995 ISBN 0-7923-3464-7
175. P. Damerow: *Abstraction and Representation.* Essays on the Cultural Evolution of Thinking. 1996 ISBN 0-7923-3816-2
176. G. Tarozzi (ed.): *Karl Popper, Philosopher of Science.* (in prep.)
177. M. Marion and R.S. Cohen (eds.): *Québec Studies in the Philosophy of Science.* Part I: Logic, Mathematics, Physics and History of Science. Essays in Honor of Hugues Leblanc. 1995 ISBN 0-7923-3559-7
178. M. Marion and R.S. Cohen (eds.): *Québec Studies in the Philosophy of Science.* Part II: Biology, Psychology, Cognitive Science and Economics. Essays in Honor of Hugues Leblanc. 1996 ISBN 0-7923-3560-0
     Set (177–178) ISBN 0-7923-3561-9
179. Fan Dainian and R.S. Cohen (eds.): *Chinese Studies in the History and Philosophy of Science and Technology.* 1996 ISBN 0-7923-3463-9
180. P. Forman and J.M. Sánchez-Ron (eds.): *National Military Establishments and the Advancement of Science and Technology.* Studies in 20th Century History. 1996 ISBN 0-7923-3541-4
181. E.J. Post: *Quantum Reprogramming.* Ensembles and Single Systems: A Two-Tier Approach to Quantum Mechanics. 1995 ISBN 0-7923-3565-1
182. A.I. Tauber (ed.): *The Elusive Synthesis: Aesthetics and Science.* 1996 ISBN 0-7923-3904-5
183. S. Sarkar (ed.): *The Philosophy and History of Molecular Biology: New Perspectives.* 1996 ISBN 0-7923-3947-9
184. J.T. Cushing, A. Fine and S. Goldstein (eds.): *Bohmian Mechanics and Quantum Theory: An Appraisal.* 1996 ISBN 0-7923-4028-0
185. K. Michalski: *Logic and Time.* An Essay on Husserl's Theory of Meaning. 1996 ISBN 0-7923-4082-5
186. G. Munévar (ed.): *Spanish Studies in the Philosophy of Science.* 1996 ISBN 0-7923-4147-3
187. G. Schubring (ed.): *Hermann Günther Graßmann (1809–1877): Visionary Mathematician, Scientist and Neohumanist Scholar.* Papers from a Sesquicentennial Conference. 1996 ISBN 0-7923-4261-5

MIX
Papier aus verantwortungsvollen Quellen
Paper from responsible sources
FSC® C105338

If you have any concerns about our products,
you can contact us on
ProductSafety@springernature.com

In case Publisher is established outside the EU,
the EU authorized representative is:
Springer Nature Customer Service Center GmbH
Europaplatz 3, 69115 Heidelberg, Germany

Printed by Libri Plureos GmbH
in Hamburg, Germany